Stem Cells for Urogynecological Diseases

Gina Torres Zambrano
Sheikh Khalifa Medical City, UAE

Juan Uría Gonzalez Tova
NMC Royal, UAE

Braulio Peramo Moyaa
Al Ain Fertility Center, UAE

Mariano Rosselló Gayáy
Instituto Médico Roselló, Spain

Fathima Mohammed Ahamed
Al Ain Fertility Center, UAE

IGI Global
Scientific Publishing
Publishing Tomorrow's Research Today

Vice President of Editorial	Melissa Wagner
Director of Acquisitions	Mikaela Felty
Director of Book Development	Jocelynn Hessler
Production Manager	Mike Brehm
Cover Design	Jose Rosado

Published in the United States of America by
IGI Global Scientific Publishing
701 East Chocolate Avenue
Hershey, PA, 17033, USA
Tel: 717-533-8845
Fax: 717-533-7115
Website: https://www.igi-global.com E-mail: cust@igi-global.com

Library of Congress Cataloging-in-Publication Data

Names: Torres Zambrano, Gina Marcela, 1979- editor
http://id.loc.gov/rwo/agents/n2024058380
Title: Stem cells for urogynecological diseases / edited by Gina Torres Zambrano, Juan Uria Gonzalez Tova, Braulio Peramo Moyaa, Mariano Rosselló Gayá, Fathima Mohammed Ahamed.
Description: Hershey, PA : IGI Global Scientific Publishing, [2026] | Includes bibliographical references and index. | Summary: "The purpose of this book is to explore the science behind stem cells, their application in urogynecology, and the real-world implications of these therapies for patients"-- Provided by publisher.
Identifiers: LCCN 2024053561 (print) | LCCN 2024053562 (ebook) | ISBN 9798369388297 h/c | ISBN 9798369388303 s/c | ISBN 9798369388310 eISBN
Subjects: LCSH: Urogynecology
http://id.loc.gov/authorities/subjects/sh91000494 | Stem cells--Therapeutic use | Women--Diseases--Treatment
http://id.loc.gov/authorities/subjects/sh2005002380
Classification: LCC RG484 .S84 2025 (print) | LCC RG484 (ebook) | DDC 616.02/774--dc23/eng/20250211
LC record available at https://lccn.loc.gov/2024053561
LC ebook record available at https://lccn.loc.gov/2024053562

British Cataloguing in Publication Data
A Cataloguing in Publication record for this book is available from the British Library.

To the pioneers in the field of stem cell therapies, whose unwavering commitment to scientific exploration and innovation has paved the way for revolutionary advancements in healthcare. Your dedication and passion have given hope to millions around the world, and this book stands as a tribute to your remarkable contributions.

And to our families, whose boundless love and support have been our foundation throughout this journey. Your encouragement has been the driving force behind every word written in these pages.

This book is dedicated to all those who believe in the power of science to transform lives and to create a healthier and brighter future for generations to come. To Dr. Carmen González Enguita, Chief of the Urology Department at the University Hospital Fundación Jiménez Díaz, Madrid, for her tireless commitment to clinical excellence, medical education, and scientific advancement in urology. Her appointment as president of the Spanish Urological Association not only honors an impeccable professional journey but also inspires a generation of professionals who continue to believe in principled, visionary leadership.

With admiration and gratitude.

Table of Contents

Foreword

In recent decades, regenerative medicine has emerged as one of the most promising frontiers in the treatment of diseases that, until recently, lacked effective solutions. Among the various therapeutic strategies, stem cells have stood out for their ability to renew and repair damaged tissues. In the fields of urology and gynecology, advancements in the application of stem cells have opened new perspectives for treating complex conditions, ranging from erectile dysfunction and Peyronie's disease to urinary incontinence and fertility disorders.

The purpose of this book is to provide a comprehensive overview of the current and future applications of stem cells in urology and gynecology. This field combines cellular biology, tissue engineering, immunology, and clinical practice to address health problems that affect millions of people worldwide. The chapters of this work not only present the scientific and clinical foundations of stem cell use but also explore the challenges and opportunities that this technology presents in managing urological and gynecological diseases.

The rapid development of stem cell therapies has been driven by the need to overcome the limitations of conventional treatments. For example, in urology, current therapies for erectile dysfunction or tissue regeneration in Peyronie's disease focus primarily on symptomatic management rather than resolving the underlying causes. In this context, stem cells offer an alternative that not only aims to improve symptoms but also regenerates the affected tissues, marking a paradigm shift in the treatment of these conditions.

In gynecology, stem cells are revolutionizing the treatment of ovarian insufficiency, endometriosis, and other reproductive disorders that impact the quality of life of many women. The ability of stem cells to differentiate into specific tissue cells, along with their modulatory effect on inflammation and healing, positions these therapies as a powerful tool in the personalized medicine of the future.

The advances presented here are the result of the dedication and effort of researchers and clinicians who, with tenacity and vision, are transforming theory into practice. This work captures not only the achievements but also the lessons learned and the ethical challenges facing these innovative therapies. The book is grounded in clinical and preclinical studies that support the use of stem cells, highlighting both the successes and the remaining challenges on the path to safe and effective clinical implementation.

Finally, this book not only celebrates the achievements but also calls for continued collaboration and scientific rigor, emphasizing the need to keep exploring the potential of stem cells to offer more effective and personalized treatments. The authors of this work, from diverse disciplines, invite us to envision a future where regenerative medicine radically transforms patient care and brings us ever closer to truly personalized medicine.

François Peinado Ibarra
Hospital Ruber Juan Bravo, Universidad Europea, Madrid, Spain

Preface

INTRODUCTION

In recent years, the field of urogynecology has been on the precipice of a major transformation. The integration of stem cell therapies into this specialty heralds a new era in the treatment of pelvic floor disorders, bringing with it the promise of more effective, personalized, and regenerative medical interventions. This book is a testament to the incredible potential of stem cells and a call to action for the medical community to embrace these advancements in the pursuit of improving women's and men's health.

Urogynecology, by its very nature, deals with some of the most sensitive and complex aspects of female health, including pelvic organ prolapse, urinary incontinence, and pelvic pain. These conditions, while often underreported, significantly impact the quality of life for millions of women worldwide. Traditional treatment options, including surgery, physical therapy, and pharmacological interventions, have their limitations—often focusing on managing symptoms rather than addressing the underlying causes of the disorders. Moreover, the recurrence rates for many conditions remain high, leaving patients in a cycle of ongoing discomfort and treatment. For male pathologies, stem cells have great potential in the treatment of erectile dysfunction, incontinence due to sphincter injuries following pelvic surgeries, and urethral stricture treatment, among others. For both sexes, they offer an alternative therapeutic approach for renal conditions.

Stem cell therapy offers a revolutionary alternative. By leveraging the body's innate ability to heal itself, stem cells have the potential to regenerate damaged tissues, restore normal function, and provide long-term relief from chronic conditions. This approach aligns with the principles of regenerative medicine, which aims to replace or regenerate human cells, tissues, or organs to restore or establish normal function. For urogynecology, this could mean a paradigm shift—from managing symptoms to curing diseases.

The purpose of this book is to explore the science behind stem cells, their application in urogynecology, and the real-world implications of these therapies for patients. We delve into the different types of stem cells, including embryonic stem cells, adult stem cells, and induced pluripotent stem cells, each with its own unique characteristics and potential applications. The book also examines the current state of research, including preclinical studies, clinical trials, and emerging therapies that are pushing the boundaries of what we thought was possible in urogynecology.

A critical aspect of this exploration is understanding the mechanisms by which stem cells exert their effects. Stem cells can differentiate into various cell types, release bioactive molecules that promote tissue repair, and modulate the immune response—all of which are crucial for addressing the complex pathophysiology of pelvic floor disorders. The book provides detailed discussions on these mechanisms, supported by the latest scientific evidence, to give readers a comprehensive understanding of how stem cells could revolutionize urogynecological care.

However, the road to integrating stem cell therapies into clinical practice is fraught with challenges. Ethical considerations, particularly regarding the use of embryonic stem cells, remain a topic of intense debate. The book addresses these ethical dilemmas head-on, offering a balanced view that considers both the potential benefits and the moral responsibilities of using stem cells in medicine. Furthermore, the regulatory landscape for stem cell therapies is complex and varies significantly across different regions. Navigating these regulatory challenges is essential for ensuring the safe and effective use of stem cells in urogynecology, and this book provides insights into the current regulations and guidelines that govern this emerging field.

In addition to ethical and regulatory challenges, the translation of stem cell research from the laboratory to the clinic requires rigorous clinical trials. This book highlights the ongoing clinical trials that are exploring the safety and efficacy of stem cell therapies in urogynecology. We discuss the design of these trials, the outcomes they seek to measure, and the potential impact they could have on clinical practice. By examining these trials, readers will gain an appreciation for the meticulous process involved in bringing a new therapy to market and the hurdles that must be overcome to ensure that these treatments are both safe and effective for patients.

One of the most exciting aspects of stem cell research in urogynecology is the potential for personalized medicine. Stem cells can be derived from the patient's own body, reducing the risk of rejection and other complications associated with donor cells or tissues. This personalized approach not only enhances the safety profile of stem cell therapies but also opens the door to customized treatments tailored to the specific needs of each patient. The book explores this potential, offering case studies and examples of how personalized stem cell therapies could transform the way we treat pelvic floor disorders.

ORGANIZATION OF THE BOOK

Chapter 1: Introduction to Regenerative Treatments

In this foundational chapter, we introduce readers to the remarkable world of stem cells and their expanding role in clinical medicine. We begin by outlining stem cell classification, core biological principles, and their translational potential across medical disciplines, particularly in urology and gynecology. Emphasizing the need for ethical oversight, we draw on recommendations from the International Society for Stem Cell Research (ISSCR) to highlight the importance of rigorous standards. Central to the chapter is the discussion on preclinical studies and the ethical complexities tied to traditional animal testing, alongside a critical examination of translational pipelines in stem cell therapy. This introduction serves as both a conceptual primer and a bridge to the practical applications explored throughout the book.

Chapter 2: Andrology

This chapter explores the therapeutic potential of stem cells in andrology, focusing on conditions such as erectile dysfunction and Peyronie's disease. We review the latest research, drawing from both human and animal studies, to evaluate how stem cell interventions can improve vascular structures like the corpus cavernosum. By presenting evidence of enhanced treatment outcomes and functional recovery, we underscore the clinical relevance of regenerative therapies. Our aim is to equip andrology specialists with current knowledge on the integration of stem cell-based approaches into patient care, setting the stage for broader clinical adoption.

Chapter 3: Urinary Incontinence

Urinary incontinence remains a pervasive issue affecting patients across gender and age groups. In this chapter, we dissect the various types—urge, stress, and male incontinence—while identifying their underlying pathologies. We explore the emerging use of stem cells as an innovative alternative to conventional treatments, offering the promise of bladder function restoration through regeneration of urothelial, vascular, and neural components. With a strong focus on preclinical advancements, we aim to present a compelling case for regenerative medicine as a transformative modality in incontinence therapy.

Chapter 4: Other Bladder Conditions

Beyond incontinence, several debilitating bladder conditions—such as underactive bladder, detrusor underactivity, interstitial cystitis, and bladder pain syndrome—present substantial therapeutic challenges. This chapter delves into the clinical presentation and pathophysiology of these disorders, exploring their intricate ties to pelvic pain and impaired detrusor function. By shedding light on diagnostic barriers and current therapeutic gaps, we advocate for a regenerative framework to guide more effective and holistic treatment strategies, ultimately enhancing patient quality of life.

Chapter 5: Vaginal and Urethral Conditions

In this chapter, we address a wide range of vaginal and urethral disorders—ranging from menopause-induced atrophy and pelvic organ prolapse to chronic infections and sexual dysfunction. We focus on the mechanisms and therapeutic possibilities of stem cell interventions, particularly using mesenchymal, endometrial, and induced pluripotent stem cells. Highlighting both preclinical and clinical studies, we examine how regenerative techniques can restore tissue integrity and function. Through this lens, we aim to present stem cell therapy as a promising avenue to significantly improve women's urogenital health and well-being.

Chapter 6: Stem Cell Therapy in Reproductive Medicine

Reproductive medicine stands to be transformed by regenerative therapies, particularly in the realm of infertility. In this chapter, we explore breakthroughs in in vitro gametogenesis and spermatogenesis, and the potential of stem cell interventions to restore fertility in men and women. We discuss testicular failure and ovarian dysfunction, while weighing scientific progress against necessary ethical safeguards. With a focus on translating promising data into responsible clinical applications, we advocate for a balanced approach that considers both innovation and integrity in reproductive health.

Chapter 7: Stem Cell Therapy for Renal Diseases

Kidney disease remains a major global health burden with few regenerative treatment options. In this chapter, we examine how stem cell therapies—especially those using mesenchymal, embryonic, and induced pluripotent stem cells—offer a novel strategy to address both acute and chronic kidney injuries. We analyze their mechanisms of action, including cellular replacement, immunomodulation, and paracrine signaling. We also reflect on experimental successes and ongoing hurdles, such as standardization and safety. This chapter presents a forward-looking view on how stem cells could reshape nephrology through next-generation therapies and precision interventions.

Chapter 8: Stem Cell Applications in Pediatric Urology

The application of stem cell therapies in pediatric urology represents an exciting frontier. Here, we explore current and potential regenerative treatments for conditions such as vesicoureteral reflux, hypospadias, and neurogenic bladder. Emphasizing the unique regenerative needs of pediatric patients, we highlight how stem cells can aid tissue repair, control inflammation, and support normal development. While current data is promising, we acknowledge the need for rigorous clinical trials to assess safety and long-term outcomes. This chapter serves as both a review and a call to action for continued innovation in pediatric urologic care.

CONCLUSION

As we look to the future, the integration of stem cell therapies into urogynecology holds immense promise. The potential to regenerate damaged tissues, restore normal function, and improve the quality of life for patients is within our reach. However, realizing this potential requires a concerted effort from researchers, clinicians, regulatory bodies, and patients alike. This book is not just a guide to the science and application of stem cells in urogynecology; it is a call to action for the medical community to embrace these advancements and to work together to overcome the challenges that lie ahead.

We would like to acknowledge the contributions of the many researchers, clinicians, and patients who have played a role in advancing the field of stem cell research in urogynecology. Their dedication and perseverance have paved the way for the therapies that are discussed in this book. We are also grateful to the patients who have participated in clinical trials and shared their experiences, helping to shape the future of urogynecology.

In conclusion, this book is intended for a diverse audience—clinicians who are looking to integrate stem cell therapies into their practice, researchers who are pushing the boundaries of what is possible, and patients who are eager to learn about the latest advancements in urogynecology. We hope that the knowledge contained within these pages will inspire you to explore the potential of stem cells and to consider how these therapies could revolutionize the care you provide or receive.

Thank you for joining us on this journey. The future of urogynecology is bright, and together, we can make a difference in the lives of millions of women around the world.

Gina Torres Zambrano
Sheikh Khalifa Medical City, Abu Dhabi, UAE

Juan Uría González-Tova
NMC Royal, Abu Dhabi, UAE

Mariano Roselló Gayáy
Instituto Médico Roselló, Palma de Mallorca, Spain

Braulio Peramo Moyaa
Al Ain Fertility Center, Abu Dhabi, UAE

Fathima Mohammed Ahameda
Al Ain Fertility Center, Abu Dhabi, UAE

Acknowledgement

Writing a book is a collaborative effort that involves the contribution and support of numerous individuals. We are deeply grateful to all those who have played a part in bringing this book on stem cell therapies to fruition.

First and foremost, we would like to express my heartfelt gratitude to the scientist, researches and medical professionals who have dedicate their lives to advancing in the field of stem cell therapy. Your ground-breaking work has paved the way for new possibilities in healthcare and has inspirited the content of this book.

Lastly, we want to express our gratitude to the readers of this book. Your curiosity and interest in stem cell therapy motivate authors and researches alike to continue exploring and sharing the wonders of science and medicine.

In conclusion, this book would not have been possible without the collective efforts of countless individuals who have contributed their time, expertise and enthusiasm. Thank you all for being a part of this journey to uncover the potential of stem cell therapy and its impact on the future of healthcare

Introduction

In recent years, the field of regenerative medicine has witnessed remarkable advancements. As the frontier of medical science continues to expand, stem cell therapy has emerged as a transformative paradigm, particularly in the fields of urology and gynecology. Regenerative treatments represent a groundbreaking frontier in medical science, offering unprecedented potential for the treatment of various urological and gynecological conditions. This book delves into the transformative impact of stem cell therapies, exploring their applications, benefits, and the future they promise in these specialized fields of medicine, and their impact in treating a range of conditions that affect millions of individuals worldwide.

Stem cells, with their unique ability to differentiate into various cell types and their capacity for self-renewal, hold immense potential for tissue repair and regeneration. In the context of urological and gynecological disorders, where tissue damage often leads to functional impairment, stem cell-based approaches offer hope for restoring organ function and improving patients' quality of life.

The application of stem cell therapies in urology and gynecology spans a wide range of conditions. From addressing female reproductive health challenges such as pelvic pain and infertility to tackling urological disorders like urinary incontinence and bladder dysfunction, stem cell therapies offer novel approaches that go beyond traditional treatment modalities.

The authors aim to provide a comprehensive overview of the current state of stem cell research and its clinical applications in urology and gynecology. Through detailed discussions of ongoing clinical trials, case studies, and expert insights, readers will gain a deeper understanding of how stem cell therapies are poised to revolutionize patient care in these fields.

In urology, stem cell therapy has shown remarkable promise in addressing conditions such as urinary incontinence, erectile dysfunction, and chronic kidney disease. There is also potential for its use in fertility. The ability of stem cells to home in on injured sites, stimulate tissue regeneration, and modulate immune responses has opened new avenues for treating these challenging conditions (Wiafe et al., 2015).

Similarly, in gynecology, stem cell therapies are being explored for their potential to treat conditions like pelvic organ prolapse, endometriosis, fertility, and menopause. The regenerative capabilities of stem cells offer hope for restoring normal function and improving the quality of life for patients suffering from these debilitating conditions (Boerner, 2024).

Within these pages, we will explore a curated selection of the most relevant articles and meta-analyses that have shaped our current understanding of stem cell research and its clinical applications. We will delve into the mechanisms by which stem cells can repair and regenerate damaged tissues, highlighting the latest advancements in research methodologies and therapeutic techniques. Moreover, we will discuss the ethical considerations and regulatory landscapes that govern the use of stem cells, including animal welfare, ensuring that practitioners and researchers alike are informed about the complexities surrounding this evolving field.

As we embark on this exploration, we will not only examine the promising clinical outcomes reported in scientific literature but also the challenges that remain in translating these findings into everyday practice. By fostering a comprehensive understanding of stem cell therapies in urology and gynecology, this book aims to serve as an essential resource for healthcare professionals, researchers, and students dedicated to advancing the frontiers of regenerative medicine. Together, let us pave the way for a future where stem cell therapies can provide hope and healing for those afflicted by urological and gynecological conditions.

Within these pages, we will review the different sources for stem cell retrieval, their particular advantages and challenges, and the choice of stem cell type, depending on the specific therapeutic application and the balance between ethical considerations, availability, and potential for differentiation and regeneration.

Additionally, we will discuss the process for isolating and culturing stem cells, including steps to ensure their viability and functionality, such as bone marrow aspiration, peripheral blood collection, adipose tissue harvesting, culture techniques, organoids, and 3D cultures, among other techniques.

These advancements are paving the way for more effective and precise stem cell therapies, bringing us closer to realizing their full potential in treating a wide range of diseases and conditions.

Scaling up stem cell production for therapeutic use involves several significant challenges, such as standardization and quality control to ensure consistent quality and functionality of stem cells across large batches. Automation and efficiency are also crucial, as current stem cell production often relies on manual processes, which are labor-intensive and not scalable. Developing automated systems that can handle the complexity of stem cell culture and differentiation is essential and will reduce costs, as current production at a scale suitable for widespread clinical use is expensive, including costs associated with specialized equipment, reagents, and skilled labor (Cytiva, s. f.).

Meeting the stringent regulatory requirements for stem cell therapies can be challenging. This includes ensuring the safety, efficacy, and quality of the stem cells produced. Moreover, the use of certain types of stem cells, such as embryonic stem cells, raises ethical and legal concerns that can impact production and distribution. Moving from laboratory research to clinical application involves rigorous testing and validation. This process can be lengthy and requires significant investment.

Despite these challenges, advancements in automation, bioprocessing technologies, and regulatory frameworks are gradually addressing these issues, paving the way for more scalable and accessible stem cell therapies (Bender, 2021). Ongoing research and technological advancements continue to push the boundaries of what is possible with stem cell therapies.

By compiling the latest research, clinical data, and expert insights, we hope to offer a valuable resource for clinicians, researchers, and students in the fields of urology, gynecology, and regenerative medicine. As we stand on the cusp of a new era in regenerative medicine, this book seeks to illuminate the path forward, highlighting both the immense potential and the challenges that lie ahead in harnessing the power of stem cells to transform patient care in urology and gynecology.

REFERENCES

Bender, E. (2021). Stem-cell start-ups seek to crack the mass-production problem. *Nature*, *597*(7878), S20–S21. DOI: 10.1038/d41586-021-02627-y

Boerner, C. (2024, marzo 11). Phase 3 CELLEBRATE trial testing regenerative stem cell-based therapy to improve urinary control for women. *VUMV*. https://news.vumc.org/2024/03/11/phase-3-cellebrate-trial-testing-regenerative-stem-cell-based-therapy-to-improve-urinary-control-for-women-2/

Cytiva. (s. f.). The Emerging Key to Smart Cell Therapy Manufacturing. *Nature portfolio*. Recuperado 6 de septiembre de 2024, de https://www.nature.com/articles/d42473-021-00282-5

Wiafe, B., Metcalfe, P. D., & Adesida, A. B. (2015). Stem Cell Therapy: Current Applications and Potential for Urology. *Current Urology Reports*, *16*(11), 77. DOI: 10.1007/s11934-015-0551-5 PMID: 26385812

Chapter 1
Introduction to Regenerative Treatments

ABSTRACT

This chapter provides a comprehensive introduction to stem cells, focusing on their classification, clinical translation, and the recommendations of the International Society for Stem Cell Research (ISSCR). It begins with a fundamental overview of stem cell biology, including the various types and their potential applications in the medical practice. The argument emphasizes the importance of adhering to established standards and guidelines to ensure ethical practices and scientific rigor in stem cell research. Key topics include translation of stem cell therapies, the role of preclinical studies, highlighting both their importance and the ethical concerns associated with traditional animal testing, and the critical evaluation of current standards to guide future research. Aimed primarily at urologists and gynecologists, this chapter provides essential background information to bridge the gap between basic stem cell science and its practical implications, setting the stage for the subsequent content of the book and offering insights into emerging opportunities and challenges

A GLIMPSE INTO THE FUTURE

SC therapy represents a groundbreaking frontier in medical science, offering unprecedented potential for the treatment of various urological and gynecological conditions. In urology, SC therapy holds immense promise for revolutionizing the treatment of urological conditions. With the ability to differentiate into various cell types, stem cells offer a potential solution for a wide range of urological disorders, and has shown remarkable promise in addressing conditions from bladder dysfunction to prostate cancer, such as stress urinary incontinence, erectile dysfunction, and chronic kidney disease. The ability of SC to home in on injured sites, stimulate

DOI: 10.4018/979-8-3693-8829-7.ch001

tissue regeneration, and modulate immune responses has opened new avenues for treating these challenging conditions (Wiafe et al., 2015).

SC could be used to repair damaged bladder tissue, offering hope for patients with conditions like interstitial cystitis or neurogenic bladder. SC could be used to repair damaged kidney, testicle, bladder or penile tissue, slowing the progression of the disease or even reversing it in some cases. (González Enguita et al., 2024; Pérez-Aizpurua et al., 2023)

Regenerative therapy might be used to regenerate urethral tissue, addressing urethral strictures and improving urinary flow. By repairing damaged sphincter muscles, SC therapy could potentially treat urinary incontinence.

Overcoming Challenges

Despite the promising potential, significant challenges must be overcome before SC therapies become standard treatments.

- **Cell Source:** Determining the optimal source of stem cells (embryonic, induced pluripotent stem cells, or adult stem cells) remains a critical issue.
- **Tumorigenicity:** Preventing the formation of tumors from transplanted stem cells is a major concern.
- **Immune Rejection:** The body's immune system may reject transplanted stem cells, requiring immunosuppressive therapy.
- **Ethical Considerations**: The use of embryonic stem cells raises ethical concerns, while the collection of adult stem cells can be invasive.
- **Efficacy and Safety:** Rigorous clinical trials are needed to establish the efficacy and safety of stem cell therapies for urological conditions.

The Road Ahead

While the journey to widespread adoption of stem cell therapies in urology is still long, the potential benefits are substantial. Continued research and development are essential to address the challenges and unlock the full potential of this groundbreaking technology.

As our understanding of stem cells grows, we can expect to see increasing numbers of clinical trials and, ultimately, the development of effective stem cell-based treatments for urological disorders. The future of urology is bright, with stem cell therapy at the forefront of innovation.

STEM CELLS

SCs are unspecialized and undifferentiated cells that have the ability to differentiate into many different cell types (such as in the skin, muscle, or blood), and are capable of self-renewal to produce more of the same type of stem cells, and differentiate in other types, giving them a potential to regenerate diseased organs, and that therefore contribute to the renewal and repair of damaged tissues. (Alwaal et al., 2015)

It is important to emphasize that tissue regeneration is highly dependent on the surrounding environment, including organ architecture and apocrine effects. (Lahlil et al., 2018)

SCs act as a repair system, replenishing damaged tissues via cell asymmetric cell division to give rise to progeny tors, and SCs are expected to be relatively quiescent and be activated under stress conditions to regenerate diseased tissues and thereby maintain homeostasis in the body organs.

Nowadays, we can treat several diseases that were incurables before, such as leukemia, which become possible through SCs transplantation with exciting results. The first therapy using SCs was a bone marrow transplant performed by French oncologist Georges Mathé in 1958 on five workers at the Vinča Nuclear Institute in Yugoslavia who had been affected by a criticality accident. However, the use of stem cells for solid tissue transplantation has not been met with similar success. (Lahlil et al., 2018)

Because the SCs can be obtained after a blood donation of the same patients, these cells therapy is an autologous treatment, avoiding immunologic reactions to this preparation. The products have a very strict evaluation for quality and safety. The adverse effect is rare but may occur fainting and dizziness during the collection procedure. (Torres Zambrano et al., 2020)

In the field of urology, there are different studies using SCs for regenerating bladder and wall architecture, creation of mesh to repair vaginal prolapse, regeneration of the kidney and urethra, promote restoration of spermatogenesis, recovery of urinary continence, reparation of Peyronies's disease (PD) and erectile dysfunction (ED). For example, the combination of stimulation by humoral factors and coculture with primary urothelial cells resulted in a further significant increase of smooth-muscle-specific gene expression in the treated tissue.

There are 2 ways of classifying the SCs, which could be according to their differentiation potential or their source of isolation.

Classification According to Differentiation Potential

Totipotent SCs: (zygote) These SCs have the highest potential and can differentiate into any tissue type, regardless of origin. The zygote and morula are examples of this SC type.

Pluripotent SCs: (embryonic SCs) These SCs can differentiate into cells from the 3 different germ cell layers (ectoderm, mesoderm, and endoderm), and gonadal ridge but not into extra-embryonic tissues. An example is embryonic SCs (ESCs), which are a derivative of the inner cell mass of the blastocyst.

Multipotent SCs: These SCs are capable of self-renewal and can differentiate into organ-specific cell types. Examples include hematopoietic SCs, mesenchymal SCs, and neural SCs.They can be found in fully developed organs and can be differentiated into cells belonging to their own germ layer.

Unipotent SCs: These SCs can give rise to only one defined cell type, epithelial cells.

Induced pluripotent SCs: These SCs are "reprogrammed" cells, i.e., differentiated cells that are manipulated in the laboratory to express genes that are normally present in ESCs and that therefore behave like ESCs. Induced pluripotent SCs can differentiate into cells of all organs and tissues. (Alwaal et al., 2015)

Classification According to the Source of Isolation

SC can be sourced from various tissues and at different stages of development:

Embryonic Stem Cells (ESCs): Derived from early-stage embryos, these cells are pluripotent, meaning they can differentiate into almost any cell type. However, their use raises ethical concerns and regulatory challenges.

Adult Stem Cells (ASCs): Found in various tissues like bone marrow, fat, and blood, these cells are multipotent, meaning they can develop into a limited range of cell types related to their tissue of origin. Examples include hematopoietic stem cells (from bone marrow) and mesenchymal stem cells (from fat tissue).

Induced Pluripotent Stem Cells (iPSCs): Created by reprogramming adult cells (like skin cells) to revert to a pluripotent state. iPSCs offer the versatility of ESCs without the associated ethical issues, but they require careful handling to ensure safety and efficacy.

Perinatal Stem Cells: Sourced from tissues associated with birth, such as the umbilical cord, placenta, and amniotic fluid. These cells are less controversial and have a high potential for differentiation.

Each source has its advantages and challenges, and the choice of stem cell type depends on the specific therapeutic application and the balance between ethical considerations, availability, and potential for differentiation and regeneration.

Figure 1. Stem cell classification

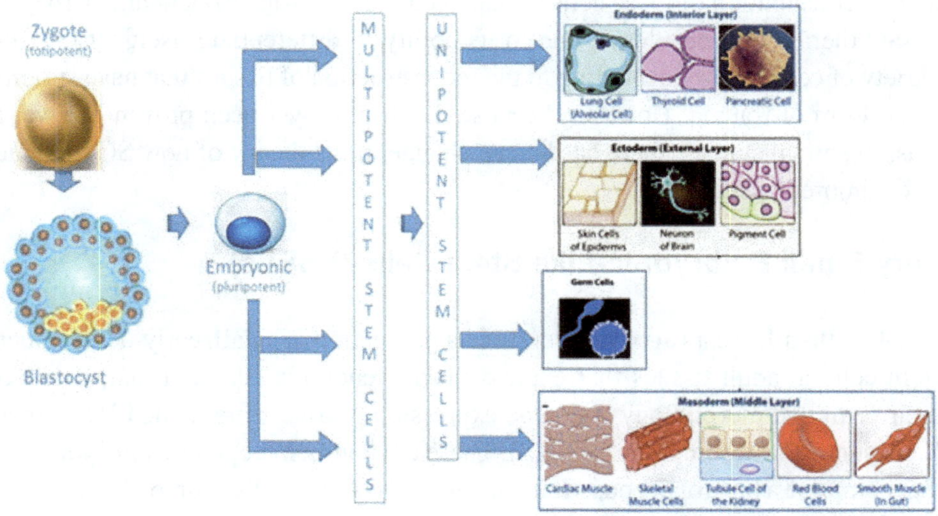

SCs can be recovered from the placenta, amniotic fluid, adipose tissue, liver, muscle, bone marrow, and umbilical cord blood. Matured SC resides in a specialized microenvironment called the stem cell niche. These niches are divided into two categories, stromal and epithelial:

Stromal niche is located in isolated anatomical locations, with SCs anchored to the niche by cell adhesion and extracellular matrix molecules. This structure allows the communication of SCs through the secretion of cytokines and growth factors. A common example is the vascular wall, which contains adipose tissue, neural, intestinal, and testicular SC.

The epithelial niche rests on the basal membrane, surrounded by mature progeny. An example of such a structure is SCs derived from muscles.

Progenitor cells show the least potential for differentiation (up to two types of cells).

Related to this classification, the use of SCs to employ in regenerative medicine is ethically controversial and they have technical problems, such as the risk of teratoma formation and life-threatening arrhythmias and their potential histocompatibility with unrelated recipients. In response to these problems, a solution for obtaining ethically acceptable pluripotent stem cells has been proposed: generating induced pluripotent stem cells (iPSCs) by genetic modification of adult cells. However, these cells have also been found to be at risk of teratoma formation and immunological rejection and demonstrate genomic instability. Moreover, the current results of clinical applications of iPSCs have demonstrated only paracrine effects of therapy and no contribution of these cells to damaged organs. (Ratajczak et al., 2014)

SC therapy has developed due to its ability to differentiate itself into a wide variety of cells, thus contributing to the reconstruction of tissues that have suffered some form of trauma. However, because this has not yet been proven in vivo, at least in animal studies, there has been a change in the theory of how SC therapies affect injured tissue.

Very Small Embryonic-Like Stem Cells (VSELs)

More than 15 years ago was identified a population of small, early-development stem cells in adult tissues that express pluripotency markers and that, based on their primitive morphology and gene expression profile, were named Very Small Embryonic-Like Stem Cells (VSELs). These are non-hematopoietic pluripotent ST that comprise a heterogeneous group of cells. (Dr. Antonio Bencomo Hernandez, s. f.) VSELs were first reported in mouse bone marrow and then in human cord blood. (Bhartiya et al., 2016; Kucia et al., 2006, 2007)

The existence of these cells across germ layers' differentiation was subsequently confirmed by at least 20 other independent groups. VSELs are quiescent small cells (measure ≈5 to 7 µm in humans), corresponding in size to the cells in the inner cell mass of the blastocyst. Thus, they are slightly smaller than red blood cells and bigger than platelets, therefore, require a special gating strategy during fluorescence-activated cell sorting. Transmission electron microscopy analysis revealed that they are spherical in shape with a high nucleo-cytoplasmic ratio, have large nuclei containing open euchromatin, and a thin rim of cytoplasm enriched in spherical mitochondria, which are characteristic of early development cells. (Bhartiya et al., 2016; Dr. Antonio Bencomo Hernandez, s. f.; Ratajczak et al., 2019)

It has been proposed that VSELs originate from cells related to the germline, being an overlapping population of the epiblast derived Primordial germ cells (PGCs) and are deposited in various developing adult organs during embryogenesis including gonads, and serve as a backup pool for tissue committed adult SCs 'progenitors'. They have asymmetric cell divisions to self-renew and give rise to the slightly bigger adult SCs, which in turn divide rapidly, undergo clonal expansion, and differentiate into tissue-specific cell types—thereby maintaining life-long homeostasis in various body organs. (Bhartiya et al., 2016)

Figure 2. Representative images with Lionheart TM FX Automated Microscope of NHPBSC population, with the presence of Erythrocytes contamination, after 24 h of isolation from COVID-19 infected peripheral blood samples. A) Bright field Microscopic image of NHPBSC morphology, indicated by the white arrow. B) DAPI (5µg/ml) DNA staining showing the nucleus. Bars in B) and C) indicate 200µm (original magnification, 10x)

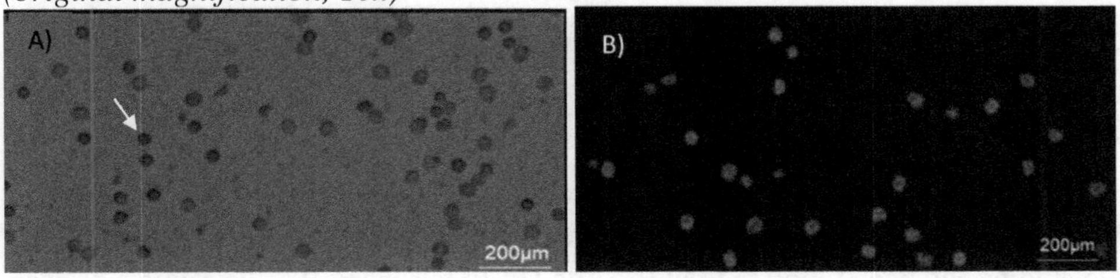

They express pluripotent and PGC-specific markers, do not divide in vitro nor form teratoma but have the potential to differentiate into three germ layers and germ cells in vitro. VSELs express some ESC markers, such as stage-specific antigen (SSEA), nuclear Oct-4A (octamer-binding transcription factor 4A), Nanog, and Rex1.

The studies show that VSELs, Lin-CD34+CD45- expressing CD133+ or NANOG+ have the same expansion and differentiation capacities towards the mesodermal and endodermal pathways, especially those VSELs expressing the Nanog gene would be the most representative of pluripotent cells and then extended to them the molecular studies. Contrary, VSELs which express the CXCR4 marker, seem to have less ability to proliferate and differentiate in different media. (Lahlil et al., 2018)

Issues with safety and production efficiency have retarded the progress and clinical applications of SCs therapy because of the fear of gene mutation and tumor development. VSELs isolated from adult tissues may solve several problems with the use of controversial ESCs and somehow problematic iPSCs in regenerative medicine. (Ratajczak et al., 2019; Suman et al., 2019)

Due to their biology, VSELs remain an excellent alternative to "natural" SCs source, because their isolation, proliferation, and differentiation are mainly controlled, and develop their potential application, by finding a way to simplify their expansion and variation. (Lahlil et al., 2018)

VSELs accepted within the adult stem cell compartment in the bone marrow and the number of these cells decreases with age. (Ratajczak et al., 2019) They are quiescent but are activated and actively mobilized from the bone marrow into the circulation under stress conditions including stroke, myocardial infarction, burn injury, after partial pancreatectomy pulmonary diseases, and post-cytotoxic treatments. They follow in direction of the stromal cell-derived factor 1 (SDF-1) gradient created by damaged tissue and play a role in tissue regeneration or regaining homeostasis and few reports have demonstrated their ability to regenerate in vivo. (Bhartiya et al., 2016; Lahlil et al., 2018)

Several papers have been published showing the benefits of the injected purified VSELs to hematopoiesis, osteogenesis, and angiogenesis and be specified to cardiomyocytes liver, pulmonary alveolar epithelium, neurons, and both in vitro and in vivo animal models. The well-demonstrated presence of chimerism in several organs indicates the potential of these cells to differentiate across germ layers. However, the precise molecular mechanism of how nascent VSELs control their pluripotency and differentiation potential remains to be determined. (Lahlil et al., 2018; Ratajczak et al., 2019)

They can be obtained through mobilization from the bone marrow, increasing its baseline levels, and could be used in patients with low concentrations of these cells. No studies had been found to treat Erectile Dysfunction (ED) or Peyronie's disease (PD), but taking into account their high regenerative potential, we can speculate that its use for these therapies is part of its potential use in the future, and prove that they can reestablish three dimensional fully functional tissue structures, which will be crucial to justify their potential application in the clinic. Some further deep sequencing analysis is needed to evaluate the genomic stability of VSELs-derived cells after current expansion strategies employing small molecular DNA modifying agents. (Ratajczak et al., 2019)

In summary, we propose that VSELs isolated from mobilization from the bone marrow should be studied for the treatment of andrology conditions, as they may provide a path forward that solves several problems with the use of regenerative medicine. (Ratajczak et al., 2019)

RECOMMENDATIONS OF THE INTERNATIONAL SOCIETY FOR STEM CELL RESEARCH (*GUIDELINES FOR STEM CELL RESEARCH AND CLINICAL TRANSLATION. - SEARCH RESULTS - PUBMED*, S. F.)

Laboratory-based Human Embryonic Stem Cell Research, Embryo Research, and Related Research Activities

SC and embryo research offers critical insights into early human development and disease, targeting causes of miscarriage, epigenetic/genetic/chromosomal disorders, and human reproduction. Derivation of certain SC lines depends on the use of human embryos. Many countries consider research on human embryos and embryonic SC lines ethically permissible when conducted under strict scientific and ethical oversight, in line with guidance from American Society for Reproductive

Medicine (ASRM, 2020), European Society of Human Reproduction and Embryology (ESHRE, s. f.; *Guidelines — International Society for Stem Cell Research*, 2021), the American College of Obstetricians and Gynecologists (ACOG Committee on Ethics, 2006), and the UK Human Fertilisation and Embryology Authority HFEA, 2019 (HFEA, 2019).

Only a few jurisdictions allow creation of embryos specifically for research, which is vital for validating and improving In Vitro Fertilization (IVF) methods (including mitochondrial replacement and in vitro–derived gametes) and for revealing the earliest stages of human development.

All research that (a) involves preimplantation stages of human development, human embryos, or embryo-derived cells or (b) entails the production of human gametes in vitro when such gametes are tested by fertilization or used for the creation of embryos shall be subject to review, approval, and ongoing monitoring by a specialized human embryo research oversight (EMRO) process capable of evaluating the unique aspects of the science. The derivation of human pluripotent stem cells from somatic cells via genetic or chemical means of reprogramming (for example, induced pluripotent stem cells or iPSCs) requires human subjects review but does not require specialized EMRO as long as the research does not generate human embryos or entail sensitive aspects of the research use of human totipotent or pluripotent stem cells as outlined in this section.

The EMRO process should be conducted by qualified scientists, ethicists, and community members who are not directly engaged in the research under consideration.

To ensure that human embryo and embryonic stem cell research is proceeding with due consideration, to ensure consistency of research practices among scientists globally, and to specify the nature of scientific projects that should be subject to review, research review and oversight should use the three categories of review described in this section.

The International Society for Stem Cell Research (ISSCR) supports laboratory-based research that entails modifying the nuclear genomes of gametes, zygotes, and/or preimplantation human embryos, performed under a rigorous EMRO process. Such research will enhance fundamental knowledge and is essential to inform any thoughtful deliberations about the potential safety and use of nuclear genome modification in strategies aimed at preventing the transmission of genetic disorders. Until further clarity emerges on both scientific and ethical fronts, the ISSCR holds that any attempt to modify the nuclear genome of human embryos for the purpose of human reproduction is premature and should be prohibited at this time.

Research that entails incorporating human totipotent or pluripotent cells into animal hosts to achieve chimerism of either the central nervous system or germline requires specialized research oversight. Such oversight should utilize available baseline animal data grounded in rigorous scientific knowledge or reasonable inferences and involve a diligent application of animal welfare principles.

Procurement of Biomaterials

Access to human gametes, embryos, fetal tissues, and somatic cells underpins embryo and stem cell research. Their acquisition must comply with internationally recognized ethical standards as well as all applicable laws and regulations in the jurisdiction where the research takes place.

Rigorous review must be performed prior to the procurement of all gametes, embryos, or somatic cells that are destined for use in human embryo and stem cell research.

Explicit and contemporaneous informed consent for the provision of all biomaterials for embryo and embryonic stem cell research is necessary, including from all gamete donors. Informed consent should be obtained at the time of the proposed transfer of any biomaterials to the research team or during the time that biomaterials are collected and stored for future research use.

Review of procurement protocols must ensure that biomaterials donors are adequately informed about the specific aspects of their voluntary research participation.

Research oversight bodies must authorize all proposals to reimburse, compensate, or provide valuable considerations of any kind to providers of embryos, gametes, or somatic cells.

For the provision of oocytes for research, when oocytes are collected outside the course of clinical treatment, compensation for non-financial burdens should not constitute an undue inducement.

Informed consent for research donation must be kept distinct from informed consent for clinical treatment.

The informed consent process and study design of human biomaterials procurement should be robust.

Derivation, Banking and Distribution of Human Pluripotent Stem Cell Lines

Proposals for derivations of new HUMAN EMBRYONIC STEM CELLS (hESC) lines should be scientifically justified and executed by scientists with appropriate expertise. Hand-in-hand with the privilege to perform these derivations is the obligation to distribute the cell lines to the research community.

A clear, detailed outline for banking and open access to the new lines should be incorporated into derivation proposals. New pluripotent stem cell lines should be made generally available as soon as possible following derivation and first publication.

Researchers and repositories should develop a policy that states whether and how incidental findings will be returned to research subjects. This policy must be explained during the informed consent process and potential subjects should be able to choose which types of incidental findings they wish to receive if any. Reporting findings with relevance to public health may be required by law in certain jurisdictions.

The ISSCR encourages the establishment of national and international repositories that are expected to accept deposits of newly derived stem cell lines and to distribute them on an international scale.

Documentation of the provenance of stem cell lines is critical if the cell lines are to be widely employed in the research community. Provenance must be easily verifiable by access to relevant informed consent documents and raw primary data regarding genomic and functional characterization.

Institutions engaged in human stem cell research, whether public or private, academic or non-academic, should develop procedures whereby research scientists are granted, without undue financial constraints or bureaucratic impediment, unhindered access to research materials for scientifically sound and ethical purposes, as determined under these guidelines and applicable laws.

Mechanisms for Enforcement

These ISSCR guidelines should be upheld and enforced through standards of academic, professional, and institutional self-regulation.

CLINICAL TRANSLATION OF STEM CELLS

Cell Processing and Manufacture, Sourcing Material

Donor Consent:

In the case of donation of cells for allogeneic use, the donor should give written and legally valid informed consent that covers, where applicable, terms for potential research and therapeutic uses, the return of incidental findings, the potential for commercial application, and other issues.

Donor Screening:

Donors should be screened for infectious diseases and other risk factors, as is done for blood and solid organ donation, and for genetic diseases as appropriate.

Manufacture

Quality Control in Manufacture.

All reagents and processes should be subject to quality control systems and standard operating procedures to ensure the quality of the reagents and consistency of protocols used in manufacturing. For extensively manipulated stem cells intended for clinical application, good manufacturing practices(GMP)procedures should be followed.

Processing and Manufacture Oversight

The degree of oversight and review of cell processing and manufacturing protocols should be proportionate to the risk induced by manipulation of the cells, their source and intended use, the nature of the clinical trial, and the number of research subjects who will be exposed to them.

Components in Culture or Preservation of Cells:

Components of animal origin used in the culture or preservation of cells should be replaced with human or chemically defined components whenever possible.

Release Criteria:

Criteria for release of cells for use in humans must be designed to minimize risk from culture-acquired abnormalities. The final product as well as in-process testing may be necessary for product release and should be specified during the review process.

Repositories and Databases:

Funding bodies, industry, and regulators should work to establish public repositories and databases of clinically useful lines that contain adequate information to determine the lines' utility for particular disease therapy.

PRECLINICAL STUDIES

General Considerations

Animal Welfare

Given that preclinical research into stem cell-based therapeutics makes heavy use of animal models, researchers should adhere to the principles of the three "Rs": reduce numbers, refine protocols, and replace animals with in vitro or non-animal experimental platforms whenever possible.

Preclinical Study Objectives:

Early phase human studies should be preceded by a rigorous demonstration of safety and efficacy in preclinical studies. The strength of preclinical evidence demanded trial launch should be proportionate with the risks, burdens, and ethical sensitivities of the anticipated trial.

Study Validity

All preclinical studies testing safety and efficacy should be designed in ways that support precise, accurate, and unbiased measures of clinical promise. In particular, studies designed to inform trial initiation should have high internal validity; they should be representative of clinical scenarios they are intended to model and they should be replicated.

Safety Studies

Cell Characterization:

Cells to be employed in clinical trials must first be rigorously characterized to assess potential toxicities through studies in vitro and, where possible for the clinical condition and tissue physiology to be examined, in animals.

Tumorigenicity Studies:

Risks for tumorigenicity must be rigorously assessed for any stem cell-based product, especially if extensively manipulated in culture, genetically modified, or when pluripotent.

Biodistribution Studies:

For all cell-based products, whether injected locally or systemically, researchers should perform detailed and sensitive biodistribution studies of cells.

Ancillary Therapeutic Components:

Before launching high-risk trials or studies with many components, researchers should establish the safety and optimality of other intervention components, like devices or co-interventions such as surgeries.

Long-Term Safety Studies:

Preclinical researchers should adopt practices to address long-term risks, and to detect new and unforeseen safety issues.

Potential of Stem Cells for Toxicology:

Researchers, regulators, and reviewers should exploit the potential for using stem cell-based systems to enhance the predictive value of preclinical toxicology studies.

Efficacy Studies

Efficacy Evidence for Initiating Trials:

Trials should generally be preceded by compelling preclinical evidence of clinical promise in well-designed studies. Animal models suited to the clinical condition and the tissue physiology should be used unless there is evidence of efficacy using similar products against similar human diseases.

Small Animal Studies

Small animal models should be used to assess the morphological and functional recovery caused by cell-based interventions, the biological mechanisms of activity, and to optimize implementation of an intervention.

Large Animal Studies

Large animal models should be used for stem cell research when they are believed to better emulate human anatomy or pathology than small animal models and where risks to human subjects in anticipated clinical trials are high.

Transparency and Publication

Sponsors, researchers, and clinical investigators should publish preclinical studies in full and in ways that enable an independent observer to interpret the strength of the evidence supporting the conclusions.

Clinic Al Research

Oversight

Prospective Review.

All research involving clinical applications of stem cell-based interventions must be subject to prospective review, approval, and ongoing monitoring by independent human subjects review committees.

Expert Review of Clinical Research:

The review process for stem cell-based clinical research should ensure that protocols are vetted by independent experts who are competent to evaluate (a) the in vitro and in vivo preclinical studies that form the basis for proceeding to a trial and (b) the design of the trial, including the adequacy of the planned endpoints of analysis, statistical considerations, and disease-specific issues related to human subject's protection.

Standards for Clinic Al Research Conduct

Systematic Appraisal of Evidence

Launch of clinical trials should be supported by a systematic appraisal of evidence supporting the intervention.

Risk-Benefit Analysis:

Risks should be identified and minimized, unknown risks acknowledged, and potential benefits to subjects and society estimated. Studies must anticipate a favorable balance of risks and benefits.

Research Subjects Lacking Consent Capacity:

When testing interventions in human subjects that lack the capacity to provide valid informed consent, risks from study procedures should be limited to no greater than a minor increase over minimal risk unless the risks associated with the intervention are exceeded by the prospect of therapeutic benefit.

Objectives of Trials:

A stem cell-based intervention must aim at ultimately being clinically competitive with or superior to existing therapies or meet a unique therapeutic demand. Being clinically competitive necessitates having reasonable evidence that the nature of existing treatments poses some type of burden related to it that would likely be overcome should the stem cell-based intervention prove to be safe and effective.

Subject Selection:

Individuals who participate in clinical stem cell research should be recruited from populations that are in a position to benefit from the results of this research. Groups or individuals must not be excluded from the opportunity to participate in clinical stem cell research without rational justification. Unless scientifically inappropriate, trials should strive to include women as well as men and members of racial and/or ethnic minorities.

Informed Consent

Informed consent must be obtained from potential human subjects or their legally authorized representatives. The informed consent of subjects must be obtained if substantial changes in risks or benefits of a study intervention or alternative treatments emerge over the course of the research.

Assessment of Capacity to Consent:

Prior to obtaining consent from potential adult subjects who have diseases or conditions that are known to affect cognition, their capacity to consent should be assessed formally.

Research teams must protect the privacy of human subjects.

Patient Sponsored and Pay to Participate Trials.

Patient-sponsored and pay-to-participate trials pose challenges for ensuring scientific merit, integrity, and priority as well as fairness. Accordingly, these financial mechanisms should be used only if they are approved and supervised by a rigorous independent review body that espouses the principles outlined in these guidelines regarding the integrity of the research enterprise, transparency, and patient welfare.

Issues Particular to Early Phase Trials

Consent in Early Phase Trials:

Consent procedures in any pre-licensure phase, but especially early phase trials of stem cell-based interventions, should work to dispel potential research subjects' overestimation of benefit and therapeutic misconception.

Pacing of Testing:

In general, initial tests of a novel strategy should be tested under lower-risk conditions before escalating to higher-risk study conditions even if they are more likely to confer therapeutic benefit.

Maximizing Value:

Researchers should take measures to maximize the scientific value of early phase trials.

Issues Particular to Late Phase Trials

Choice of Comparators:

Clinical research should compare new stem cell-based interventions against the best therapeutic approaches that are currently or could be made reasonably available to the local population.

Placebo and Sham Comparators:

Where there are no proven effective treatments for a medical condition and stem cell-based interventions involve invasive delivery, it may be appropriate to test them against placebo or sham comparators, assuming early experience has demonstrated feasibility and safety of the particular intervention.

Research Subject Follow- Up and Trial Monitoring

Data Monitoring:

An independent data-monitoring plan is required for clinical studies. When deemed appropriate, aggregate updates should be provided at predetermined times or on-demand. Such updates should include adverse event reporting and ongoing statistical analyses if appropriate. Data monitoring personnel and committees should be independent of the research team.

Long-term Follow-up:

Given the potential for transplanted cellular products to persist, and depending on the nature of the experimental stem cell-based intervention, subjects should be advised to undergo long-term health monitoring. Additional safeguards for ongoing research subject privacy should be provided. Subject withdrawal from the research should be done in an orderly fashion to promote physical and psychological welfare.

Autopsy:

To maximize the opportunities for scientific advance, research subjects in stem cell-based intervention studies should be asked for consent to a partial or complete autopsy in the event of death to obtain information about cellular implantation and functional consequences. Requests for an autopsy must consider cultural and familial sensitivities. Researchers should strive to incorporate a budget for autopsies in their trials and develop a mechanism to ensure that these funds remain available over long time horizons if necessary.

Transparency and Reporting of Research Results

Registration:

All trials should be prospectively registered in public databases.

Adverse Event Reporting:

Investigators should report adverse events including their severity and their potential causal relationship with the experimental intervention.

Publication:

Researchers should promptly publish aggregate results regardless of whether they are positive, negative, or inconclusive. Studies should be published in full and according to international reporting guidelines.

Stem Cell-Based Medical Innovation

Provision of Innovative Care:

Clinician-scientists may provide unproven stem cell-based interventions to at most a very small number of patients outside the context of a formal clinical trial and according to the highly restrictive provisions outlined in this section.

Clinical Application

Regulatory Approval:

The introduction of novel products into routine clinical use should be dependent on the demonstration of an acceptable balance of risk and clinical benefit appropriate to the medical condition and patient population for which new treatments are designed.

Bio- and Pharmacovigilance:

Developers, manufacturers, providers, and regulators of stem cell-based interventions should continue to systematically collect and report data on safety, efficacy, and utility after they enter clinical use.

Patient Registries:

Registries of specific patient populations can provide valuable data on safety and outcomes of stem cell-based interventions within defined populations but should not substitute for stringent evaluation through clinical trials prior to introduction into standard care.

Off-label Use:

Off-label uses of stem cell-based interventions should be employed with particular care, given uncertainties associated with stem cell-based interventions.

Access and Economics

Comparative Value for Healthcare Systems and Access Issues.

Stem cell-based interventions should be developed with an eye towards delivering economic value to patients, payers, and healthcare systems

Pricing:

Developers, funders, providers, and payers should work to ensure that cost of treatment does not prevent patients from accessing stem cell-based interventions for life-threatening or seriously debilitating medical conditions.

Communications

Public Representation of Science:

The stem cell research community should promote accurate, balanced, and responsive public representations of stem cell research.

Communications about Clinical Trials:

When describing clinical trials in the media or in medical communications, investigators, sponsors, and institutions should provide balance and not emphasize statistically significant secondary results when pre-specified primary efficacy results are not statistically significant. They should also emphasize that research is primarily aimed at generating systematic knowledge on safety and efficacy, not therapeutic care.

Communications about Clinical Care:

The provision of information to patients on stem cell-based interventions must be consistent with the primacy of patient welfare and scientific integrity

Standards in Stem Cell Research

Standards Development:

Researchers, industry and regulators should work towards developing, and implementing standards on design, conduct, interpretation, and reporting of research in stem cell science and medicine.

These guidelines should be periodically revised to accommodate scientific advances, new challenges, and evolving social priorities.

CRISPR-BASED GENE EDITING

CRISPR (Clustered Regularly Interspaced Short Palindromic Repeats), originally a bacterial adaptive-immune system, has been repurposed into a precise genome-editing tool that lets us modify DNA sequences directly within living organisms. In medicine, it enables targeted gene disruption, correction, or insertion to treat both monogenic and complex diseases far more quickly, cheaply, and efficiently than earlier techniques. (*Gene Editing*, s. f.) Inspired by bacterial antiviral defense mechanisms, CRISPR's wide-ranging applications span basic research to potential therapies for genetic disorders and cancer. Although its potential in urogynecology is considerable, we remain several steps away from clinical translation in this specialty. Its greatest potential may lie in the treatment of urogynecological cancers, which in our field are the diseases most closely associated with genetic mutations, as well as congenital defects.

Anatomical dysfunctions—such as posterior urethral valves, spinal cord maldevelopment with resultant obstructive bladder damage, renal pelvic obstruction, or vesicoureteral reflux from malformed ureterovesical junctions—underlie these conditions, and uncovering the molecular mechanisms of bladder and renal tissue injury, repair, and prevention could one day revolutionize their management.

Targeted regulation of gene expression offers great promise for tumor therapy, but exogenously delivered genetic elements are often neutralized by the intracellular innate immune response (IIR), which diminishes transgene efficiency, adding one more problem to therapies and clinical translation. As the cell's frontline defense, the IIR comprises enzymes and signaling pathways that detect and eliminate foreign DNA. In mammalian cells, DNA is normally confined to the nucleus, so any DNA in the cytoplasm triggers a kinase cascade that suppresses transgene expression. Developing compact CRISPR platforms has eased cellular delivery but often sacrifices editing efficiency, limiting their broader application. Improving CRISPR performance remains a critical challenge with profound biotechnological implications.

To overcome this barrier, some studies introduced a synthetic dual-antibody (ANAM) system, developed to boost transgenic performance by inhibiting NF-κB (Nuclear Factor kappa-light-chain-enhancer of activated B cells) and β-catenin, two key IIR mediators, by dampening these innate immune mediators, boosting CRISPR transgene expression and gene-editing efficiency, while also delivering a marked antitumor effect. Furthermore, selective blockade of other IIR components can enhance gene expression in mammalian cells to varying degrees (Zhan et al., 2024). Another study for renal cell carcinoma (RCC), using a non-integrating lentiviral vector approach can overcome this immune rejection, allow the growth of transduced cells in an immunocompetent host. This technique enables the establishment of a von Hippel-Lindau (VHL) gene knockout RENCA cell line in BALB/c mice, generating an improved model of immunocompetent, metastatic RCC. (Hu et al., 2018)

Mechanism of Action

A Guide RNA (gRNA) duplex molecule, containing a 20-nt sequence, is used as a guide or conductor for the Cas9 (CRISPR- associated protein-9 nucleases) endonuclease, or enzyme, that acts as "molecular scissors" to cut DNA, by inducing a double-strand break (DSB) at the gRNA-specified locus that is damage and need to be removed. For this binding between gRNA and Cas9, a Protospacer Adjacent Motif (PAM) is required, it is a short DNA motif (e.g., "NGG" for Cas9) adjacent to the target site.(*Gene Editing*, s. f.)

The DNA Repair Pathways can be:

- Non-Homologous End Joining (NHEJ): Error-prone ligation often causing insertions/deletions (indels) and gene knockout.
- Homology Directed Repair (HDR): Uses an exogenous donor template to introduce precise base changes or gene inserts.

The delivery methods could be Viral Vectors, such Adeno-Associated Virus (AAV) for in vivo organ targeting or Lentivirus for ex vivo editing of hematopoietic stem cells. Also, Non-Viral Systems can be used as Lipid Nanoparticles (LNPs) for mRNA/gRNA delivery or Electroporation or nucleofection for isolated cells in culture. (Zhan et al., 2024) Gene circuits can be engineered to synthesize immunomodulatory factors with high specificity inside cancer cells. Packaged into AAV vectors, these circuits move seamlessly from in vitro assembly to in vivo delivery, triggering a potent antitumor response. The therapy platform is designed with future clinical translation in mind, but several challenges remain. We must validate circuit function across additional disease models—such as patient-derived tumor cells—and rigorously assess off-target effects and potential side effects in living organisms.

Ortinski et al. (Ortinski et al., 2017) have shown that, while using integrase-deficient lentiviral vectors (NIL) to deliver the CRISPR/Cas9 system and knock out VHL in the murine RENCA renal carcinoma line, NIL matches integration-competent lentivirus in both vector production and Cas9-mediated editing efficiency. Despite the popularity of lentiCRISPR for editing cell lines, few reports describe Cas9-modified clones that can be implanted into immunocompetent mice—likely because stable Cas9 expression can trigger an immune response. (Wang et al., 2015)

To our knowledge, no prior work has demonstrated tumor regression or growth arrest in such hosts. By swapping one plasmid in the transduction workflow for NIL, we generated nonintegrated RENCA clones that successfully engraft in immunocompetent mice without genomic integration.

Beyond monotherapy, the approach could synergize with T-cell treatments, not o only with the use of CRISP, but other cell derived therapies. By driving tumor-specific expression of immune activators, gene circuits can recruit and activate T cells at the tumor site and dismantle the immunosuppressive microenvironment. Combining this strategy with neoantigen-based immunotherapy, gene circuits could selectively express tumor neoantigens to flag cancer cells for the immune system, while secreted immunomodulators enhance antigen presentation by dendritic cells, further amplifying the antitumor response.

3D BIOPRINTING IN REGENERATIVE MEDICINE

The interdisciplinary effort draws on tissue engineering, molecular medicine, biotechnology, and three-dimensional printing, and urologists have played a leading role in translating these advances into clinical possibilities. Researchers continue to refine compatible biomaterials, versatile cell sources, fabrication techniques, and bioactive molecules, all while carefully minimizing risks.

Knowing that the risk of graft rejection is a problem, and that greater genetic compatibility lowers the risk of new tissue rejection, autologous and homologous pluripotent cells have enabled significant advances. Furthermore, the introduction of new biomaterials has also been a crucial element in this progress. The main methods currently used in organ construction are based on in vitro cell expansion, which can be sourced from autologous or allogeneic stem cells. Three-dimensional (3D) printing encompasses a suite of layer-by-layer manufacturing methods that have advanced dramatically since the early 1980s. Improvements in hardware, falling costs, broader material choices, and greater accessibility have accelerated its use in biotechnology. Depending on the chosen technique, 3D printing can rapidly and cost-effectively produce intricate constructs.

Biomaterials, such as connective tissue "scaffolds," guide cell growth and mimic matrices, which is expected to help implanted cells survive; and 3D printing makes it possible to design these structures. Despite all this, despite ongoing efforts and achievements, there are enormous obstacles to regenerative medicine, and there is still a long way to go before it becomes a reality in everyday clinical practice. (García-Perdomo & Jurado-Penagos, 2022)

The key Advantages of 3D Bioprinting are the high resolution, having precisely controls porosity, pore size, nanoarrays, and surface roughness, that enhances nutrient diffusion and host-cell infiltration, improving tissue regeneration compared to mold casting or electrospinning. It also reproduces the unique extracellular matrix (ECM) layouts of urologic organs (kidney, prostate, bladder), mimics organ-specific complexity that traditional methods struggle to achieve.(Xu et al., 2022)

3D printing has rapidly expanded beyond manufacturing into sectors like education, food, fashion, transportation, and healthcare. Within the life sciences, bioprinting applies this technology to tissue engineering, enabling the fabrication of scaffolds with intricate architectures that replicate native organs and, ultimately, support clinical transplantation.(Xu et al., 2022)

Figure 3. 3D printing in urology

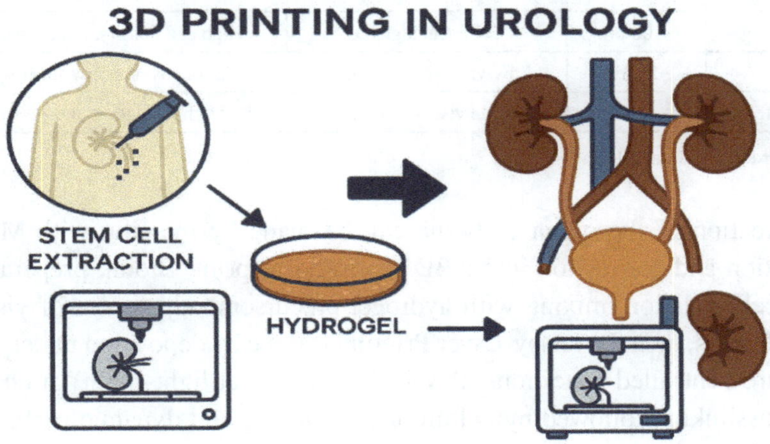

3D bioprinting fabricates living tissues by depositing cell-laden “bioinks” layer by layer according to a digital model. It aims to create functional grafts, organoids, and implantable tissues. These bioink could be a hydrogel matrix (e.g., collagen, alginate, gelatin methacryloyl), Living cells (stem cells, primary parenchymal cells) or Biochemical additives (growth factors, ECM proteins). The Scaffold Materials used for biding this bioinks could be Synthetic polymers (PLGA (poly(lactide-co-glycolide)) and PCL (polycaprolactone)) for mechanical strength or Natural polymers for biocompatibility Skin grafts for burn patients. The potential uses for this technologies in urology are the printing of vascularized organoids, such the kidney or bladder, also for drug screening and disease modeling, which is an important advance in the study of tissues and diseases, and the abolition of animal testing, making it a more ethical technique than many we currently use. (García-Perdomo & Jurado-Penagos, 2022)

Beyond creating bespoke biomaterials, 3D printing turns medical imaging—CT (computed tomography) scans and MRI (magnetic resonance imaging)—into tangible anatomical models. Surgeons can rehearse complex procedures on accurate replicas that reveal an organ’s architecture, blood supply, and nerve pathways, improving preoperative planning and patient outcomes.

Currenlty in urology, 3D printing has been applied to the male urethra recreation, mimicking a tubular structure about 18–20 cm long composed of a shifting epithelial lining, fibroblasts, and smooth muscle. Urethral reconstruction remains challenging because of its shape and cellular and immunity complexity. Approaches include grafting cultured autologous cells, implanting natural or synthetic biomaterial scaffolds, and combining both by seeding scaffolds with patient-derived cells.

Table 1. Printing technologies

Technology	Resolution	Typical Bioink	Advantages
Extrusion	~100 µm	Collagen, alginate	High cell density, multi-material
Inkjet	~50 µm	Low-viscosity hydrogels	Fast deposition, low shear stress
Laser-Assisted	~10 µm	Wide viscosity range	High resolution, nozzle-free

(García-Perdomo & Jurado-Penagos, 2022)

The creation of the organ or tissue can be made by imaging with MRI/CT segmentation and conversion into a 3D scaffold blueprint. Bioink preparation is done by cell isolation, mixing with hydrogel precursors, adjustment of viscosity and crosslinkers, then, Layer-by-Layer Printing is done by deposition under sterile, temperature-controlled conditions. Physical (Ultra violet lights light) or chemical (ionic) crosslinking followed by culture in bioreactors and dynamic perfusion or mechanical stimulation to enhance tissue maturation.

Bioprinting Modalities

Cell-Laden Bioprinting

Delivers living cells directly into the printed construct, creating tissue-like architectures with native cell distributions. One example is the use a hyaluronic acid–based scaffold printed with glioblastoma cells recreated mechanical and biological features of the human brain microenvironment for invasion studies and drug screening, another a bilayer nerve conduit, fabricated with a multi-nozzle printer, incorporated BMSC internally and PC-12 cells externally to support peripheral nerve repair.

Cell-Free Bioprinting

Prints acellular scaffolds with a wider choice of biomaterials and higher mechanical strength; cells are seeded post-print, avoiding stress to living cells during fabrication. It has bein use for printing a hierarchical hydroxyapatite/tricalcium phosphate scaffold, then seeded with NIH3T3 cells, fibroblast cell line that was isolated from a mouse NIH/Swiss embryo, demonstrating efficacy in bone regeneration. Also combining 3D printing with freeze-casting produced scaffolds featuring microscale pores that successfully supported breast cancer cell growth.

These advances underscore how 3D bioprinting is revolutionizing tissue engineering, offering unprecedented control over scaffold architecture, material composition, and patient-specific design.

Emerging Trends and Integration

By integrating CRISPR with 3D bioprinting, we can perform in situ gene editing of printed tissues to improve engraftment and immune compatibility, couple bioprinted constructs with microfluidic platforms for real-time biosensing, and ultimately achieve personalized medicine by printing autologous grafts derived from CRISPR-edited, patient-specific cells.

Figure 4. Integration of CRISP and 3D printing technologies in urology

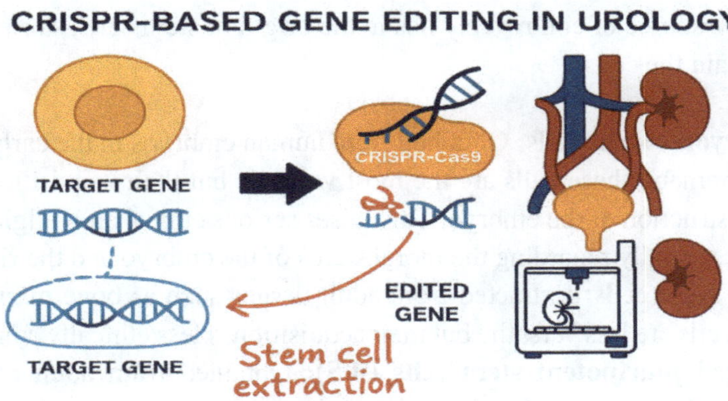

To move this vision forward, researchers are focusing on optimizing bioink formulations that support both cell viability and CRISPR delivery, refining microfluidic–bioprinter interfaces for seamless monitoring, and developing nonviral vectors—such as lipid nanoparticles—that remain functional throughout the printing process.

Because our system confines therapeutic activity to tumors, it offers a versatile platform for probing cancer biology. Moreover, synthetic gene circuits can detect and respond to disease-specific markers beyond the cancer setting. By swapping promoter inputs and effector outputs, this sensing-and-response framework can be adapted to program precise, multifactorial immune functions for a wide range of complex diseases.

A PROMISE WITH CONTROVERSY

SC have generated great excitement in the field of regenerative medicine. Their ability to transform into different types of cells makes them a promising tool for treating a wide range of diseases, from spinal cord injuries to neurodegenerative diseases like Parkinson's, that can affect the bladder and micturition quality. However, their use has been surrounded by intense controversy.

Why So Much Controversy?

The main source of controversy lies in the origin of stem cells, staring with the way we obtain the cells:

- **Embryonic stem cells:** Obtained from human embryos in the early stages of development, these cells are the most versatile but their acquisition involves the destruction of the embryo. This raises serious ethical and religious dilemmas, especially regarding the moral status of the embryo and the right to life.
- **Adult stem cells:** Extracted from adult tissues such as bone marrow or fat, these cells are less versatile but their acquisition is less ethically controversial.
- **Induced pluripotent stem cells (iPS):** Obtained from adult cells reprogrammed to acquire the characteristics of embryonic cells, these cells offer a promising alternative, but their development and safety are still under investigation.

The Main Points of Debate

- **The moral status of the embryo:** Some moral controversies could be proposed by a part of the population, such as the consideration of the human embryo as an individual, with a right to life from the moment of conception. This question could become fundamental for a certain part of the researches in determining whether research with embryonic stem cells is ethically acceptable.
- **Therapeutic potential:** Although stem cells hold great promise, the results of clinical trials so far have been mixed. The long-term efficacy and safety of stem cell treatments are not yet fully established.
- **Associated risks:** Stem cell research carries potential risks, such as tumor formation and the transmission of infectious diseases.
- **Tumorigenicity**: A significant concern is the potential for stem cells to form tumors. There have been instances where transplanted stem cells have led to

uncontrolled growth, raising safety concerns. Rigorous testing and monitoring are essential to mitigate this risk.

- **Social implications:** The development of stem cell-based technologies raises a number of social issues, such as access to these treatments, economic implications, and potential non-therapeutic applications.
- **Animal warfare:** Similar to the moral debate surrounding the status of a human embryo, the use of animals in experimentation raises significant ethical concerns. These animals, unable to provide consent, are subjected to irreparable harm and often death. Compounding this issue is the fact that a substantial majority (90%) of preclinical animal studies fail to translate effectively to human trials due to the inherent biological differences previously discussed. (Dawkins, 2012)

Specific Controversial Research Areas in Urology

- **Bladder Regeneration:** While stem cell therapy shows promise for repairing bladder tissue, there are concerns about the long-term functionality of the regenerated tissue and the potential for tumor formation.
- **Prostate Cancer:** Stem cell-based therapies could be use for the side effect of the treatments for prostate cancer such urinary incontinence, with questions about their efficacy and safety, particularly in terms of potential side effects and tumo recurrence.
- **Kidney Repair:** Stem cell-based treatments for kidney disease are also in development, but challenges include the complex structure of the kidney and the potential for immune rejection.

To progress in cell therapy development for urogynaecological conditions, a comprehensive understanding of underlying mechanisms is crucial for optimizing treatment protocols. Rigorous, standardized preclinical and clinical studies are essential to establish efficacy, durability, and overall therapeutic potential. It is crucial to approach SC research in urology with caution and ethical consideration. While the potential benefits are significant, addressing the controversies and ensuring patient safety are paramount. (Aghamir, 2020; González Enguita et al., 2024)

The Current Landscape

Despite the controversies, stem cell research continues to advance rapidly. Many countries have established specific value regulations for this area of research, seeking to balance potential therapeutic benefits with ethical considerations.

The regulatory landscape for stem cell therapies is complex and evolving. Balancing the need for innovation with patient safety is a challenge for regulatory authorities. In conclusion, stem cells represent one of the most promising and controversial areas of biomedical research today. Resolving these controversies will require open and multidisciplinary dialogue among scientists, doctors, ethicists, lawmakers, and society as a whole.

CONCLUSION

Regenerative therapy in urology presents a promising horizon for treating a range of debilitating conditions. The potential to regenerate damaged tissues and restore normal function is undeniably captivating. However, the path to realizing this potential is fraught with challenges. Ethical considerations, primarily centered on the source of SC and their potential impact on human embryos, have sparked intense debate. Animal welfare concerns are also paramount as preclinical research is essential for advancing this field, but there are promising solutions created for solving this issue, like the direct study in human cells, that in the future, can avoid the use of animals for experimentation.

Despite these hurdles, the pursuit of safe and effective stem cell therapies for urological diseases remains a compelling goal. Rigorous scientific inquiry, coupled with a steadfast commitment to ethical principles and animal welfare, will be crucial in guiding this research. As our understanding of stem cell biology deepens, we can anticipate significant advancements in the treatment of urological disorders, ultimately improving the quality of life for countless patients.

It is imperative to approach this field with a balanced perspective, acknowledging both the potential benefits and the associated challenges. By fostering collaboration between scientists, ethicists, policymakers, and the public, we can navigate the complexities of stem cell research and harness its full potential for the betterment of human health.

By compiling the latest research, clinical data, and expert insights, we hope to offer a valuable resource for clinicians, researchers, and students in the fields of urology, gynecology, and regenerative medicine. As we stand on the cusp of a new era in regenerative medicine, this book seeks to illuminate the path forward, highlighting both the immense potential and the challenges that lie ahead in harnessing the power of stem cells to transform patient care in urology and gynecology.

REFERENCES

ACOG Committee on Ethics. (2006). ACOG Committee Opinion No. 347, November 2006: Using preimplantation embryos for research. *Obstetrics and Gynecology*, *108*(5), 1305–1317. DOI: 10.1097/00006250-200611000-00057 PMID: 17077265

Aghamir, S. M. K. (2020). Introduction. In Aghamir, S. M. K. (Ed.), *Stem Cells in Urology* (pp. 1–6). Springer International Publishing., DOI: 10.1007/978-3-030-41476-4_1

Alwaal, A., Hussein, A. A., Lin, C. S., & Lue, T. F. (2015). Prospects of stem cell treatment in benign urological diseases. *Korean Journal of Urology*, *56*(4), 257–265. DOI: 10.4111/kju.2015.56.4.257 PMID: 25874038

ASRM. (2020). *Ethics in embryo research: A position statement by the ASRM Ethics in Embryo Research Task Force and the ASRM Ethics Committee (2020)* [2020]. https://www.asrm.org/practice-guidance/ethics-opinions/ethics-in-embryo-research-a-position-statement-by-the-asrm-ethics-in-embryo-research-task-force-and-the-asrm-ethics-committee-2020/

Bhartiya, D., Shaikh, A., Anand, S., Patel, H., Kapoor, S., Sriraman, K., Parte, S., & Unni, S. (2016). Endogenous, very small embryonic-like stem cells: Critical review, therapeutic potential and a look ahead. *Human Reproduction Update*, *23*(1), 1–36. DOI: 10.1093/humupd/dmw030 PMID: 27614362

Dawkins, M. (2012). *Animal Suffering: The Science of Animal Welfare*. Springer Science & Business Media.

Hernandez, A.B. (s. f.). *STEM CELLS THERAPY MANUAL*. ADSCC.

ESHRE. (s. f.). *Documents of the Task Force Ethics Law*. ESHRE Ethical Considerations. Recuperado 6 de julio de 2025, de https://www.eshre.eu/Specialty-groups/Special-Interest-Groups/Ethics-and-Law/Documents-of-the-Task-Force-Ethics-Law

García-Perdomo, H. A., & Jurado-Penagos, A. (2022). Application of regenerative medicine and 3d bioprinting in urology. *Actas Urológicas Españolas (English Ed.)*, *46*(6), 323–328. DOI: 10.1016/j.acuroe.2022.03.006 PMID: 35660078

Gene Editing. (s. f.). CRISPR Therapeutics. Recuperado 5 de julio de 2025, de https://crisprtx.com/gene-editing

González Enguita, C., Garranzo García-Ibarrola, M., Tufet, I., Jaumont, J. J., Garde García, H., González López, R., Quintana Franco, L. M., Torres Zambrano, G. M., & García-Arranz, M. (2024). Cell Therapy in the Treatment of Female Stress Urinary Incontinence: Current Status and Future Proposals. *Life (Chicago, Ill.)*, *14*(7), 7. Advance online publication. DOI: 10.3390/life14070861 PMID: 39063615

Guidelines for Stem Cell Research and Clinical Translation. - Search Results—PubMed. (s. f.). Recuperado 28 de marzo de 2021, de https://pubmed.ncbi.nlm.nih.gov/?term=Guidelines+for+Stem+Cell+Research+and+Clinical+Translation

Guidelines—International Society for Stem Cell Research. (2021). https://www.isscr.org/guidelines

HFEA. (2019). *HFEA: UK fertility regulator.* Human Fertilisation & Embryology Authority (HFEA). https://www.hfea.gov.uk/

Hu, J., Schokrpur, S., Archang, M., Hermann, K., Sharrow, A. C., Khanna, P., Novak, J., Signoretti, S., Bhatt, R. S., Knudsen, B. S., Xu, H., & Wu, L. (2018). A Non-integrating Lentiviral Approach Overcomes Cas9-Induced Immune Rejection to Establish an Immunocompetent Metastatic Renal Cancer Model. *Molecular Therapy. Methods & Clinical Development*, *9*, 203–210. DOI: 10.1016/j.omtm.2018.02.009 PMID: 29766028

Kucia, M., Halasa, M., Wysoczynski, M., Baskiewicz-Masiuk, M., Moldenhawer, S., Zuba-Surma, E., Czajka, R., Wojakowski, W., Machalinski, B., & Ratajczak, M. Z. (2007). Morphological and molecular characterization of novel population of CXCR4+ SSEA-4+ Oct-4+ very small embryonic-like cells purified from human cord blood—Preliminary report. *Leukemia*, *21*(2), 297–303. DOI: 10.1038/sj.leu.2404470 PMID: 17136117

Kucia, M., Reca, R., Campbell, F. R., Zuba-Surma, E., Majka, M., Ratajczak, J., & Ratajczak, M. Z. (2006). A population of very small embryonic-like (VSEL) CXCR4+ SSEA-1+Oct-4+ stem cells identified in adult bone marrow. *Leukemia*, *20*(5), 857–869. DOI: 10.1038/sj.leu.2404171 PMID: 16498386

Lahlil, R., Scrofani, M., Barbet, R., Tancredi, C., Aries, A., & Hénon, P. (2018). VSELs Maintain their Pluripotency and Competence to Differentiate after Enhanced Ex Vivo Expansion. *Stem Cell Reviews and Reports*, *14*(4), 510–524. DOI: 10.1007/s12015-018-9821-1 PMID: 29736843

Ortinski, P. I., O'Donovan, B., Dong, X., & Kantor, B. (2017). Integrase-Deficient Lentiviral Vector as an All-in-One Platform for Highly Efficient CRISPR/Cas9-Mediated Gene Editing. *Molecular Therapy. Methods & Clinical Development*, *5*, 153–164. DOI: 10.1016/j.omtm.2017.04.002 PMID: 28497073

Pérez-Aizpurua, X., Garranzo-Ibarrola, M., Simón-Rodríguez, C., García-Cardoso, J. V., Chávez-Roa, C., López-Martín, L., Tufet i Jaumot, J. J., Alonso-Román, J., Maqueda-Arellano, J., Gómez-Jordana, B., Ruiz de Castroviejo-Blanco, J., Osorio-Ospina, F., González-Enguita, C., & García-Arranz, M. (2023). Stem Cell Therapy for Erectile Dysfunction: A Step towards a Future Treatment. *Life (Chicago, Ill.)*, *13*(2), 502. Advance online publication. DOI: 10.3390/life13020502 PMID: 36836859

Ratajczak, M. Z., Ratajczak, J., & Kucia, M. (2019). Very Small Embryonic-Like Stem Cells (VSELs): An Update and Future Directions. *Circulation Research*, *124*(2), 208–210. DOI: 10.1161/CIRCRESAHA.118.314287 PMID: 30653438

Ratajczak, M. Z., Zuba-Surma, E., Wojakowski, W., Suszynska, M., Mierzejewska, K., Liu, R., Ratajczak, J., Shin, D. M., & Kucia, M. (2014). Very small embryonic-like stem cells (VSELs) represent a real challenge in stem cell biology: Recent pros and cons in the midst of a lively debate. *Leukemia*, *28*(3), 473–484. DOI: 10.1038/leu.2013.255 PMID: 24018851

Suman, S., Domingues, A., Ratajczak, J., & Ratajczak, M. Z. (2019). Potential clinical applications of stem cells in regenerative medicine. In *Advances in Experimental Medicine and Biology* (Vol. 1201, pp. 1–22). Springer., DOI: 10.1007/978-3-030-31206-0_1

Torres Zambrano, G. M., Valverde, C. A. V., Hernandez, A. B., Hadi, L. A., Rivero, R. A., & Carmenate, Y. V. (2020). Renal Involvement in Patients with COVID-19 Pneumonia and Outcomes After Stem Cell Nebulization. *medRxiv*, 2020.12.16.20236877. DOI: 10.1101/2020.12.16.20236877

Wang, D., Mou, H., Li, S., Li, Y., Hough, S., Tran, K., Li, J., Yin, H., Anderson, D. G., Sontheimer, E. J., Weng, Z., Gao, G., & Xue, W. (2015). Adenovirus-Mediated Somatic Genome Editing of Pten by CRISPR/Cas9 in Mouse Liver in Spite of Cas9-Specific Immune Responses. *Human Gene Therapy*, *26*(7), 432–442. DOI: 10.1089/hum.2015.087 PMID: 26086867

Wiafe, B., Metcalfe, P. D., & Adesida, A. B. (2015). Stem Cell Therapy: Current Applications and Potential for Urology. *Current Urology Reports*, *16*(11), 77. DOI: 10.1007/s11934-015-0551-5 PMID: 26385812

Xu, K., Han, Y., Huang, Y., Wei, P., Yin, J., & Jiang, J. (2022). The application of 3D bioprinting in urological diseases. *Materials Today. Bio*, *16*, 100388. DOI: 10.1016/j.mtbio.2022.100388 PMID: 35967737

Zhan, T., Li, X., Liu, J., & Ye, C. (2024). CRISPR-based gene expression platform for precise regulation of bladder cancer. *Cellular & Molecular Biology Letters*, *29*(1), 66. DOI: 10.1186/s11658-024-00569-7 PMID: 38724931

Chapter 2
Andrology

ABSTRACT

This chapter explores the application of stem cells in andrology, with a focus on their potential to address conditions such as erectile dysfunction and Peyronie's disease. It provides a thorough review of current research, including both human and animal studies, and examines how stem cell therapies are being used to enhance treatment outcomes in these areas. The chapter discusses key findings related to the improvement of vascular tissues, like the one that forms the corpus cavernosum, through regenerative treatments, highlighting the advances in stem cell technology that hold promise for future clinical applications, and evaluates how stem cell therapies are contributing to better clinical outcomes and discusses their role in enhancing treatment options for these conditions. this chapter synthesizes the main articles and studies in the field, offering insights into the evolving role of stem cells in managing and treating andrological conditions. It aims to inform specialists on the developments and benefits of incorporating stem cell-based approaches into clinical practice.

ERECTILE DYSFUNCTION

Erectile dysfunction (ED) is defined as the inability to achieve and maintain an erection sufficient for satisfactory sexual performance and is the most common sexual disorder, showing an increasing incidence. ED has a prevalence of about 50% in men between the ages of 40 and 70 years and a full DE prevalence of 15% in the same age group. The recent increase in modifiable risk factors for erectile dysfunction, such as cardiovascular disease, atherosclerosis, metabolic syndrome, diabetes mellitus, and prostate cancer, has led to an increase in the prevalence of erectile dysfunction

DOI: 10.4018/979-8-3693-8829-7.ch002

in recent years. Some predictive models show a projection of 322 million cases of ED by 2025. (Pérez-Aizpurua et al., 2023; Vakalopoulos et al., 2018)

In the absence of true causal treatment, Phosphodiesterase type 5 (PDE5) inhibitors are widely prescribed. In cases with limitations in efficacy and safety, patients tend to invasive treatment options, such as intraurethral alprostadil, intracavernous injections, vacuum devices, and penile prostheses. In recent years, the increased incidence ED has encountered an increase in new treatments. Some of these treatment modalities are considered restorative therapy because they can reverse the underlying changes that have caused the patient's ED.

The goal of many of these proposed options has been to cure or decrease symptomatology without causing unnecessary side effects. These restorative therapies include shock wave therapy, injectable platelet-rich plasma, and SC therapy (SCT). (Patel et al., 2019) SCT is a versatile new therapy, used in various specialties. Within cardiovascular disease, particularly heart failure, bone marrow-derived stem cells have been used for cardiomyocyte regeneration and repair. From a neurological point of view, SCT is under clinical study for use in the treatment of Parkinson's disease and dementia. In traumatology, SCT is used for the regeneration of tendons, bones, and cartilage. With the same underlying benefits that SCT offers in the treatment of diseases in other fields, SCT is being investigated in the treatment of ED.

It has been proposed that SC, through its paracrine effect, can secrete growth factors that favor the regeneration of the affected tissue. This is achieved by its chemotactic, anti-inflammatory, regenerative, angiogenetic, and anti-apoptotic properties of excreted chemokine.

Improved Vascular Function and Endothelium of the Cavernous Body

Impaired endothelial function is an important factor in the development and progression of ED. This endothelial dysfunction results in decreased synthesis of endothelial nitric oxide, changes in the environment, increased oxidative stress, decreased levels of growth factors, and cytokines and elevated levels of vasoconstrictors.

Each of these factors can produce erectile dysfunction on its own. Nitric oxide, a key substance, produced by endothelium cells, that mediates the relaxation of the cavernous body by forming cyclic guanosine monophosphate. Free radicals may be responsible for decreasing nitric oxide synthase in men with ED and SC could reverse this reduction by secreting various types of cell growth factors that can improve the microenvironment of the cavernous body.

Figure 1. Mechanism of erection and pathophysiology of erectile dysfunction. AMP: adenosine monophosphate. cAMP: cyclic adenosine monophosphate. ATP: adenosine triphosphate. Ca: Calcium. GPT: guanosine triphosphate. cGPT: cyclic guanosine triphosphate

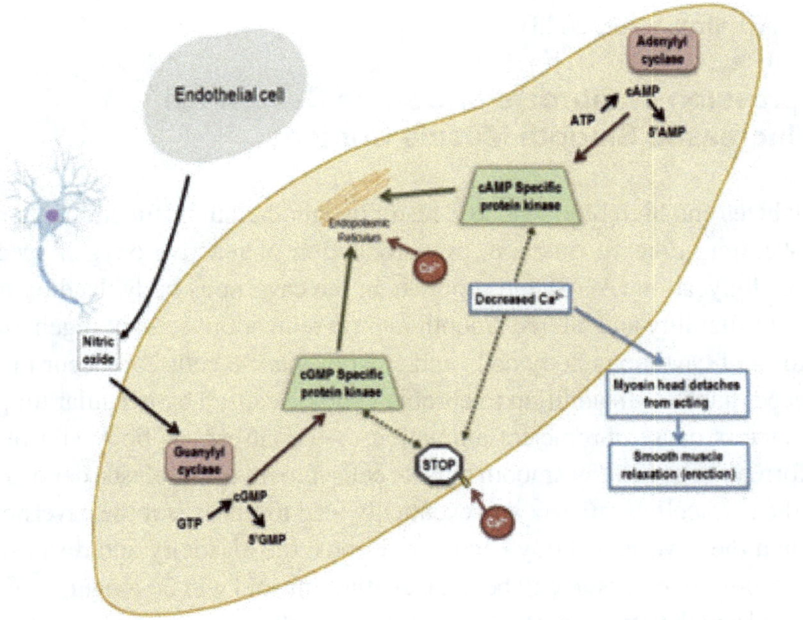

(Pérez-Aizpurua et al., 2023)

The study published by Wang et al.(X. Wang et al., 2017)(B. Wang et al., 2018) has examined SC preconditioned in hypoxia for the treatment of rats with diabetes-induced ED. Some cytokines related to vascular protection, such as vascular endothelial growth factor (VEGF) and Ang-1, have been shown to be not significantly regulated in these pre-conditioned SC compared to normoxia. The results suggested that Adipose Stem Cells (ASCs) may promote penile vascularity. Jeon et al used factor 1 derived from stromal cells that expresses genetically modified mesenchymal stem cells (MSCs) to treat rats with ED. Their results indicated that as the expression of factor 1 derived from stromal cells increased, recovery in rats with ED was improved.

After injecting rats with MSC, the expression of the endothelial cell and platelet adhesion molecule increased in the cavernous body, indicating a high level of vascularization in the damaged tissues. The results of another study showed that the intracavernosal injection (ICI) of ASC markedly increased the expression of VEGF in the penis of DMED rats compared to the non-injected control group. This could contribute to the revascularization in the cavernous body of diabetic rats injected with ASC and therefore lead to improvements in the endothelium.(Yan et al., 2020)

A literature review by Pérez-Aizpurua et al, on SCT for the treatment of ED, concluded that it has potential future applications. However, most urological societies agree that further research is required to conclusively prove its potential benefit. (Pérez-Aizpurua et al., 2023)

Suppression of Fibrosis of Corpus Cavernous and Increased Smooth Muscle Content

Diabetes and high blood glucose result in endothelial dysfunction, altered function of nitrergic nerve function, overproduction of reactive oxygen species, and advanced glycation (AGE) end products in the cavernous body, leading to reduce NO bioavailability and altered smooth muscle relaxation. Loss of regeneration and apoptosis of cavernous body cells and smooth muscle cells can occur in ED with advanced diabetes. In addition to chronic hypoxia caused by irregular tumescence, these factors cause chronic inflammation of the cavernous body and induce the transformation of healthy smooth muscle cells into myofibroblasts, deposit proteins from the extracellular matrix and eventually lead to fibrosis of the cavernous body.

When the cavernous body becomes fibrous, the elasticity and distensibility of the cavernous body tissue will be reduced, the sinusoid will no longer be effectively congested and the cavernous body will subsequently cause insufficient compression of the subtunic venules, causing blood to leak during erection. Leaks make the penis completely stiff, resulting in ED.

There are studies studying the use of MSCs and the regulation of certain key mediators in the mechanism of erection, including intracellular nitric oxide and calcium concentrations. Regulation of both nitric oxide and calcium intracellularly via the action of various transmembrane transport ionic channels is believed to be the therapeutic target of stem cell therapy in ED; however, the exact mechanism is still unknown. (Pérez-Aizpurua et al., 2023) One study reported that allogeneic ASCs improved smooth muscle recovery and reduced cavernous body fibrosis in an STZ-induced rat diabetes model and reversed ED in rats. (X. Y. Wang et al., 2015)

Another study reported that VEGF-modified ASCs could reverse diabetic ED in rats more effectively. The study found that the relationship in the cavernous body between smooth muscle and collagen, together with the expression of nitric oxide ligase, was improved in the group treated with VEGF transfected ASCs compared to the group treated with non-transfected. (G. Liu et al., 2013) Similarly, another study evaluated the effectiveness of ASCs expressing hepatocyte growth factor in a diabetic rat model. Transfected ASCs with hepatocyte growth factor increased the expression of molecule 1 adhesion of platelets and endothelial cells and Endothelial Nitric Oxide Synthase (eNOS), improved smooth muscle content, improved smooth-collagen muscle ratio, and decreased apoptosis rate, resulting in reduced fibrosis and improved erectile function. (T. Liu et al., 2015)

In general, these studies suggest that the use of ASC is an effective antifibrotic strategy for the treatment of cavernous body fibrosis. However, the precise mechanism of ASC anti-fibrosis has not been fully elucidated. Because ASCs are a type of MSC, the antifibrotic mechanism may refer to the previous Chitosan Microspheres (CSM) study. Studies have shown that MSCs can affect innate and adaptive immune cells, resident fibroblasts, and smooth muscle cells, as well as various pro-fibrotic and antifibrotic factors, limit the host's response to injury through immunomodulation, and mitigate the transition of fibroblast-producing Extracellular Matrix (ECM). In addition, stem cells can directly modulate the composition of ECM by secreting many matrix regulatory enzymes.

Neurological Improvement of the Cavernous Body

Another important factor leading to erectile dysfunction of diabetes is diabetic neuropathy. The incidence of peripheral and autonomic neuropathy in men with diabetic ED is reported to be significantly higher than in men with diabetes with normal erectile function. Diabetic ED can also occur before symptoms of neuropathy appear.

Neuropathy in patients with diabetic erectile dysfunction is mainly characterized by neurodegeneration and decreased activity of neuronal nitric oxide ligase. Neuropathy has also been reported to occur in cell bodies and nerve axons in diabetic rats. In the initial stage of diabetes, there were purely functional lesions on the nitrergic nerves. However, with the progression of diabetes, the damage became irreversible. Therefore, it is crucial to maintain the integrity of the nitrergic nerve for normal erectile function.

In the year 2004, Bochinski et al. first described the potential benefit of neural crest-derived stem cells in a rat model with ED of neurogenic origin. (Bochinski et al., 2004; Pérez-Aizpurua et al., 2023) Yang et al. study reports that ASCs promote nerve regeneration, inhibit apoptosis and eventually restore erectile function by promoting recovery of bioactive factors such as nerve growth factor, VEGF, and neurturin. (R. Yang et al., 2015) In 2016, Lu et al found that overexpression of the pigment epithelium-derived factor (PEDF) in a high glucose environment can induce the neuroprotection of ASCs by transfecting PEDF into ASC. (Lu et al., 2016)

A 2017 Hou et al published a meta-analysis on the use of adipose tissue-derived stem cells (ADSCs) in rat models of erectile dysfunction, revealing that ADSCs can regenerate erectile tissues thanks to their paracrine effect, including neurotrophic factors, such as brain-derived, vascular endothelial growth factor (VEGF) and hepatocyte growth factor. (Hou et al., 2017; Pérez-Aizpurua et al., 2023)

In 2017, Zhou et al noticed that ASC's traditional single-celled injection strategy had a low ASC retention rate due to cell loss and low survival rate.(Zhou et al., 2017) So, they turned ASCs into ASC-based micro tissues (MTs) using the drop pendant method. In investigating the difference between the effectiveness of ASC and MT treatments for diabetes mellitus erectile dysfunction (DMED), they found that the high rate of ASC-based MT retention can improve the neuroprotection of ASCs in the treatment of DMED.

Reducing Inflammation and Apoptosis

Elevated levels of apoptosis and inflammatory response in erectile tissue have been linked to diabetes mellitus. Zhou et al used the diabetic rat model to adapt ASC combined with an insulin treatment strategy and found that the combined use of insulin and ASC can further regulate inflammation and reduce the accumulation of end products of sugar metabolism in the penis. Combined therapy is more effective at improving erectile function than a single treatment strategy. (Zhou et al., 2016)

In 2010, García et al randomly divided 22 male diabetic obese rats into 2 groups, all of whom had bilateral cavernous nerve injury. (Garcia et al., 2010) The experimental group was injected with ASC into the cavernous body and the control group was injected with phosphate-buffered saline. The results of apoptosis showed that the apoptosis rate of the experimental group was significantly lower than that of the control group. The mechanism may not be the direct differentiation of ASCs in endothelial cells, but through indirect pathways, including paracrine effects.

According to a study by Lu et al, tunnel staining of apoptotic cells in the cavernous body revealed that the significant increase in diabetes-induced apoptosis improved by injecting ASC, and further improved with PEDF-transfected ASCs. (Lu et al., 2016) Similar results are described in others meta-analysis, like he one of Park et al. in published in 2019 (analyzing 19 studies with the aim of evaluating the effects of ADSC) and the one from Wani et al. (including 29 animal studies and three human trials), with a significant improvement in histological and molecular parameters, with a better response to nitric oxide synthase (NOS), and measures of cyclic guanosine monophosphate (cGMP) and smooth muscle-to-collagen ratio. (Park et al., 2019; Pérez-Aizpurua et al., 2023)

All these data have prompted the study of augmented therapies: SC is incubated in vitro with growth factors or drugs for ED to enhance their action. Combination therapy, which uses SC and low-energy shock waves (LESW), has proven to be an effective treatment option. LESW is thought to activate not only exogenously administered SCs, but also endogenous SCs of the penis, which are present in subtunic and perisinusoidal spaces.

Animal Studies

Numerous models have been conducted that simulate different types of human erectile dysfunction. The most widely applied models are chronic DE induced by diabetes mellitus and acute ED induced by cavernous nerve injury (CNI). In short, the most used technique was a direct injection of SC in experimental models (single-celled therapy). Stromal vascular factor (SVF) and ADSC are equally effective at regaining erectile function in rats with CNI. Their findings provided a new option for SC therapies, as ADSC cultivation has several disadvantages, such as high cost, risk of contamination, and the need for advanced facilities. MTs seem to improve ADSC retention in the penis and improve erectile function even more than traditional ADSC injection treatments, possibly due to their high retention rate or induction.

Feng et al. observed a reduction in oxidative stress and iron levels in tissues in diabetic rats treated with UCSC. In addition to an overexpression of ferroptosis inhibitor genes SLC7A11 and GPX4, concluding that CMs could exert an attenuation of ferroptysis induced by diabetes. (Feng et al., 2022; Pérez-Aizpurua et al., 2023) Likewise, Luo et al. published a study with diabetic rats treated with ADSC, observing a suppression of pyroptosis mediated by NLRP3, which could be one of the causes of ED secondary to DM. (Luo et al., 2022; Pérez-Aizpurua et al., 2023)

The downregulation of markers such as TGF-beta, alpha-SMA and collagen has also been seen with the use of intraperitoneal UCSC in a preclinical study of streptozotozin-induced diabetic rats. (Mukti et al., 2022; Pérez-Aizpurua et al., 2023)

Table 1. Animal studies

First author	Experimental model	Cell type	Outcome
Albersen (Albersen et al., 2010)	CNI	ADSC or ADSC lysate	Improved ICP/MAP
Bivalacqua (Bivalacqua et al., 2007)	ED (age)	BMSC ± eNOS	Improved ICP/MAP
Cengiz (Cengiz et al., 2017)	DM	HUC SC	Improved ICP/MAP
Chen(X. Chen et al., 2016)	CNI	ADSC	Improved ICP/MAP
Chen (F. Chen et al., 2017)	DM	ADSC or ADSC exosomes	Improved ICP/MAP
Gokce(Gokce et al., 2014a)	PD	ADSC	Improved ICP/MAP
Huang(Huang et al., 2010)	ED (tobacco)	ADSC	Improved ICP/MAP

continued on following page

Table 1. Continued

First author	Experimental model	Cell type	Outcome
Jeon(Jeon et al., 2016)	CNI	ADSC ± LISW	Improved ICP/MAP
Kendirci (Kendirci et al., 2010)	CNI	BMSC	Improved ICP/MAP in p75 group
Lin (H. Lin et al., 2016)	CNI	ADSC ± magnetic nanoparticles	No significant difference in nanoparticle group
Martinez-Salamanca(Martínez-Salamanca et al., 2016)	CNI	BMSC ± PDE5	Improve ICP/MAP in combined group
Qiu (Qiu et al., 2012)	CNI (radiation)	ADSC	Improved ICP/MAP
Ryu (Ryu et al., 2016)	CNI	BMSC	Improved ICP/MAP
Ryu 15(Ryu et al., 2016)	DM	BMSC	Improved ICP/MAP
Shan (Shan et al., 2017)	DM	BMSC ± LISW	Improve ICP/MAP in combined group
Song(Song et al., 2014)	CNI	SVF	Improved ICP/MAP
Sun (Sun et al., 2017)	DM	B cell lymphom-2 BMSC	Improved ICP/MAP
Woo (Woo et al., 2011)	CNI	MDSC	Improved ICP/MAP
Xu (Xu et al., 2014)	CNI	ADSC micro tissue	Improved ICP/MAP
Yang (Q. Yang et al., 2016)	CNI	Urine-derived SC	Improved ICP/MAP
Yang(R. Yang et al., 2015)	CNI	ADSC	Improved ICP/MAP
You(You et al., 2015)	CNI	SVF or ADSC	Improved ICP/MAP

Vakalopoulos et al., 2018

Human Studies

Of the 23 studies evaluating the safety and effectiveness of CM therapy in the treatment of ED registered on Clinical-Trials.gov since 2010, 9 have been completed. Four studies were opened. None of the studies provided information on the cost of treatment for patients included in each trial. The main SCs used include bone marrow-derived cells (BMSCs) and ADSCs, followed by umbilical cord (UCSCs) and placenta (PSCs)-derived MSCs. In most of the published studies, the main used route was the direct ICI of SC, either by autologous or allogeneic transplantation. Alternatively, SVF has been used, which is isolated from the SC obtained.

Most of the studies are phase I/II pilot clinical trials, the main objective of was to evaluate the tolerability and safety profile of regenerative therapies and their overall effect on erectile function. In general, no serious adverse effects derived from the ICI of MSCs were found. (Pérez-Aizpurua et al., 2023)

These studies included a total of 61 patients, with an average age of between 50 and 70 years. The primary endpoints of the 5 studies included the tolerability, safety, and efficacy of SCT for ED. Most studies showed improvement in erectile function due to SCT in patients, including improvements in penile vascular flow, International Index of Erectile Function-15 (IIEF-15) item and Erectile Hardness Scale scores. The 5 studies reported that no serious adverse effects occurred during the follow-up period in each study and 2 of the studies that focused on patient tolerability reported that patients were able to tolerate stem cell collection and injection well.

According to current data from human trials focusing on the efficacy and safety of SCT in the treatment of erectile dysfunction, the treatment modality looks promising. Many of the findings in the included studies indicated that SCT had modest efficacy in the treatment of drug-resistant ED. In all clinical trials investigated in this narrative, SCT did not cause any serious adverse events. Although the available studies make SCT, look promising, they are not free of limitations. None of these studies had large cohorts. These 5 studies collectively investigate only 61 patients in total. This is partly because the 5 studies used strict inclusion and exclusion criteria. 4 of the 5 studies were Phase I trials and only 1 of the studies was a Phase II clinical trial. All 5 studios were opened and only one was simple blind. None of the studies mentioned whether enrollment or treatment in the trial required payment from patients or whether patients were compensated for their enrollment. Payment or compensation could have introduced a response bias, which is not mentioned in these studies. In most of these studies, the main assessment criterion is safety and tolerability rather than erectile function. In addition, studies are difficult to standardize as they use stem cells from different sources. This makes widespread security assumptions difficult to conclude without further corroborated studies on each type of SCT. Only one of the studies contained a placebo control arm. Studies on post-prostatectomy patients do not contain a postoperative spontaneous recovery control arm of erectile function.

This makes effectiveness findings challenging to validate, especially in this specific population. Certain ethical concerns surrounding the use of placenta-derived stem cells and umbilical cord for TCM, and neither of the two studies investigating the use of these sources addresses this controversy. Finally, only 2 of the studies that met the inclusion criteria for this systematic review were published in moderate impact journals. Ultimately, the 5 studies concluded that more research on stem cell transplantation for erectile dysfunction should be conducted before it can become a routinely offered treatment modality.

There is also a large discrepancy between the number of SCT studies in animal ED models and SCT studies in human DE models. This is further emphasized with the absence of phase II clinical trials of SCT in human models of erectile dysfunction, although animal trials have been very successful. The success of TCM with animal allograft may not result in success in larger-scale human trials. This is because, in animal models, stem cells are allografts of young healthy animals, while surgical practice in the United States and other countries focuses on self-grafting stem cells directly from patients. These patients with ED generally have comorbidities and moderately advanced age, which can affect the quality of stem cells used in SCT.

In addition to the differences between the animal model and the human model of SCT, stem cells from various sources may also be different depending on safety and efficacy. This may be especially true for diabetic and/or elderly patients whose bone marrow stem cells may contain damage and defects. SCT is still relatively new in nature and multiple clinical trials are currently being conducted to investigate its safety and efficacy. Currently, there are 22 trials listed on the website ClinicalTrials.gov. The use of SCT in cancer patients has also been of concern, as the potential of MMC to act as tumor promoters has been investigated. Previous research has shown that MMC can infiltrate prostate cells, including prostate cancer tissue. Infiltration of MMC into prostate cancer tissue can also contribute to carcinogenesis.

Therefore, SCT is not part of the recommended routine treatment for erectile dysfunction. Current American Urological Association (AUA) guidelines for erectile dysfunction management consider intracavernous therapy with Stem Cells in the research phase (Conditional Recommendation; Evidence Level: Grade C). The European Urology Association (EUA) similarly recognizes the therapeutic potential of stem cells but states that SCT requires more research in large-scale randomized clinical trials.

Regardless of the experimental nature of SCT for erectile dysfunction by the AUA and UAE, clinics offering stem cell therapy have grown exponentially over the past decade, like the rapid expansion of other experimental therapies for erectile dysfunction. It is important to remember that the Food and Drug Administration (FDA) has not approved SCT and has issued warnings to patients regarding clinical use and its safety. These warnings have reached the media; however, it seems that the popularity of alternative therapies for erectile dysfunction is increasing, and stem cell clinics are expanding to many cities in the United States.

These clinics that offer unused treatments for erectile dysfunction should educate patients about the novel nature of treatment and should expect more research before marketing SCT as a recommended treatment option for erectile dysfunction.

Table 2. Human studies

Study	Date	Study design	Comorbidity	Type of cells	Sample size (n)	Age of patients	Follow-up period	Primary end point	Key findings	Serious adverse events
Haahr et al (Haahr et al., 2018) NCT02240823	2018	Open label phase I	Post-radical prostatectomy	ADRC	21	Mean: 60.2 years	12 months	Adverse events	53% of erections sufficient for intercourse. Not seen in incontinent or pre-op ED men.	None
Levy et al (Levy et al., 2016) NCT02398370	2016	Open label phase I	COED	PM-MSCs	8	40 70 years	6 months	Penile vascular changes	3/8 patients achieved erections at 3 mo. Significant improvements only in PSV	None
Bahk et al (Bahk et al., 2010)	2010	Single-blind phase I	DM	UC-MSCs	10	Mean: 69.5 years	12 months	Efficacy	3/7 patients had morning erections at 3 mo. At 6 mo f/u 2/7 patients sufficient erection with PDE-5 inhibitor	None
Yiou et al (Yiou et al., 2016, 2017) NCT01089387	Phase I: 2016	Phase I & II clinical	Post-radical Prostatectomy	BM-MNSCs	Phase I: 12	Mean: 63.6 years	Phase I: 12 months	Tolerance	All patients tolerated injections well. IIEF-15 and EHS significant improvements. No pCa recurrences at f/u	None
	Phase II: 2017				Phase II: 12 + 6		Phase 2: 62.1 ± 11.7 months			
Al Demour et al (Al Demour et al., 2018) NCT02945462	2018	Open label phase I	DM	BM-MSCs	4	Mean: 54.75 years	2 years for safety, 12 months for efficacy	Safety and tolerability	4/4 good tolerance to procedure. 3/4 patients had significant improvements in IIEF-15 and all 4 in EHS.	None
Ichim et al (Ichim et al., 2013)	2013	Case report	COED	BM-MNSCs	1/18	35 year	18 months	Safety and tolerability	18 months after procedure, the patient still reported improved sexual function	None
Garber et al(Garber MG & Carlos ND, 2015)	2015	Pilot study	Type II DM	ADSC	6/12	55-81 years	9 months	Efficacy	1 returned for prosthesis, 1 returned to a non-erectile condition, and 4 maintained erection sufficient for coitus with medication until the 9th month	None
Protogerou et al (Protogerou et al., 2019)	2019	Open Label/Pilot study	Type II DM, AHT, DL, Peyronie D	AD-MSCs/ PLP	8/3	40 - 70 years	3 months	Efficacy	All patients were characterized by improved penile triplex and increased morning erections.	None
Ory et al (Ory et al., 2020)	2019	Retrosp. Cohort	Cardiomyopathy-ED	Trans endocardial hMSC	36/12	median age was 66.5 years	12 months	Efficacy	EF may improve after systemic delivery of SCT in men with ischemic cardiomyopathy and at least mild ED.	None

continued on following page

Table 2. Continued

Study	Date	Study design	Comorbidity	Type of cells	Sample size (n)	Age of patients	Follow-up period	Primary end point	Key findings	Serious adverse events
Protogerou et al (Protogerou et al., 2020a)	2020	Phase I/Pilot Study. Open Label	Type II DM, AHT, DL, Peyronie D	AD-MSCs/ PLP	5/6		12 months	Efficacy	SC + PLP appears to show some improvement in EF and has minimal side effects in the short term.	None
Zasieda et al (Zasieda, 2020)	2020	Prosp. Cohort	Metabolic syndrome-ED	MSC-DE/ LISWT	38/3	45-60 years	12 weeks	Efficacy	EF treatment with combination of mesenchymal stem cell derived exosomes and focused lowintensive shock wave therapy.	None
Bieri et al. (Bieri et al., 2020) NCT3699943	2020	Phase I/ Dose escalation	COED	ABMC	40/12	Mean age: low dose 36±5y high dose 52±12y register 57±15	12 months	safety and efficacy	No long-term adverse events were noted related to the intervention.	None
Al Demour et al, (Al Demour et al., 2021) NCT2945449	2021	Phase I/ Open Label	Type II DM	WJ-MSCs	22/12	25-75 years	12 months	Tolerability, safety, and efficacy	No patient-reported serious adverse effects. There were significant improvements in IIEF-5, EHS, peak systolic velocity (PSV) basal, and 20-min PSV	None
You et al. (You et al., 2021) NCT23448499	2021	Phase I/ Open Label	DM/Post-RP	BM-MSCs	10	Mean age: 62.0 +/- 13.0 years	49.0 +/- 44.8 months	Efficacy	This trial confirmed the safety and potential efficacy of autologous BMSC in patients with ED	None
Mirzaei et al (Mirzaei et al., 2021)	2021 (58)	Rand. Single-blinded	DM-ED	Oral mucosa-MSCs	20/6	Mean age: 63.8 (7.4) years	6 months	Efficacy	Intracavernosal injection of SC improved sexual function and PSV and RI indices of penile arteries in diabetic patients.	None
Koga et al (Koga & Horiguchi, 2022)	2022 (59)	Open Label/ Pilot study	COED	SHED-CM	38/8	31–79 years old	54 months	Efficacy	SHED-CM repair vascular damages of the corpus cavernosum.	None
Fode et al (Fode et al., 2023)	2022 (60)	Prosp. case series	COED	ADSC	10/6		6 months	Investigate feasibility and safety	The study confirms the feasibility and safety of this minimally invasive, delivery of ADSC.	None

(Lokeshwar et al., 2020)

Umbilical Cord-(UCSC) and Placenta-Derived (PDSC) Stem Cells

Bahk et al. used allogeneic umbilical cord (UCSC) and placental (PDSC) derived SCs for the treatment of ED secondary to DM. ICI with e performed 1.5x 10^7 cells in seven men aged between 57 and 83 years, finding recovery spontaneous morning erections in the period of one month, maintaining the effect during the 6 months of follow-up. In addition, a decrease in blood glucose levels was also observed after two weeks of starting treatment, which opens a window of potential benefit of MSCs in the treatment of diabetes. (Bahk et al., 2010; Pérez-Aizpurua et al., 2023) The safety profile of the treatment was not measured in this study.

Subsequently, in a phase I clinical trial by Levy et al., (Levy et al., 2016) with eight PDEi non-responders patients, they received ICI of PDSC, without observing serious adverse events and an improvement in mean penile arterial flow., which was maintained 6 months after starting treatment.

Bone Marrow-Derived Stem Cells (BMSC)

Yiou et al. published a phase I/II clinical trial (INSTIN Study) in which included a total of 18 patients (12 in the first phase of the study) after radical prostatectomy, (Pérez-Aizpurua et al., 2023; Yiou et al., 2016, 2017) using dose-escalating ICI of BMSC, finding that a single autologous BMSC ICI was safe and effective in patients with vasculogenic ED, with improvement in perceived erectile function and sexual satisfaction, maintained up to 1-year after the injection. No treatment-derived adverse were found during a period of 62.1 months. However, a slight worsening in erectile function was observed after the first year following treatment, mentioned the requirement of repeating injection for maintaining the therapeutic effect.

Bieri et al. confirmed the safety profile and efficacy of BMSC using an FDA-approved medical device for cellular delivery in a phase I clinical trial that included patients who did not respond to PDEi. (Bieri et al., 2020) In the field of ED arising from diabetes mellitus, Al Demour et al. they published two consecutive clinical trials in diabetic patients treated with autologous BMSCs and allogeneic Wharton's gelatin-derived SCs (WJSC) to treat ED. In the first trial, they described that the use of BMSC is safe and effective, with a significant improvement in scores on the index of erectile function (IIEF-15) and erection hardness score (EHS) questionnaires. (Al Demour et al., 2018) In the second, the use of two consecutive ICIs of WJSC was explored in 22 diabetic patients with ED, with good tolerability and safety, in addition to the efficacy measured by the IIEF-15 and EHS questionnaires and penile duplex Doppler in the follow-up of 12 months.(Pérez-Aizpurua et al., 2023)

Adipose-Derived Stem Cells (ADSC)

ADSCs are cells that play a primary role in revascularization of damaged tissue, inhibition of apoptosis and immunomodulatory processes, with unique self-regenerative and multipotential capacity, similar to BMSC, they are easy to obtain and in large quantities. (G. Lin et al., 2009) In a 2015 study by Garber et al., 1.5x 10^7 ADSCs were injected into six diabetic patients. (Garber MG & Carlos ND, 2015) Four of the six patients recovered spontaneous morning erections in a period of one month, being able to maintain sexual intercourse up to twelve months after ICI with the help of PDEi during the last month. Similarly, the Danish group led by Haahr, which demonstrated that ICI treatment of autologous ADSC was safe and effective in patients with ED after RP, (Haahr et al., 2016) without observing adverse events after 12 months, maintained improvement in the score from IIEF-15 up to 12 months. (Haahr et al., 2018) It is important to note that improvement was observed in patients with normal preoperative erectile function and urinary continence at the time of inclusion. The same group is currently developing a phase III study with a much larger sample of post-prostatectomy patients.

On the other hand, Protogenau et al. have devised a combined administration technique of ADSC with the substances present in SVF, suspended in platelet-rich plasma (PRP). They demonstrated improvement with treatment after one, three and six months without describing adverse effects. (Protogerou et al., 2019, 2020b) A prospective study in 10 men by Fode et al. (Fode et al., 2023) is evaluating the the feasibility and safety of the myStem® X2 kit, a new minimally invasive method of autologous ADSC transplantation. (Pérez-Aizpurua et al., 2023)

PEYRONIE'S DISEASE

Up to date the data about plasma rich platelets (PRP) and SC in the treatment of Peyronie's disease (PD)(Castiglione et al., 2013) is still poor and lacking clear scientific evidence, we may only find 45 articles in PubMed related to PD (Jan2021) and 5 for PRP. They generally all comment on their supposed benefits together with the fact that the series of patients are normally small, and the majority of studies use animal models the methodology of each study is peculiar regarding how to assess "positive result", some focus on erection quality, others on plaque size other on Doppler ultrasound results. We have selected published articles with consistent methodology and data to clarify the current status of these therapies.

Ideally, we should be able to determine which PD patients may benefit from SC or PRP therefore diagnosis of the pathology is mandatory for a correct selection of patients that may be candidates to these treatment modalities. Ultimately we should know and understand what these treatments may offer and what mechanisms are involved in reverting the pathologic situation. Unfortunately, the data regarding PRP, and SC and their benefits are limited although some papers are promising in their findings.

In February 2016 we may find "Research highlights on stem cell therapy for the treatment of Peyronie's disease". (Sangkum, 2016) In this first review the author points out the four relevant published articles to that date. (Castiglione et al., 2013; Gokce et al., 2014b; Hellstrom et al., 2006; Sangkum, 2016)After revising these four preliminary papers the authors conclude that there is a potential role of adipose derived stem cells (ADSC) for both prevention and treatment of PD. Up to that date no studies report significant improvement in plaque size or penile curvature, which is to be considered the main objective of PD treatment and there is a general recommendation for further animal and clinical studies to validate therapeutic benefits of ADSCs for PD.So, four years ago, urge for more basic science was already mentioned in this field. The authors comment that SC have shown efficacy in preclinical models of injury and chronic disease but there isn't definitive evidence that SC act by engrafting into the diseased tissue and differentiation in specialized cells in these defective tissues, whereas paracrine modulation of the environment is an increasingly recognized mechanism of action of SC therapy.

There is an increasing evidence for the role of mesenchymal derived stem cells (MSCs) as a potential treatment strategy to alleviate fibrosis [5], and that the exact mechanisms of the antifibrotic effects of the MSCs remained to be proved. Most preclinical studies suggest that the MSC work through immunomodulation, thereby limiting the host response to injury and preventing the onset of fibrosis. (Jackson et al., 2012; Schwarzer et al., 2001)

To that date March 2016 only 3 studies had evaluated mesenchymal stem cell local injection as therapy for PD in a rat model, Castiglione et al (Castiglione et al., 2013)tested the efficacy of human ADSCs in preventing fibrosis using a rat model of PD based on TGF-B1 injection in tunica albuginea. At 1 month after TGF-b1 injection the rats displayed extensive tunica-corporeal fibrosis and elastosis proved at the injection site with a significant increase of collagen III and elastin deposition. The rats showed a mild erectile dysfunction visible only when cavernous nerve was stimulated at submaximal voltage but not when maximal voltage was applied. The injection of 1 million xenogeneic human ADSCs 1 day after the TGF-b1 injection was able to prevent the development of this extensive fibrosis preserving the overall structure, elastin, and collagen III contents of the corpus cavernosum. In addition, human ADSCs treatment prevented development of erectile dysfunction.

Gokce et al. (Gokce et al., 2014b, 2015) had tested the local injection of allogeneic ADSCs and ADSCs expressing human interferon a-2b for the prevention and treatment of erectile dysfunction and fibrosis in the same animal model of Peyronie's disease used by Castiglione et al. The ADSC treated groups received tunica albuginea injections with 0.5 million rat ADSCs on day 0 (prevention) or day 30 (treatment) from the TGF-b1 injection. Forty-five days following TGF-b1 injection, in both prevention and treatment groups, tunica albuginea injection of both types of ADSCs resulted in significant lower tunica albuginea fibrosis and in a better erectile function response, also for high voltage, compared with the rats treated only with TGF-b1. In addition, local injection of ADSCs prevented and reduced fibrosis of tunica albuginea by decreasing it.

Although human data on SC therapy is emerging and repeatedly seen in the literature references it is still significantly inconsistent for human healthcare. Bahk et al. (Bahk et al., 2010) implanted umbilical cord stem cells in penis of seven men with diabetes-related erectile dysfunction. This study reported increased rigidity insufficient for penetration without addition of PDE5 inhibitors. Various cellular products including MSCs, SVF, and BMSCs are currently being studied in clinical trials (NCT02087397, NCT02240823, NCT02414308, NCT02344849, NCT01089387, NCT02398370, NCT01953523, NCT02472431, NCT01983709, NCT01601353) (Ichim et al., 2013) Ichim et al. (Ichim et al., 2013) reported a phase 1/2 trial of intracavernous bone marrow mononuclear cells in after radical prostatectomy erectile dysfunction. This study enrolled 12 participants with penile arterial insufficiency and/or venoocclusive dysfunction at 6 months to 3 years after treatment for localized prostate cancer. There were no significant adverse reactions and no signs of cancer recurrence. At 6 months increased erectile function by measures of erectile hardness score (EHS) and Doppler ultrasound parameters were more pronounced in combination with pharmacotherapy. This terminology is vague to decide or know what the authors mean by "parameter was more pronounced in combination with pharmacotherapy".

As seen these last studies are not specifically about Peyronie's disease patients but they try to search finding answers to what exactly SC do in the cavernous tissue and tunica albuginea. It has been proven about the beneficial effects of SC on restoration of erectile function through paracrine effects and cellular mechanisms. But research clinical trials in various disease states are required to define the role of SC therapy in the treatment line-up.

In February 2019, Uros Milenkovic et al. (Milenkovic et al., 2019) published a review about "The mechanisms and potential of stem cell therapy for penile fibrosis. In this review the authors attempt to answer the key points of cell therapy because as they point out "nonetheless, these hypothesis have not been proven and additional studies focusing on the mechanisms of MSC function are ongoing", regarding the fact that MSCs could cause some sort of immune response and exert their antifibrotic functions through immune modulation or attenuation of the profibrotic phenotypic changes of resident fibroblast into the contractile and ECM-producing myofibroblasts.

The complete list of preclinical studies on cell therapy for penile fibrosis, together with tunica albuginea as well as urethral stricture is listed reaching to the conclusion that up to that date (Oct 2018), the clinical application and investigation of conventional antifibrotic therapies has yielded limited results. However, the use of SC in translational research has the potential to exert antifibrotic functions on several levels by modulating the host response. Despite the amount of research on stem cells and penile fibrosis, the field is still in its infancy and is subject to many limitations.

Most preclinical research regarding stem cells in penile fibrosis has focused on corpora cavernosa fibrosis owing to its clear pathophysiology (iatrogenic postprostatectomy ED and corpora cavernosa fibrosis) and representative animal models. Conversely, PD and urethral stricture disease research have been limited by poor disease models and invalidated findings. Thus, the treatment of corpora cavernosa fibrosis with stem cells seems to be the closest to potential clinical application given additional studies evaluating the efficacy, dosage, timing and route of administration of stem cells or stromal vascular elements.

There is one study that focuses on PRP and PD (Culha et al., 2019) on rats' penises, unfortunately they finally concluded that PRP is an inexpensive way of obtaining experimental animal models with PD as there was no efficacy in the PD treated rats and what is most important is that the injection of PRP had effect which were histologically like PD.

On the other hand, other authors Epifanova M.V et al. (Epifanova et al., 2020) comment that the available data shows lack of adverse reaction with PRP treatment, and that the studies found were limited by small groups, and that is why data on safety "should be taken carefully. However, it is important to mention that PRP therapy has the potential for treating male sexual dysfunction and may be useful in andrology". Due to these "scientific" contradictions we should at least study more in depth what exactly is the cause of Peyronie's disease, perhaps then we will be able to come up with a medical treatment that reduces fibrosis and restitutes size, angle and erection to those males affected by PD. If we can prove results in a systematic and protocolised way which is easily learnt, we will then be of use for all those PD patients which are silently suffering their condition.

As the precise aetiology of PD is still unknown and under investigation its treatment is still "blind" and although cell therapy regarding SC or PRP have had good results in other pathologies, we are still lacking knowledge of what, where, how and most importantly who can benefit from these therapeutic options. The fact that we can find studies that present results on animal model or even in humans might give encouraging results, but it should not lead us to assume that they are current therapeutic options. In depth, the findings lack information of how to obtain de SC or the PRP in a protocolled way and to which patients offer it. Although the perspective of the evidence reviewed may seem discouraging it is precisely for this reason that better study of the disease and its aetiology is needed before thinking in better study design and clear patient inclusion criteria that must be agreed amongst opinion leaders and researchers to obtain scientific validated data that gives us the key to what we should inject, how and to who.

REFERENCES

Al Demour, S., Adwan, S., Jafar, H., Rahmeh, R., Alhawari, H., & Awidi, A. (2021). Safety and Efficacy of 2 Intracavernous Injections of Allogeneic Wharton's Jelly-Derived Mesenchymal Stem Cells in Diabetic Patients with Erectile Dysfunction: Phase 1/2 Clinical Trial. *Urologia Internationalis*, *105*(11-12), 935–943. DOI: 10.1159/000517364 PMID: 34384079

Al Demour, S., Jafar, H., Adwan, S., AlSharif, A., Alhawari, H., Alrabadi, A., Zayed, A., Jaradat, A., & Awidi, A. (2018). Safety and Potential Therapeutic Effect of Two Intracavernous Autologous Bone Marrow Derived Mesenchymal Stem Cells injections in Diabetic Patients with Erectile Dysfunction: An Open Label Phase I Clinical Trial. *Urologia Internationalis*, *101*(3), 358–365. DOI: 10.1159/000492120 PMID: 30173210

Albersen, M., Fandel, T. M., Lin, G., Wang, G., Banie, L., Lin, C. S., & Lue, T. F. (2010). Injections of adipose tissue-derived stem cells and stem cell lysate improve recovery of erectile function in a rat model of cavernous nerve injury. *Journal of Sexual Medicine*, *7*(10), 3331–3340. DOI: 10.1111/j.1743-6109.2010.01875.x PMID: 20561166

Bahk, J. Y., Jung, J. H., Han, H., Min, S. K., & Lee, Y. S. (2010). Treatment of diabetic impotence with umbilical cord blood stem cell intracavernosal transplant: Preliminary report of 7 cases. *Experimental and Clinical Transplantation*, *8*(2). PMID: 20565373

Bieri, M., Said, E., Antonini, G., Dickerson, D., Tuma, J., Bartlett, C. E., Patel, A. N., & Gershman, A. (2020). Phase I and registry study of autologous bone marrow concentrate evaluated in PDE5 inhibitor refractory erectile dysfunction. *Journal of Translational Medicine*, *18*(1), 24. DOI: 10.1186/s12967-019-02195-w PMID: 31937310

Bivalacqua, T. J., Deng, W., Kendirci, M., Usta, M. F., Robinson, C., Taylor, B. K., Murthy, S. N., Champion, H. C., Hellstrom, W. J. G., & Kadowitz, P. J. (2007). Mesenchymal stem cells alone or ex vivo gene modified with endothelial nitric oxide synthase reverse age-associated erectile dysfunction. *American Journal of Physiology. Heart and Circulatory Physiology*, *292*(3), H1278–H1290. Advance online publication. DOI: 10.1152/ajpheart.00685.2006 PMID: 17071732

Bochinski, D., Lin, G. T., Nunes, L., Carrion, R., Rahman, N., Lin, C. S., & Lue, T. F. (2004). The effect of neural embryonic stem cell therapy in a rat model of cavernosal nerve injury. *BJU International*, *94*(6), 904–909. DOI: 10.1111/j.1464-410X.2003.05057.x PMID: 15476533

Castiglione, F., Hedlund, P., Van Der Aa, F., Bivalacqua, T. J., Rigatti, P., Van Poppel, H., Montorsi, F., De Ridder, D., & Albersen, M. (2013). Intratunical injection of human adipose tissue-derived stem cells prevents fibrosis and is associated with improved erectile function in a rat model of Peyronie's disease. *European Urology, 63*(3), 551–560. DOI: 10.1016/j.eururo.2012.09.034 PMID: 23040209

Cengiz, T., Kaya, E., Oral, D. Y., Ozakca, I., Bayatli, N., Karabay, A. Z., Ensari, T. A., Karahan, T., Yilmaz, E., & Gur, S. (2017). Intracavernous Injection of Human Umbilical Cord Blood Mononuclear Cells Improves Erectile Dysfunction in Streptozotocin-Induced Diabetic Rats. *Journal of Sexual Medicine, 14*(1), 50–58. DOI: 10.1016/j.jsxm.2016.11.314 PMID: 28065360

Chen, F., Zhang, H., Wang, Z., Ding, W., Zeng, Q., Liu, W., Huang, C., He, S., & Wei, A. (2017). Adipose-Derived Stem Cell-Derived Exosomes Ameliorate Erectile Dysfunction in a Rat Model of Type 2 Diabetes. *Journal of Sexual Medicine, 14*(9), 1084–1094. DOI: 10.1016/j.jsxm.2017.07.005 PMID: 28781215

Chen, X., Yang, Q., Zheng, T., Bian, J., Sun, X., Shi, Y., Liang, X., Gao, G., Liu, G., & Deng, C. (2016). Neurotrophic effect of adipose tissue-derived stem cells on erectile function recovery by pigment epithelium-derived factor secretion in a rat model of cavernous nerve injury. *Stem Cells International, 2016*(1), 5161248. Advance online publication. DOI: 10.1155/2016/5161248 PMID: 26783403

Culha, M. G., Erkan, E., Cay, T., & Yücetaş, U. (2019). The Effect of Platelet-Rich Plasma on Peyronie's Disease in Rat Model. *Urologia Internationalis, 102*(2), 218–223. DOI: 10.1159/000492755 PMID: 30317233

Epifanova, M. V., Gvasalia, B. R., Durashov, M. A., & Artemenko, S. A. (2020). Platelet-Rich Plasma Therapy for Male Sexual Dysfunction: Myth or Reality? En *Sexual Medicine Reviews* (Vol. 8, Número 1, pp. 106-113). Elsevier B.V. DOI: 10.1016/j.sxmr.2019.02.002

Feng, H., Liu, Q., Deng, Z., Li, H., Zhang, H., Song, J., Liu, X., Liu, J., Wen, B., & Wang, T. (2022). Human umbilical cord mesenchymal stem cells ameliorate erectile dysfunction in rats with diabetes mellitus through the attenuation of ferroptosis. *Stem Cell Research & Therapy, 13*(1), 450. Advance online publication. DOI: 10.1186/s13287-022-03147-w PMID: 36064453

Fode, M., Nadler, N., Lund, L., & Azawi, N. (2023). Feasibility of minimally invasive, same-day injection of autologous adipose-derived stem cells in the treatment of erectile dysfunction. *Scandinavian Journal of Urology, 57*(1-6), 110–114. DOI: 10.1080/21681805.2022.2162117 PMID: 36586416

Garber, M. G., & Carlos, N. D. (2015). Intracavernous administration of adipose stem cells: A new technique of treating erectile dysfunction in diabetic patient, preliminary report of 6 cases. *MOJ Cell Science & Report*, *2*(1). Advance online publication. DOI: 10.15406/mojcsr.2015.02.00018

Garcia, M. M., Fandel, T. M., Lin, G., Shindel, A. W., Banie, L., Lin, C. S., & Lue, T. F. (2010). Treatment of erectile dysfunction in the obese Type 2 diabetic ZDF rat with adipose tissue-derived stem cells. *Journal of Sexual Medicine*, *7*(1 PART 1), 89–98. DOI: 10.1111/j.1743-6109.2009.01541.x PMID: 20104670

Gokce, A., Abd Elmageed, Z. Y., Lasker, G. F., Bouljihad, M., Braun, S. E., Kim, H., Kadowitz, P. J., Abdel-Mageed, A. B., Sikka, S. C., & Hellstrom, W. J. (2015). Intratunical Injection of Genetically Modified Adipose Tissue-Derived Stem Cells with Human Interferon α-2b for Treatment of Erectile Dysfunction in a Rat Model of Tunica Albugineal Fibrosis. *Journal of Sexual Medicine*, *12*(7), 1533–1544. DOI: 10.1111/jsm.12916 PMID: 26062100

Gokce, A., Abd Elmageed, Z. Y., Lasker, G. F., Bouljihad, M., Kim, H., Trost, L. W., Kadowitz, P. J., Abdel-Mageed, A. B., Sikka, S. C., & Hellstrom, W. J. (2014a). Adipose tissue-derived stem cell therapy for prevention and treatment of erectile dysfunction in a rat model of Peyronie's disease. *Andrology*, *2*(2), 244–251. DOI: 10.1111/j.2047-2927.2013.00181.x PMID: 24574095

Gokce, A., Abd Elmageed, Z. Y., Lasker, G. F., Bouljihad, M., Kim, H., Trost, L. W., Kadowitz, P. J., Abdel-Mageed, A. B., Sikka, S. C., & Hellstrom, W. J. (2014b). Adipose tissue-derived stem cell therapy for prevention and treatment of erectile dysfunction in a rat model of Peyronie's disease. *Andrology*, *2*(2), 244–251. DOI: 10.1111/j.2047-2927.2013.00181.x PMID: 24574095

Haahr, M. K., Harken Jensen, C., Toyserkani, N. M., Andersen, D. C., Damkier, P., Sørensen, J. A., Sheikh, S. P., & Lund, L. (2018). A 12-Month Follow-up After a Single Intracavernous Injection of Autologous Adipose-Derived Regenerative Cells in Patients with Erectile Dysfunction Following Radical Prostatectomy: An Open-Label Phase I Clinical Trial. *Urology*, *121*, 203.e6–203.e13. DOI: 10.1016/j.urology.2018.06.018 PMID: 29958973

Haahr, M. K., Jensen, C. H., Toyserkani, N. M., Andersen, D. C., Damkier, P., Sørensen, J. A., Lund, L., & Sheikh, S. P. (2016). Safety and Potential Effect of a Single Intracavernous Injection of Autologous Adipose-Derived Regenerative Cells in Patients with Erectile Dysfunction Following Radical Prostatectomy: An Open-Label Phase I Clinical Trial. *EBioMedicine*, *5*, 204–210. DOI: 10.1016/j.ebiom.2016.01.024 PMID: 27077129

Hellstrom, W. J. G., Kendirci, M., Matern, R., Cockerham, Y., Myers, L., Sikka, S. C., Venable, D., Honig, S., McCullough, A., Hakim, L. S., Nehra, A., Templeton, L. E., & Pryor, J. L. (2006). Single-Blind, Multicenter, Placebo Controlled, Parallel Study to Assess the Safety and Efficacy of Intralesional Interferon α-2b for Minimally Invasive Treatment for Peyronie's Disease. *The Journal of Urology*, *176*(1), 394–398. DOI: 10.1016/S0022-5347(06)00517-9 PMID: 16753449

Hou, Q., Ge, M., Zhang, C., Tian, D., Wang, L., Tian, H., Wang, W., & Zhang, W. (2017). Adipose tissue-derived stem cell therapy for erectile dysfunction in rats: A systematic review and meta-analysis. *International Urology and Nephrology*, *49*(7), 1127–1137. DOI: 10.1007/s11255-017-1590-2 PMID: 28417342

Huang, Y. C., Ning, H., Shindel, A. W., Fandel, T. M., Lin, G., Harraz, A. M., Lue, T. F., & Lin, C. S. (2010). The effect of intracavernous injection of adipose tissue-derived stem cells on hyperlipidemia-associated erectile dysfunction in a rat model. *Journal of Sexual Medicine*, *7*(4 PART 1), 1391–1400. DOI: 10.1111/j.1743-6109.2009.01697.x PMID: 20141586

Ichim, T. E., Warbington, T., Cristea, O., Chin, J. L., & Patel, A. N. (2013). Intracavernous administration of bone marrow mononuclear cells: A new method of treating erectile dysfunction? En *Journal of Translational Medicine* (Vol. 11, Número 1). J Transl Med. DOI: 10.1186/1479-5876-11-139

Jackson, W. M., Nesti, L. J., & Tuan, R. S. (2012). Mesenchymal stem cell therapy for attenuation of scar formation during wound healing. *Stem Cell Research & Therapy*, *3*(3), 20. Advance online publication. DOI: 10.1186/scrt111 PMID: 22668751

Jeon, S. H., Shrestha, K. R., Kim, R. Y., Jung, A. R., Park, Y. H., Kwon, O., Kim, G. E., Kim, S. H., Kim, K. H., & Lee, J. Y. (2016). Combination therapy using human adipose-derived stem cells on the cavernous nerve and low-energy shockwaves on the corpus cavernosum in a rat model of post-prostatectomy erectile dysfunction. *Urology*, *88*, 226.e1–226.e9. DOI: 10.1016/j.urology.2015.10.021 PMID: 26522972

Kendirci, M., Trost, L., Bakondi, B., Whitney, M. J., Hellstrom, W. J. G., & Spees, J. L. (2010). Transplantation of nonhematopoietic adult bone marrow stem/progenitor cells isolated by p75 nerve growth factor receptor into the penis rescues erectile function in a rat model of cavernous nerve injury. *The Journal of Urology*, *184*(4), 1560–1566. DOI: 10.1016/j.juro.2010.05.088 PMID: 20728109

Koga, S., & Horiguchi, Y. (2022). Efficacy of a cultured conditioned medium of exfoliated deciduous dental pulp stem cells in erectile dysfunction patients. *Journal of Cellular and Molecular Medicine*, *26*(1), 195–201. DOI: 10.1111/jcmm.17072 PMID: 34845823

Levy, J. A., Marchand, M., Iorio, L., Cassini, W., & Zahalsky, M. P. (2016). Determining the feasibility of managing erectile dysfunction in humans with placental-derived stem cells. *The Journal of the American Osteopathic Association*, *116*(1), e1–e5. DOI: 10.7556/jaoa.2016.007 PMID: 26745574

Lin, G., Banie, L., Ning, H., Bella, A. J., Lin, C. S., & Lue, T. F. (2009). Potential of adipose-derived stem cells for treatment of erectile dysfunction. *The journal of sexual medicine, 6 Suppl 3*(Suppl 3), 320-327. DOI: 10.1111/j.1743-6109.2008.01190.x

Lin, H., Dhanani, N., Tseng, H., Souza, G. R., Wang, G., Cao, Y., Ko, T. C., Jiang, H., & Wang, R. (2016). Nanoparticle Improved Stem Cell Therapy for Erectile Dysfunction in a Rat Model of Cavernous Nerve Injury. *The Journal of Urology*, *195*(3), 788–795. DOI: 10.1016/j.juro.2015.10.129 PMID: 26519654

Liu, G., Sun, X., Bian, J., Wu, R., Guan, X., Ouyang, B., Huang, Y., Xiao, H., Luo, D., Atala, A., Zhang, Y., & Deng, C. (2013). Correction of Diabetic Erectile Dysfunction with Adipose Derived Stem Cells Modified with the Vascular Endothelial Growth Factor Gene in a Rodent Diabetic Model. *PLoS One*, *8*(8), e72790. Advance online publication. DOI: 10.1371/journal.pone.0072790 PMID: 24023647

Liu, T., Peng, Y., Jia, C., Fang, X., Li, J., & Zhong, W. (2015). Hepatocyte growth factor-modified adipose tissue-derived stem cells improve erectile function in streptozotocin-induced diabetic rats. *Growth Factors (Chur, Switzerland)*, *33*(4), 282–289. DOI: 10.3109/08977194.2015.1077825 PMID: 26339935

Lokeshwar, S. D., Patel, P., Shah, S. M., & Ramasamy, R. (2020). A Systematic Review of Human Trials Using Stem Cell Therapy for Erectile Dysfunction. En *Sexual Medicine Reviews* (Vol. 8, Número 1, pp. 122-130). Elsevier B.V. DOI: 10.1016/j.sxmr.2019.08.003

Lu, J., Xin, Z., Zhang, Q., Cui, D., Xiao, Y., Zhuo, J., Sun, F., Xia, S., & Shao, Y. (2016). Beneficial effect of PEDF-transfected ADSCs on erectile dysfunction in a streptozotocin-diabetic rat model. *Cell and Tissue Research*, *366*(3), 623–637. DOI: 10.1007/s00441-016-2494-7 PMID: 27655233

Luo, C., Peng, Y., Zhou, X., Fan, J., Chen, W., Zhang, H., & Wei, A. (2022). NLRP3 downregulation enhances engraftment and functionality of adipose-derived stem cells to alleviate erectile dysfunction in diabetic rats. *Frontiers in Endocrinology*, *13*, 913296. Advance online publication. DOI: 10.3389/fendo.2022.913296 PMID: 35937790

Martínez-Salamanca, J. I., Zurita, M., Costa, C., Martínez-Salamanca, E., Fernández, A., Castela, A., Vaquero, J., Carballido, J., & Angulo, J. (2016). Dual Strategy With Oral Phosphodiesterase Type 5 Inhibition and Intracavernosal Implantation of Mesenchymal Stem Cells Is Superior to Individual Approaches in the Recovery of Erectile and Cavernosal Functions After Cavernous Nerve Injury in Rats. *Journal of Sexual Medicine*, *13*(1), 1–11. DOI: 10.1016/j.jsxm.2015.12.001 PMID: 26755080

Milenkovic, U., Albersen, M., & Castiglione, F. (2019). The mechanisms and potential of stem cell therapy for penile fibrosis. En *Nature Reviews Urology* (Vol. 16, Número 2, pp. 79-97). Nature Publishing Group. DOI: 10.1038/s41585-018-0109-7

Mirzaei, M., Bagherinasabsarab, M., Pakmanesh, H., Mohammadi, R., Teimourian, M., Farsinejad, A., & Jahani, Y. (2021). The Effect of Intracavernosal Injection of Stem Cell in the Treatment of Erectile Dysfunction in Diabetic Patients: A Randomized Single-blinded Clinical Trial. *Urology Journal*, *18*(6), 675–681. DOI: 10.22037/UJ.V18I.6503 PMID: 34655071

Mukti, A. I., Ilyas, S., Warli, S. M., Putra, A., Rasyid, N., Munir, D., Siregar, K. B., & Ichwan, M. (2022). Umbilical Cord-Derived Mesenchymal Stem Cells Improve TGF-β, α-SMA and Collagen on Erectile Dysfunction in Streptozotocin-Induced Diabetic Rats. *Medical archives (Sarajevo, Bosnia and Herzegovina), 76*(1), 4-11. DOI: 10.5455/medarh.2022.76.4-11

Ory, J., Saltzman, R. G., Blachman-Braun, R., Dadoun, S., DiFede, D. L., Premer, C., Hurwitz, B., Hare, J. M., & Ramasamy, R. (2020). The Effect of Transendocardial Stem Cell Injection on Erectile Function in Men With Cardiomyopathy: Results From the TRIDENT, POSEIDON, and TAC-HFT Trials. *Journal of Sexual Medicine*, *17*(4), 695–701. DOI: 10.1016/j.jsxm.2020.01.003 PMID: 32059936

Park, H. J., Jeong, H., Park, Y. H., Yim, H. W., Ha, U. S., Hong, S. H., Kim, S. W., Kim, N. J., & Lee, J. Y. (2019). Adipose Tissue-Derived Stem Cell Therapy for Cavernous Nerve Injury-Induced Erectile Dysfunction in the Rat Model: A Systematic Review and Meta-Analysis Using Methodological Quality Assessment. *International Journal of Stem Cells*, *12*(2), 206–217. DOI: 10.15283/ijsc18122 PMID: 31022999

Patel, D. P., Pastuszak, A. W., & Hotaling, J. M. (2019). Emerging Treatments for Erectile Dysfunction: A Review of Novel, Non-surgical Options. En *Current Urology Reports* (Vol. 20, Número 8). Current Medicine Group LLC 1. DOI: 10.1007/s11934-019-0908-2

Pérez-Aizpurua, X., Garranzo-Ibarrola, M., Simón-Rodríguez, C., García-Cardoso, J. V., Chávez-Roa, C., López-Martín, L., Tufet i Jaumot, J. J., Alonso-Román, J., Maqueda-Arellano, J., Gómez-Jordana, B., Ruiz de Castroviejo-Blanco, J., Osorio-Ospina, F., González-Enguita, C., & García-Arranz, M. (2023). Stem Cell Therapy for Erectile Dysfunction: A Step towards a Future Treatment. En *Life* (Vol. 13, Número 2). MDPI. DOI: 10.3390/life13020502

Protogerou, V., El Beshari, S., Michalopoulos, E., Mallis, P., Chrysikos, D., Samolis, A. A., Stavropoulos-Giokas, C., & Troupis, T. (2020a). The Combined Use of Stem Cells and Platelet Lysate Plasma for the Treatment of Erectile Dysfunction: A Pilot Study–6 Months Results. *Medicines (Basel, Switzerland)*, *7*(3), 14. DOI: 10.3390/medicines7030014 PMID: 32197323

Protogerou, V., El Beshari, S., Michalopoulos, E., Mallis, P., Chrysikos, D., Samolis, A. A., Stavropoulos-Giokas, C., & Troupis, T. (2020b). The Combined Use of Stem Cells and Platelet Lysate Plasma for the Treatment of Erectile Dysfunction: A Pilot Study–6 Months Results. *Medicines (Basel, Switzerland)*, *7*(3), 14. DOI: 10.3390/medicines7030014 PMID: 32197323

Protogerou, V., Michalopoulos, E., Mallis, P., Gontika, I., Dimou, Z., Liakouras, C., Stavropoulos-Giokas, C., Kostakopoulos, N., Chrisofos, M., & Deliveliotis, C. (2019). Administration of Adipose Derived Mesenchymal Stem Cells and Platelet Lysate in Erectile Dysfunction: A Single Center Pilot Study. *Bioengineering (Basel, Switzerland)*, *6*(1), 21. DOI: 10.3390/bioengineering6010021 PMID: 30841525

Qiu, X., Villalta, J., Ferretti, L., Fandel, T. M., Albersen, M., Lin, G., Dai, Y., Lue, T. F., & Lin, C. S. (2012). Effects of Intravenous Injection of Adipose-Derived Stem Cells in a Rat Model of Radiation Therapy-Induced Erectile Dysfunction. *Journal of Sexual Medicine*, *9*(7), 1834–1841. DOI: 10.1111/j.1743-6109.2012.02753.x PMID: 22548750

Ryu, J. K., Kim, D. H., Song, K. M., Ryu, D. S., Kim, S. N., Shin, D. H., Yi, T., Suh, J. K., & Song, S. U. (2016). Intracavernous delivery of clonal mesenchymal stem cells rescues erectile function in the streptozotocin-induced diabetic mouse. *Andrology*, *4*(1), 172–184. DOI: 10.1111/andr.12138 PMID: 26711324

Sangkum, P. (2016). Research highlights on stem cell therapy for the treatment of Peyronie's disease. *Translational Andrology and Urology*, *5*(3), 363–365. DOI: 10.21037/tau.2016.03.14 PMID: 27298783

Schwarzer, U., Sommer, F., Klotz, T., Braun, M., Reifenrath, B., & Engelmann, U. (2001). The prevalence of Peyronie's disease: Results of a large survey. *BJU International*, *88*(7), 727–730. DOI: 10.1046/j.1464-4096.2001.02436.x PMID: 11890244

Shan, H. T., Zhang, H. B., Chen, W. T., Chen, F. Z., Wang, T., Luo, J. T., Yue, M., Lin, J. H., & Wei, A. Y. (2017). Combination of low-energy shock-wave therapy and bone marrow mesenchymal stem cell transplantation to improve the erectile function of diabetic rats. *Asian Journal of Andrology*, *19*(1), 26–33. DOI: 10.4103/1008-682X.184271 PMID: 27427555

Song, K. M., Jin, H. R., Park, J. M., Choi, M. J., Kwon, M. H., Kwon, K. D., Batbold, D., Yin, G. N., Kim, W. J., Koh, G. Y., Ryu, J. K., & Suh, J. K. (2014). Intracavernous delivery of stromal vascular fraction restores erectile function through production of angiogenic factors in a mouse model of cavernous nerve injury. *Journal of Sexual Medicine*, *11*(8), 1962–1973. DOI: 10.1111/jsm.12597 PMID: 24902866

Sun, X., Luo, L. H., Feng, L., Li, D. S., & Zhong, K. Z. (2017). B Cell Lymphoma-2-Modified Bone Marrow-Derived Mesenchymal Stem Cells Transplantation for the Treatment of Diabetes Mellitus-Induced Erectile Dysfunction in a Rat Model. *Urologia Internationalis*, *98*(3), 358–366. DOI: 10.1159/000452253 PMID: 27894122

Vakalopoulos, I., Memmos, D., Mykoniatis, I., Toutziaris, C., & Dimitriadis, G. (2018). Stem cell therapy in erectile dysfunction: Science fiction or realistic treatment option? En *Hormones* (Vol. 17, Número 3, pp. 315-320). Springer. DOI: 10.1007/s42000-018-0050-4

Wang, B., Zhou, J., Banie, L., Reed-Maldonado, A. B., Ning, H., Lu, Z., Ruan, Y., Zhou, T., Wang, H. S., Oh, B. S., Wang, G., Qi, S. L., Lin, G., & Lue, T. F. (2018). Low-intensity extracorporeal shock wave therapy promotes myogenesis through PERK/ATF4 pathway. *Neurourology and Urodynamics*, *37*(2), 699–707. DOI: 10.1002/nau.23380 PMID: 28763567

Wang, X., Liu, C., Xu, Y., Chen, P., Shen, Y., Xu, Y., Zhao, Y., Chen, W., Zhang, X., Ouyang, Y., Wang, Y., Xie, C., Zhou, M., & Liu, C. (2017). Combination of mesenchymal stem cell injection with icariin for the treatment of diabetes-associated erectile dysfunction. *PLoS One*, *12*(3), e0174145. Advance online publication. DOI: 10.1371/journal.pone.0174145 PMID: 28350842

Wang, X. Y., Liu, C. L., Li, S. D., Xu, Y., Chen, P., Liu, Y., Ding, Q., Wahafu, W., Hong, B. F., & Yang, M. H. (2015). Hypoxia precondition promotes adipose-derived mesenchymal stem cells based repair of diabetic erectile dysfunction via augmenting angiogenesis and neuroprotection. *PLoS One*, *10*(3), e0118951. Advance online publication. DOI: 10.1371/journal.pone.0118951 PMID: 25790284

Woo, J. C., Bae, W. J., Kim, S. J., Kim, S. D., Sohn, D. W., Hong, S. H., Lee, J. Y., Hwang, T. K., Sung, Y. C., & Kim, S. W. (2011). Transplantation of muscle-derived stem cells into the corpus avernosum restores erectile function in a rat model of cavernous nerve injury. *Korean Journal of Urology*, *52*(5), 359–363. DOI: 10.4111/kju.2011.52.5.359 PMID: 21687398

Xu, Y., Guan, R., Lei, H., Li, H., Wang, L., Gao, Z., Song, W., & Xin, Z. (2014). Therapeutic potential of adipose-derived stem cells-based micro-tissues in a rat model of postprostatectomy erectile dysfunction. *Journal of Sexual Medicine*, *11*(10), 2439–2448. DOI: 10.1111/jsm.12636 PMID: 25042722

Yan, H., Ding, Y., & Lu, M. (2020). Current Status and Prospects in the Treatment of Erectile Dysfunction by Adipose-Derived Stem Cells in the Diabetic Animal Model. En *Sexual Medicine Reviews* (Vol. 8, Número 3, pp. 486-491). Elsevier B.V. DOI: 10.1016/j.sxmr.2019.09.006

Yang, Q., Chen, X., Zheng, T., Han, D., Zhang, H., Shi, Y., Bian, J., Sun, X., Xia, K., Liang, X., Liu, G., Zhang, Y., & Deng, C. (2016). Transplantation of human urine-derived stem cells transfected with pigment epithelium-derived factor to protect erectile function in a rat model of cavernous nerve injury. *Cell Transplantation*, *25*(11), 1987–2001. DOI: 10.3727/096368916X691448 PMID: 27075964

Yang, R., Fang, F., Wang, J., & Guo, H. (2015). Adipose-derived stem cells ameliorate erectile dysfunction after cavernous nerve cryoinjury. *Andrology*, *3*(4), 694–701. DOI: 10.1111/andr.12047 PMID: 26198799

Yiou, R., Hamidou, L., Birebent, B., Bitari, D., Le Corvoisier, P., Contremoulins, I., Rodriguez, A. M., Augustin, D., Roudot-Thoraval, F., de la Taille, A., & Rouard, H. (2017). Intracavernous Injections of Bone Marrow Mononucleated Cells for Postradical Prostatectomy Erectile Dysfunction: Final Results of the INSTIN Clinical Trial. *European Urology Focus*, *3*(6), 643–645. DOI: 10.1016/j.euf.2017.06.009 PMID: 28753830

Yiou, R., Hamidou, L., Birebent, B., Bitari, D., Lecorvoisier, P., Contremoulins, I., Khodari, M., Rodriguez, A. M., Augustin, D., Roudot-Thoraval, F., De La Taille, A., & Rouard, H. (2016). Safety of Intracavernous Bone Marrow-Mononuclear Cells for Postradical Prostatectomy Erectile Dysfunction: An Open Dose-Escalation Pilot Study. *European Urology*, *69*(6), 988–991. DOI: 10.1016/j.eururo.2015.09.026 PMID: 26439886

You, D., Jang, M. J., Kim, B. H., Song, G., Lee, C., Suh, N., Jeong, I. G., Ahn, T. Y., & Kim, C.-S. (2015). Comparative Study of Autologous Stromal Vascular Fraction and Adipose-Derived Stem Cells for Erectile Function Recovery in a Rat Model of Cavernous Nerve Injury. *Stem Cells Translational Medicine*, *4*(4), 351–358. DOI: 10.5966/sctm.2014-0161 PMID: 25792486

You, D., Jang, M. J., Song, G., Shin, H. C., Suh, N., Kim, Y. M., Ahn, T. Y., & Kim, C. S. (2021). Safety of autologous bone marrow-derived mesenchymal stem cells in erectile dysfunction: An open-label phase 1 clinical trial. *Cytotherapy*, *23*(10), 931–938. DOI: 10.1016/j.jcyt.2021.06.001 PMID: 34326007

Zasieda, Y. (2020). ERECTILE DYSFUNCTION TREATMENT WITH COMBINATION OF MESENCHYMAL STEM CELL DERIVED EXOSOMES AND FOCUSED LOW-INTENSIVE SHOCK WAVE THERAPY. *Men's Health. Gender and Psychosomatic Medicine*, *1-2*, 70–78. DOI: 10.37321/UJMH.2020.1-2-08

Zhou, F., Hui, Y., Xin, H., De Xu, Y., Lei, H. E., Yang, B. C., Guan, R. L., Li, M., Hou, J. Q., & Xin, Z. C. (2017). Therapeutic effects of adipose-derived stem cells-based microtissues on erectile dysfunction in streptozotocin-induced diabetic rats. *Asian Journal of Andrology*, *19*(1), 91–97. DOI: 10.4103/aja.aja_29_20 PMID: 27345005

Zhou, F., Hui, Y., Xu, Y., Lei, H., Yang, B., Guan, R., Gao, Z., Xin, Z., & Hou, J. (2016). Effects of adipose-derived stem cells plus insulin on erectile function in streptozotocin-induced diabetic rats. *International Urology and Nephrology*, *48*(5), 657–669. DOI: 10.1007/s11255-016-1221-3 PMID: 26820518

Chapter 3
Urinary Incontinence

ABSTRACT

This chapter, addresses the complexities of the different types of urinary incontinence, such us the urge, stress and male incontinence. It aims to raise awareness that these conditions, often associated with females but that can also affect the male patients, can be effectively treated with the evolution of the regenerative therapies. The chapter explores the latest studies and interventions, many of which are in the preclinical phase, emphasizing their potential to significantly enhance the quality of life for those affected. By targeting the urothelium, the goal is to restore bladder function and alleviate symptoms, through the intervention in the pathologies that affects the involved tissues, including the vascular and neural. Urinary incontinence, a prevalent condition affecting both men and women, significantly impacts quality of life. While traditional treatments such as surgery (e.g., urethral sling) offer relief, they often present limitations. Stem cell therapy emerges as a promising alternative, capitalizing on the regenerative potential of these cells.

BACKGROUND

Pelvic floor dysfunction (PFD) is a highly prevalent urogynaecology disorder among middle-aged and elderly women, with an incidence of 30% to 50%, with a group of pathologies that had an impact in the quality of life, like pelvic organ prolapse (POP), stress urinary incontinence (SUI), fecal incontinence, and overactive bladder syndrome (OAB). There are many risk factors for developing these conditions, but the most important causes are related with childbirth-associated pelvic floor injuries.

The prevalence of PFD varies in different geographical regions. Wu et al. predicted that the number of women with one or more of this conditions will increase by 55.8% from 2010 to 2050. (M. Lin et al., 2022; J. M. Wu et al., 2009)

DOI: 10.4018/979-8-3693-8829-7.ch003

Urinary incontinence (UI) is defined by the International Continence Society (ICS) as the "complaint of involuntary loss of urine". (D'Ancona et al., 2019) Is a problem that affects women and men frequently, about 200 million people worldwide are facing this problem, and according to demographic predictors, it will increase to 40 million in the next two decades, with a profound impact on the quality of life, their subjective health status, levels of depression and need for care. In addition to causing much distress and embarrassment, it also has significant costs, both for individuals and for societies. Despite the differences in consensus regarding the definition of incontinence and the population studied, and therefore the variations in estimate prevalence, there is universal agreement on the importance of the problem in terms of human suffering and economic cost.

Although the symptom is the same, which is the uncontrolled loss of urine with different severity, it is important to distinguish if it is due to muscle weakness or stress incontinence, or related to disorders in bladder function and/or architecture, which may cause urinary urgency, or by overflow, when the bladder has lost its contractility.

For stress incontinence for example, treatment is scaled starting with physical therapy, lifestyle changes, and especially control of aggravating or triggering diseases such as diabetes, hypertension, and overweight. When these measures fail, surgical treatment with synthetic slings is chosen, which mid-urethral slings are the gold standard, with a success rate of 85–90%, and failure rates up to 15–10% (Hakim et al., 2015), requiring reoperations, without counting the complications added by the type of mesh used, flow obstruction, infection, exposure of the mesh, etc.

Short-term responses have been seen with therapies using injectable bulking agents such as Teflon, bovine collagen, silicone particles, and carbon beads; conversely, these bulking agents can cause chronic inflammation, foreign body giant cell response, periurethral abscess, erosion of the urethra, obstruction of the lower urinary tract with resultant urinary retention, and migration to inner organs and pulmonary embolism, and the success rates are inferior compared to the surgical treatment. (Carr et al., 2008; Hillary et al., 2020)

Recommendations from the European Association of Urology conclude that bulking agents should not be used as first-line treatment for patients seeking a permanent cure. (Burkhard et al., 2018) Likewise, the American Urological Association establishes that bulking agents are viable treatments; Nevertheless, patients should be advised that the results may not be as expected and that there may be long-term complications. (Kobashi et al., 2017)

On the other hand, urge incontinence, which shares some of the triggers of stress, also begins with changes in lifestyle, diet, and stress management. Then treatments are started with substances that regulate detrusor contraction such as antimuscarinics or beta 3-agonists, and when this fails or tolerance is created to them, the injection of botulinum toxin A, sacral neuromodulators, is opted for. More severe cases, partial or total cystectomy with bowel neobladder.

Being a disabling problem, due to the shame, the failure rate of the treatments, the number of re-interventions, among others, new techniques for its treatment are being studied all over the world, including regenerative medicine.

REGENERATIVE TREATMENTS FOR STRESS URINARY INCONTINENCE (SUI)

Stress urinary incontinence (SUI) has been described as the most common type of urinary incontinence. SUI is defined by the International Continence Society as a "Complaint of involuntary loss of urine on effort or physical exertion including sporting activities, or on sneezing or coughing." (D'Ancona et al., 2019) The most relevant risk factors are obesity and increasing parity, principally in patients younger than 40 years (relative risk (RR) = 3.3). (Hillary et al., 2020; Rortveit et al., 2001)

Female SUI often has a multifactorial cause with functional defects of the urinary tract bladder sphincter and morphological nerve damage. The underlying mechanisms for the onset and progression of SUI are not completely understood, but it is well known that the integrity of the external and internal urinary sphincters is necessary to maintain the continent function. Furthermore, somatic and autonomic nerves, fascial and ligaments structures, and urethral mucosa play an important role for the urinary continence mechanism. In male SUI is related to prostate resection or radical prostatectomy (RP) for example.

Blaivas and Olsson created a classification, based on the urethral hypermobility and intrinsic sphincter deficiency (ISD). The first one might be the results in a lack of equal pressure transmission to the bladder and the urethra, while ISD results in weakness in all the smooth and striated muscle components, essentials for the function of the female sphincter mechanism, as well as the submucosal vascular cushions. ISD seems to lead to a more symptomatic form of SUI than hypermobility. (Blaivas & Olsson, 1988; Hillary et al., 2020)

Following the 2019 ban by the U.S. Food and Drug Administration (FDA) on polypropylene mesh for female pelvic floor repair due to complications such as erosion and tissue exposure, tissue engineering has emerged as a promising alternative. But, despite its therapeutic potential, the integration of tissue engineering strategies in female pelvic floor reconstruction remains insufficiently examined. (X. Wu et al., 2020)

Regenerative medicine is based on regeneration of the damaged rhabdosphincter, to improve the function of the external (striated) muscles and internal (smooth) muscles, as well as the sphincter blood circulation (Barakat et al., 2020a; Cao et al.,

2020; Furuta et al., 2007; Goldman et al., 2012; Vaegler et al., 2012; H. J. Wang et al., 2011).

Figure 1. SC therapy for urinary incontinence and vaginal conditions

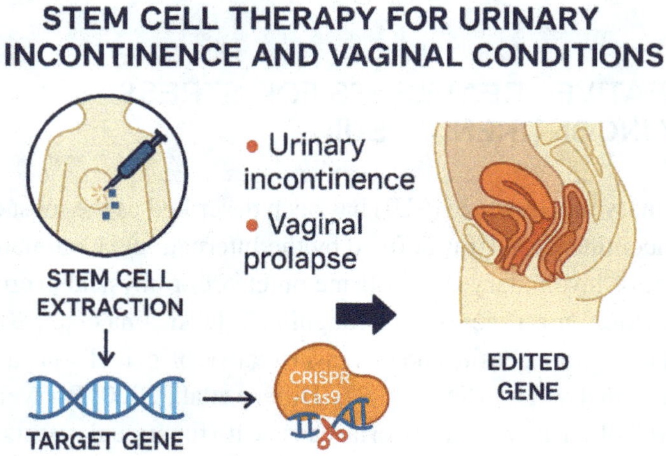

The stem cells therapeutic potential is derived by their plasticity, but it is now known that different bioactive factors secreted by stem cells also play an important role, by mediating their plasticity or the recruitment for replacement cells from the host at the injury site. A secretome is defined as all the proteins and components of the secretory pathway secreted by a cell into the extracellular space. In humans, this subset of the proteome encompasses 13-20% of all proteins, including growth factors, chemokines, cytokines, adhesion molecules, proteases, and receptors (Uhlén et al., 2015; Zambon et al., 2019).

The autocrine/paracrine effects of these bioactive molecules have been recognized as key regulators of several mechanisms, including immunomodulation, angiogenesis, inflammation, apoptosis, and tissue repair.(Dissaranan et al., 2014) The use of secretomes, with or without stem cells, can represent a promising approach to the treatment of SUI. (Zambon et al., 2019)

Preclinical trials have shown a better understanding of the mechanism of action on how externally injected cells can restore the container function. Multipotent mesenchymal stem cells (M-MSCs) from different tissues have been used in different clinical trials. Adipocytes have been found to have a tendency to become myocytes, so their regenerative potential on the artificial sphincter is high, as well as nerve regeneration. Not only stem cells have regenerative potential, but also the growth factors found in cocktails, such as hypoxia-induced factor-1 (HIF-1), vascular endothelial growth factor (VEGF), basic fibroblast growth factor (BFGF), and insulin-like growth factor (ILGF), exerting a paracrine effect for the modulation of inflammation and cell differentiation.

Most studies use mesenchymal stem cells (MSCs), like skeletal muscle-derived (SkMSCs: myoblasts, satellite cells, muscle progenitor cells and muscle-derived stem cells (MDSCs)), bone marrow stem cells (BMSC), human umbilical mononuclear cells (HUCBs), adipose-derived stem cells (ADSC), and modified/sorted stem cells which show various efficacy.(Hakim et al., 2015; Hillary et al., 2020) The first preclinical incontinence model was developed by Lin et al. in 1998. They observed possible differentiation of ADSC into the smooth muscle in a vaginal dilatation (VD) model, causing a significantly decreased periurethral muscle function and injured neurons in the pelvic plexus, then treated with intravenous ADSC injection by the appearance of Edu (5-ethynyl-2′-deoxyuridine)-positive nuclei localized within cells that stained positive for alpha-smooth muscle actin (α-SMA). (A. S. Lin et al., 1998)

However, preclinical tests have well-known problems. The first is that vaginal distension, the section made of the external sphincter muscle, or in the pudendal nerve is acute, while in the patient, this problem is chronic and due to other factors, with which reproducibility is questioned. It is well known that 90% of animal studies are not reproducible in humans.(Hakim et al., 2015)

Choosing a suitable preclinical model is difficult. Models ranging from the rat to the canine have been described (Eberli et al., 2009; Hakim et al., 2015). Other models such as non-human primates could be closer to the human subject, but their use is obviously limited. Despite these limitations, Badra et al. (Badra et al., 2013) demonstrated that striated muscle-derived stem cells integrated well into the sphincter system after an injury to the pudendal nerve in cynomolgus monkeys and that this also resulted in better sphincter function.

The second problem is the poor response of cell therapy in a chronic injury, as opposed to an acute injury model, because the natural healing makes these animal models generally not durable because of the tendency to recover spontaneously in a short period, unlike SUI in women. Some authors developed a more durable model by combining injury methods, longer injury time, and/or use additional substances to prolong the natural recovery of these models. (Hakim et al., 2015; G. Wang et al., 2011; Zambon et al., 2019)

In addition, the success tests were performed histologically, in sections where the striated muscle is found, but this does not imply an improvement in functionality. Finally, the injection route of the cells is different in animals, since it is usually periureteral, while in humans the transurethral route is used. Knowing the potential of autologous stem cells, without the problems of graft rejection, the minimal manipulation they require, and the clearly demonstrated safety, makes it logical to think that human trials will soon give us a new tool for SIU treatment, with accurate data about its efficacy

The main problem we face is the fear of patients to participate in clinical trials, in contrast to those who suffer from heart disease, it may be because although it greatly affects quality of life, this disease does not affect life itself. It also has a problem of cost of treatment, which for a disease with less relevance in society, many study groups are not willing to assume. Moreover, due to the lack of standardization around stem cell-based therapy, with different doses of cells in each study, it is difficult to reach a conclusion on what is the ideal cell concentration, or if it is required to be individualize according to the condition and the patient tissues.

Muscle-Derived Stem Cells (MDSCs)

MDSCs are the most common type of cell used, especially autologous, and they can be obtained under local anesthesia, resulting in low morbidity. In vitro, these cells can be differentiated into different mesenchymal lines, but also into neuronal and endothelial lineages, which is vital for sphincter regeneration and functional neural pathways. (J. Y. Lee et al., 2000; Qu-Petersen et al., 2002) Several studies have found good tissue regeneration, improved VLPP, and recovered contractile urethral sphincter function in animal models of SUI. (Cannon et al., 2003; Gul et al., 2020) However, isolated autologous MDSCs must first be expanded in vitro and need to be cultured for a certain time before injection into the urethral sphincter. (Barakat et al., 2020a; Y. Li et al., 2010; Zhou et al., 2016) The advantage of MD-SCs is that cells injected into the rhabdosphincter, administered transurethrally or periurethrally, contains myotubes and myofibers, which are essential for sphincter function and also can work as a blocking agent.(Chancellor et al., 2000; Seval & Koyuncu, 2022) Barak et al published a study on 11 clinical trials (637 patients), which used autologous MDSC directly into the rhabdosphincter via the transurethral or periurethrally route, nine included female patients (n = 352) and two studies included male patients (n = 285) (Table 1). (Barakat et al., 2020a)

The first MDSC clinical study (42 women and 21 men) with autologous myoblasts and fibroblasts was carried out by Strasser et al., reporting an improvement in the quality of life and contractility of the urethral sphincter, with success rate of >90% for women and >50% for men between 2002 and 2004. (Strasser et al., 2007) Fibroblasts were injected in the submucosa of the urethra, while myoblasts were implanted in the rhabdosphincter. However, this Austrian study was controversial because it was withdrawn due to scientific irregularities, ethics violations, and protocol irregularities.

The interpretation and comparison between the clinical trials with MDSC was difficult due to variability in surgical techniques and the number of cells injected. The median rate of postoperative improvement was 77% after MDSC injection (improvement rate: 1–2 pads).

The methodology used by Chancellor et al. (Chancellor et al., 2000) in a preclinical trial with cell therapy using an intraurethral injection of MDSCs to enhance the urethral wall closing and to possibly improve the urinary sphincter function for SUI, is now being followed by other researchers. They described that the regenerative response of MDSCs has been shown to be superior to myoblasts in a myocardial infarct model, because of their ability to undergo long-term proliferation, self-renewal, and multipotent differentiation, including differentiation toward endothelial and neuronal lineages. (J. Y. Lee et al., 2000; Qu-Petersen et al., 2002)

Mitterberger et al. published the largest series of patients a study with 123 women treated with autologous injections of collagen matrix combined with MDSCs and fibroblasts into the rhabdosphincter and into the urethral submucosa, and after a 1-year, they observed a subjectively complete improvement in 79% of women between 36–84 years of age, 13% of substantial improvement and 8% of slight improvement following. Each patient received a large number of injections (40–50 in total) and, the collagen matrix vehicle could have acted as a bulking material, particularly with such a large injection volume. (Gul et al., 2020; Mitterberger, Pinggera, et al., 2008) This group, however, had to withdraw their publications because of ethical issues. The composite continence score they developed to show this success rate is no longer used and the highly favorable results cannot be trusted. (Hakim et al., 2015)

Some other studies showed a cure rate between 40% and 75%. (Carr et al., 2008; C. N. Lee et al., 2010; Sèbe et al., 2011) Sèbe et al. (Sèbe et al., 2011) Published a study with 12 women with persistent SUI after failed surgery, treated with MDSCs injection. They reported quality of life improvement in 50% at 3 months' time, 3 patients became dry, 7 improved and 2 worsened. Another study from Gerullis et al. using autologous MDSCs in 222 patients, (Gerullis et al., 2012a) reported 12% of patient's continent, 42% improved, and 46% had chronic urinary incontinence after a 6- to 12-month follow-up.

Other studies did not reach these cure rates, like the one from Stangel-Wojcikiewicz et al. with 16 women informing an improvement in 25% of them based on clinical and urodynamic outcomes. (Stangel-Wojcikiewicz et al., 2014) Sharifiaghdas et al. conducted a prospective cohort research using MDSCs for the treatment of SUI. (Sharifiaghdas et al., 2016) Three out of ten patients regained full continence after 3 years of follow-up, assessed with a cough stress test, a 1-h pad test, and questionnaires. Three patients did not respond to the medication, whereas four individuals shown great improvement. (Seval & Koyuncu, 2022)

Intrasphincteric injection of autologous adult muscle-derived cells (AMDC) in high doses seemed to be safe. Additionally, efficacy data suggested a trend toward greater efficacy with doses of 32 × 10 6 AMDC or greater. (Hakim et al., 2015)

Carr et al. (Carr et al., 2008) first published a study with 8 women with SUI of whom 3 withdrew from the trial early. Of the 5 remaining patients, one achieved continence while the others showed only some improvement. Importantly, they noted that the onset of improvement was only after 3–8 months. The same author also issued 5 years later a study regarding the different doses. (Carr et al., 2013) A total of 38 women underwent intra-sphincter injection of low doses (1, 2, 4, 8 468 or 16 × 10 469 6) or high doses (32, 64 or 128 × 106) of AMDC, obtained by a large biopsy of skeletal muscle (5–30 g), which was associated with donor site pain, bruising and bleeding. The study finalized with 33 patients, demonstrating significant improvements in pad weight testing following AMDC injection therapy ($p<0.001$). Compared with those receiving a low dose, a greater proportion of patients treated with a high cell concentration reported 0–1 leak over 3 days at their 18-month follow-up (88.9% versus 33.3%).

In 2014, Peters et al. published another of the largest series to date of AMDC injection for SUI, including 80 female patients, and 72 completed diaries and pad tests at 12-month follow-up. The investigators pooled the data from two trials of autologous MDSC injections at a dose range of 10 × 106–200 × 106, derived from biopsies of each patient's quadriceps femoris. After 1 year, all treatment groups described subjective improvement that was maintained throughout the study period. The higher dose AMDC group had significant improvements in mean pad weight testing ≥50% over a 12-month period ($p<0.05$), curiously, this group also had the greatest mean baseline pad weight. (Peters et al., 2014)

Blaganje et al. (Blaganje & Lukanović, 2012) injected AMDC in 38 women with SUI. Initially, the women were treated with electrical stimulation alone for 6 weeks. Then the stem cells were injected, and another 6 weeks of electrical stimulation were given. At the end of this 6-week period, 5 women were cured and 29 improved. Unfortunately, there was no control group. Kuismanen et al. (Kuismanen et al., 2014) injected 5 women with ADSC of whom 1 became continent after 6 months and 2 more at 12 months, using the cough stress test as an outcome measure. So far, no randomized controlled data are available, nor are there any long-term data.

Garcia-Arranz et al. designed two prospective, nonrandomized phase I-IIa clinical trials of urinary incontinence involving 9 men (8 treated) and 10 women using AMDC, obtained from liposuction containing, through endoscopic intraurethral injection, doing the follow-up with pad test, quality-of-life and incontinence questionnaires, and additionally, urethrocystoscopy and urodynamic studies on the first and last visit. They found an objective improvement of >50% ($P < .05$) and a subjective improvement of 70% to 80% from baseline in 3 men (37.5%) and 5 women (50%). (Garcia-Arranz et al., 2020)

A meta-analysis published by Pokryczynska et al. enrolled 320 female patients who underwent periurethral injection of AMDC. When they reach 12 months' follow-up, the cure and recurrence rates reported were 20–40% and 50–70%, respectively. No major adverse events occurred during the study timeframe. (Pokrywczynska et al., 2016)

The regenerative potential that these cells have is clear, in addition to their own components inherent to them. However, MDSC injection has certain disadvantages including cell migration and uptake, as occurs with other commercial injectable agents, which requires the development of new strategies, such as tissue engineering, to improve the retention and survival of transplanted stem cells. (Barakat et al., 2020a)

Table 1. Clinical studies using MDSCs for SUI

Author	Patients, n	n Mean follow-up	months Continent (no pads) and improvement (1–2 pads) (n/N)	Injection site Periurethra or Submucosa (o'clock positions)	Functional outcomes
Stangel-Wojcikiewicz et al., 2014[38]	16 Females	24, 48	8/16 (50) continent 4/16 (25) improvement 4/16 (25) persistent SUI	3, 9, 12	NR
Sharifiaghdas et al., 2016	10 Females	1, 6, 12, 36	3/10 continent 4/10 Improvement 3/10 persistent SUI	2, 5, 7	MUCP: 15.6 cmH2O 12 months: 26.1 cm H2O
Peters et al., 2014[34]	80 Females	1, 3, 6, 12	40/80 (50) Improvement		No adverse events; all groups had subjective improvements; significant improvements in pad weight ($P < 0.01$) and bladder diary leaks ($P < 0.001$) by >50% for highest dose versus baseline.
Cornu et al., 2014[40]	11 Female same patients as Sebe et al. [32]	72	3/11 (27) improvement	3, 12	NR
Gräs et al., 2014[41]	20 Females with complicated SUI 15 Females with uncomplicated SUI		5/20 (25) continent 1/20 (7) improvement 13/20 (63) persistent SUI	3, 9, 12	Reduction in mean number of leaks from 2.8 to 1.5 per day for complicated SUI ($P < 0.01$) versus 4.1 to 3.3 per day ($P < 0.01$) for uncomplicated SUI; total subjective improvements in 63% for complicated SUI versus 57% for uncomplicated SUI.
Carr et al., 2013 [42]	38 Female	1, 3, 6, 12, 18	29/38 (76) improvement	External urethral sphincter	>50% reduction in pad weight in 88.9% for high dose versus 61.5% for low dose; >50% reduction in number of leakages in 77.8% for high dose versus 53.3% for low dose (all $P < 0.01$).

continued on following page

Table 1. Continued

Author	Patients, n	n Mean follow-up	months Continent (no pads) and improvement (1–2 pads) (n/N)	Injection site Periurethra or Submucosa (o'clock positions)	Functional outcomes
Blaganje et al., 2012 [35]	38 Female	1, 1/2	5/38 (13) continent 29/38 (76) improvement 3/38 (7) persistent SUI	Two levels	Stress test negative in 78.4%; 13.5% subjective cure, 78.4% subjective improvement.
Sebe et al., 2011[31]	12 Females	1, 3, 6. 12	3/12 continent 7/12 improvement 2/12 worsening	3,9	QOL improved in 66.7%; improvement on pad testing in 58.3%; subjective cure in 25%.
Mitterberger et al., 2007[43]	123 Females	12	94/119 (79) continent 16/119 (13) improvement 9/119 (8) slight improvement	Mid-urethra	NR
Gerullis et al., 2012[44]	222 Males	6	26/222 (11) continent 94/222 (42) improvement 102/222 (45) persistent SUI	Around urethra	VLPP/MUCP/MBC/Qmax/ Pdet-max/MRU/24-h voiding diary/pad test: (~5)/QOL (~50) score.
Mitterberger et al., 2008[45]	63 Males	12	41/63 (65) continent 17/63 (26) improvement 5/63 (8) persistent SUI	Around urethra. Around submucosa	40% improvement in MUCP; 79% subjective cure per I-QOL.
Garcia-Arranz et al. (2020)	8 males and 10 females	4, 12, 24 weeks, and 1 year	subjective improvement of 70%–80% from baseline	Endoscopic intraurethral injection	3B36:G45/8 (37.5%) 5/10 (50%)

Kaufman's comprehensive review of autologous muscle-derived cell injections for urethral sphincter regeneration (AMDC-USR) synthesized safety and efficacy data from initial clinical studies. No long-term adverse events were reported. Patients receiving higher-dose AMDC-USR injections achieved at least a 50% reduction in stress urinary leakage and pad weight at 12-month follow-up. All dosage cohorts demonstrated statistically significant improvements in incontinence-specific quality of life at one year. Pooled analyses confirm that AMDC-USR is safe across a broad dose range and reveal a dose-dependent efficacy, with higher doses eliciting greater patient response rates (Kaufman, 2020).

Adipose-Derived Stem Cells (ADSCs)

ADSCs are one of the most widely used stem cells for autologous transplants, due to the easy recovery of adipose tissue, being able to recover large amounts of cells safely with minimal morbidity. In addition, ADSCs have the ability to proliferate rapidly, even at low serum levels. ADSCs are an alternate source of autologous adult stem cells that is obtainable in large quantities by suction-assisted lipectomy under local anesthesia, with minimal discomfort.

They have the potential for mesodermal tissue differentiation, they could be induced to differentiate into adipocytes, smooth muscle cells, and neuron-like cells, the latter of which expressed neuronal markers, which is applicable for tissue regeneration and revascularization, using substances like Isobutylmethylxanthine (IBMX), indomethacin (INDO), and insulin were the active ingredients in a previously established neural induction medium (NIM). (Ning et al., 2006; Zuk et al., 2001). Some studies suggest that ADSCs hold substantial potential for restoring urethral sphincter muscle function. Successful regeneration of the sphincter depends not only on enhanced myotube formation but also on the concurrent reestablishment of the vascular network and the integrity of the ECM—both of which are vital for the restoration of normal urethral physiology. (Knoll et al., 2024)

Rodriguez et al. (Rodríguez et al., 2006) showed that ADSCs can be obtained from processed liposuctions and can differentiate into functional urothelium and smooth muscle tissue, with adequate contractility under cholinergic stimuli, being a cellular source for the treatment of injured smooth muscle tissues. Thus, ADSCs can be used for both the male and female UI. Gotoh et al. (Gotoh et al., 2014), Yamamoto et al. (Yamamoto et al., 2012), and Choi et al. (Choi et al., 2016a) reported evidence that periurethral injection of Autologous ADSCs are a safe and technically feasible treatment option for men with moderate SUI after Radical Prostatectomy (RP) and Holmium Laser Enucleation of the Prostate. Arjmand et al. used ADSC injection in 10 female patients with SUI and demonstrated a substantial improvement at 2–24 weeks after the procedure. (Arjmand et al., 2017) (Table 2).

The high volumes injected might also contribute to a bulking effect. Kuismanen et al. gave transurethral ADSC injections to 5 women with SUI. Cells were harvested from the abdominal wall, isolated, expanded for 3-4 weeks, and suspended in a bovine collagen gel and saline matrix for injection. No clinically significant morbidity resulted from the cell harvest. At 1 year, 3 of 5 patients had a negative cough test with the bladder filled to 500 ml, 2 of them were satisfied with the treatment and did not wish further treatment for SUI. The authors conclude that the feasibility and efficacy of the treatment were not optimal (Kuismanen et al., 2014).

Table 2. Clinical studies using ADSCs for SUI

Author	Patients, n	n Mean follow-up	months Continent (no pads) and improvement (1–2 pads) (n/N)	Injection site Periurethral or Submucosa (o'clock positions)	Functional outcomes
Gotoh et al., 2014 (Gotoh et al., 2014)	13 Males	1/2, 1, 3, 6, 9, 12, 48	8/11 improvement 3/11 no improvement	5, 7 4, 6, 8	MUCP ↑ (~9.2 cmH2O) FPL ↑ (~5.6 mm) PVR ↑ (~4.5 mL)
Choi et al., 2016 (Choi et al., 2016b)	6 Males	3	6/6 improvement	5, 7 4, 6, 8	
Arjmand et al., 2017 (Arjmand B, Safavi M, Heidari R, s. f.)	10 Females	1/2, 1,5, 6	10/10 improvement	2, 10	2 weeks ($P < 0.001$) 24 weeks ($P = 0.018$)
Kuismanen et al., 2014 (Peters et al., 2014)	5 Females	3, 6, 12	3/5 continent 2/5 improvement	3, 9	Negative cough test in 60% at 1 year ; improvements in pad weight testing in those with negative pad testing; trend towards subjective improvement on UISS.

A clinical review from Barakat et al, found in a total of five clinical trials (40 patients) using adult ADSCs (three in men and two in women). The median post-operative improvement rate was 88% after ADSC injection (improvement rate: 1–2 pads). They clarify that it is difficult to assess the long-term safety of the results due to insufficient data, since the published studies were carried out in small populations and had short-term results. More scientific studies with a longer follow-up period and more patients are required. (Table 2).

The injection route has also been investigated. For instance, Li et al. compared periurethral and intravenous administration of ADSCs in a rat model of stress urinary incontinence. Their study assessed differences in urethral sphincter recovery, cytokine secretion levels, and ADSC retention within urethral tissues. However, no significant differences were observed between the two delivery methods (G. Li et al., 2022). In vivo studies have demonstrated that neurally induced ADSCs (NI-ADSCs) facilitate nerve regeneration by promoting and sustaining the repair-associated phenotype of Schwann cells. In a rat model of SUI, intraurethral injection of NI-ADSCs significantly upregulated the expression of neural markers such as S100β and demyelination indicators like GFAP, alongside improving urodynamic parameters. Conversely, administration of DAPT—a Notch signaling pathway inhibitor—attenuated these neural repair effects (M. Liu et al., 2025).

However, giving the potential of the ADSCs, additional researches are needed to develop SUI injection therapies, probably with a different type of matrix or count of cells.

Human Umbilical Cord Blood Stem Cell (HUCBs

HUCB are an alternative source to get large numbers of hematopoietic cells. HUCBs can be extracted from human umbilical cord vein blood by puncture without invasive measures, and are not considered embryonic cells, which is a great advantage compared to other cell sources. Additionally, there is a low risk of graft-vs.-host disease and virus contamination with HUSCs, and probably requiring that the matching with the HLA types be less rigorous. Stem cells from the elderly have a low differentiation potential compared to stem cells from the young. (Barakat et al., 2020b; Seval & Koyuncu, 2022)

The first transplantation of HUCB cells was successfully performed in 1988 to treat Fanconi's anemia, (Gluckman et al., 1989; Kong, 2004) accumulating extensive experience in the use of HUCB cells to reconstitute the hematopoietic system, and their proliferative capacity seems to be superior to that of cells from blood, (Gluckman et al., 1997) and also with a reduced risk of graft-versus-host disease.(Rocha et al., 2000) Some authors confirm their differentiation into nonhematopoietic cells such neurons or glia,(Sanchez-Ramos et al., 2001) hepatic, endothelial, or muscle cells.(Ishikawa et al., 2003; Pesce et al., 2003)

Lee et al. (C. N. Lee et al., 2010) used peri-urethral injections of HUCB for the treatment of SUI in 39 women after failure of conservative treatment and surgical therapy. Once the blood was collected, it was stored in an aseptic bag with 23 ml of citrate-phosphate-dextrose-adenine (CPDA-1) at –70°C. HUCB injection was done under local anesthesia in the submucosal area of the proximal urethra at the 4 and 8 o'clock positions of the urethra ($430 \pm 190 \times 106$ cells/2mL). No postoperative complications were reported. Using the Patient's Satisfaction tool, they indicated subjective improvement in 72% of cases after 12 months of treatment: 13 patients (36%) were completely satisfied and 13 patients (36%) reported a significant improvement of their symptoms (Table 3). However, 10 patients (27%) in this study remained incontinent. They also noted a progressive improvement over time since more patients were cured at 12 months than at 3 or 6 months.

Table 3. Clinical studies using other types of stem cells, HUCBs and TNCs

Author	Patients, n	n Mean follow-up	months Continent (no pads) and improvement (1–2 pads) (n/N)	Injection site Periurethral or Submucosa (o'clock positions)	Functional outcomes
Lee et al., 2010(C. N. Lee et al., 2010)	39 Females only 36 completed follow-up Heterologous HUCB	1, 3,12	13/36 (36) continent 13/36 (36) improvement 10/36 (27) persistent SUI	4, 8	Improvement in MUCP by 120% for ISD and by 62.9% for MUI; subjective cure in 36.1%, no improvement in 27.8%.
Shirvan et al., 2013(Shirvan et al., 2013)	9 Females TNCs/ platelets	1, 3,6	8/9 improvement 1/9 persistent SUI	1.5, 3, 4.5, 6, 7.5, 9, 12	↑ MUCP (from <30 cmH2O)

Bone Marrow-Derived Stem Cells (BMDSC)

The bone marrow contains multiple cell populations, such as erythrocytes, macrophages, endothelial cells, fibroblasts, adipocytes, and hematopoietic cells/ mesenchymal SC (Barakat et al., 2020a; Salem & Thiemermann, 2010) The first BMSCs were described as osteogenic progenitor cells in 1966. They were later used for tissue engineering.

BMDSC had promising regenerative properties in preclinical studies, and they have a capacity of differentiating into various cell types including skeletal muscles, smooth muscles, and neurons, (Kinebuchi et al., 2010) but the main disadvantage is the biopsy required in humans to obtain this cells for clinical use, since it requires aspiration under local anesthesia, which is painful and often only a small number of suitable BMDSC are isolated because the stem cell population is very small. (Gunetti et al., 2012; Kinebuchi et al., 2010) The alternative could be mobilization from the bone marrow to the blood, obtaining a greater number of cells, and with an easier method of extraction through the vein. The isolated BMSCs it could also be expanded in vitro and cultured for a shortly before the final injection. (Barakat et al., 2020a)

No trials in humans on the effect of BMDSC injection for the treatment of SUI have been reported to date. (Hillary et al., 2020) Some studies in animal models have shown that injected BMSCs were able to restore damaged urethral sphincter tissue and significantly improve UI. (Gunetti et al., 2012) However, the results of other animal studies were very contradictory to these findings.

Kinebuchi et al. (Kinebuchi et al., 2010) issued a preclinical study where GFP-labelled BMDSC were injected into the urethral sphincter of Sprague–Dawley rats, which were rendered incontinent by urethrolysis. After 13 weeks, transplanted cells survived and differentiated into striated muscle cells and peripheral nerve cells. Fluorescent imaging was used for demonstrating skeletal muscle regeneration, and they evidenced a greater proportion in the BMSC group compared to the CFM group, as well as the proportions of peripheral nerves in the urethra. Conversely, the VLPP of the urethra was not improved. (Hillary et al., 2020) In a similar study, Gunetti et al. injected GFP-labelled BMDSCs into rats injured by urethrolysis, they also demonstrated the persistence of cells after 3-month, located within neuromuscular sites. (Gunetti et al., 2012)

Unfortunately, there is a lack of compelling evidence demonstrating long-term efficacy, as well as a study of malignant transformation. (Pokrywczynska et al., 2016; Zambon et al., 2019). To try to improve results, studies have been done with Low-Intensity Extracorporeal Shock Wave Therapy (Li - ESWT) to stimulate the growth of striated muscle and promote tissue repair, due to the mobilization and activation of stem cells at the site of injury, and thus stimulation of the neoangiogenesis and nerve plasticity. (B. Wang et al., 2018; Zambon et al., 2019)

Urine-Derived Stem Cells (UDSCs)

UDSCs can be easily obtained from human urine and be expanded in vitro, which means that invasive measures are not required for their recovery, which translates into a reduction in economic costs. Barak et al found few experimental studies using UDSC for the treatment of UI, which are one by Zhang et al. (Barakat et al., 2020a; Y. Zhang et al., 2008) with the collection of 55 urine samples of 15 healthy individuals and eight patients with VUR, its isolation and expansion in vitro being effective. The authors demonstrated that individual clones of UDSC have the ability to differentiate into cell lines that express urothelium, smooth muscle, endothelial and interstitial cell markers (Barakat et al., 2020a; Gunetti et al., 2012; Salem & Thiemermann, 2010; Y. Zhang et al., 2008)

Total Nucleated Cells (TNCs)

There is the possibility of obtaining Multipotent Endothelial cells (EPC) from peripheral blood to produce autologous TNC for the treatment of UI. In addition, during this process growth factors and platelet-rich plasma are obtained. Some of the most important factors are: platelet-derived growth factor, transforming growth factor β1, fibroblast growth factors, vascular endothelial growth factors, etc.). (Barakat et al., 2020a; Roobrouck et al., 2008) In a clinical study using mixed platelets TNC for the treatment of severe female UI 120 ml of peripheral blood was drawn to obtain platelets and nucleated cells. After centrifugation, the total volume of platelets and TNC was equal. (Barakat et al., 2020a; Shirvan et al., 2013) A 10 mL mixture was then injected into the periurethral tissue. Eight of the nine patients had a clinically significant improvement according to the International Incontinence-Urinary Incontinence Questionnaire (ICIQ-UI) and ICIQ-QOL at 3 and 6 months of follow-up. One patient had a UI improvement, but not cured.

Biological Scaffolds

In natural physiological conditions, cellular behavior and the structural development of tissues are regulated by the nanoscale architecture of the ECM. Due to the anatomical characteristics of the vaginal wall, for example for treatment of UI, direct stem cell injection is not a viable therapeutic option for tissue repair, as most mammalian cells require an adhesive substrate to survive and function. Consequently, researchers are increasingly turning to biological scaffolds that mimic the ECM topography specific to vaginal tissues. These scaffolds provide a three-dimensional matrix conducive to cell adhesion, proliferation, and differentiation with high efficiency.

To be suitable for tissue engineering applications in pelvic floor dysfunction, scaffold materials must meet the following criteria: (1) exhibit excellent biocompatibility, promoting cell adhesion and proliferation without toxicity or immunogenicity; (2) be biodegradable; (3) possess sufficient mechanical strength to support tissue regeneration; and (4) present appropriate porosity and aperture size to facilitate nutrient diffusion and cellular integration. Scaffold materials employed in pelvic floor reconstruction fall into two primary categories: synthetic and natural biomaterials.(X. Wu et al., 2020)

Two of the most common used synthetic polymers utilized materials in tissue engineering, are polyhydroxy acids such as polylactic acid (PLA), polyglycolic acid (PGA), and their copolymer polylactic-co-glycolic acid (PLGA), (Okamoto & John, 2013; Pan & Ding, 2012). These polymers are inherently non-toxic and undergo chemical hydrolysis, producing lactic and glycolic acids that are naturally metabolized by the body. Their degradation is independent of enzymatic activity within the local tissue microenvironment, allowing for predictable and controllable resorption rates in vivo (Atala et al., 2006). These materials have demonstrated efficacy in applications such as urethral reconstruction and bladder replacement (H.-K. Lin et al., 2015; Raya-Rivera et al., 2011; Ribeiro-Filho & Sievert, 2015).

Synthetic polymer-based hydrogels also serve as platforms for the sustained release of bioactive molecules, thereby enhancing stem cell differentiation and tissue integration within three-dimensional scaffolds (Nguyen et al., 2014). Another hydroxyalkanoate polymer, the Polycaprolactone (PCL), has an excellent biocompatibility, low immunogenicity, and ability to undergo hydrolysis under physiological conditions (Boffito et al., 2015; Z. Li & Tan, 2014), but certain polyhydroxy esters, can produce acidic degradation byproducts, potentially altering the local pH and affecting cellular behavior, viability, and eliciting unwanted inflammatory responses (L. Lu et al., 2000; Vacanti et al., 2012). Moreover, synthetic polymers generally lack intrinsic bioactive domains, limiting their capacity to support direct cell adhesion.

To overcome these limitations, advanced fabrication strategies have emerged. These include biofunctionalization techniques such as collagen-rich serum coatings or covalent conjugation of ECM-derived peptides, which facilitate cell attachment and ECM deposition on synthetic scaffolds(Coburn et al., 2012; X. Li et al., 2008; Zhu, 2010). The incorporation of biological activators onto scaffold surfaces further promotes cell–matrix interactions and enhances lineage-specific stem cell differentiation (X. Liu et al., 2012; Rossi & van Griensven, 2014).

An ideal scaffold material should minimize immunogenicity while simultaneously offering robust mechanical integrity and biological functionality. It must support cellular signaling, cytokine integration, and recreate a microenvironment conducive to tissue regeneration (Rice et al., 2013). Indeed, scaffold parameters—including geometry, porosity, surface topology, and mechanical profile—exert critical influence on host cell behavior and overall biomaterial performance once deployed in vivo.

Table 4. Summary of materials commonly used in female pelvic floor tissue engineering

Materials	Application	Advantages
PLA	Urogynecological alternative implant in vitro study (Mangera et al., 2013), ascorbic-acid releasing biomaterials for pelvic floor repair (Mangera et al., 2013; Roman et al., 2016; Vashaghian et al., 2017)	Phenotypical morphology and functionality maintained, increased cell metabolic activity and proliferation, matrix deposition, and collagen production (Mangır et al., 2016)
PLGA/PCL	Implant for pelvic floor (Vashaghian et al., 2017)	Comparative strength to native tissue, cell adhesion and growth, migration through the scaffold (Vashaghian et al., 2017)
SILK	Urethra reconstruction (Xie et al., 2013),Tissue engineering mesh for pelvic floor reconstruction in abdominal wall rat model (Horan et al., 2009), biocompatibility, no inflammation, tissue growth (Xie et al., 2013), tissue ingrowth, degradation (Horan et al., 2009)	Cell migration, adhesion and proliferation (Xie et al., 2013), endothelial and smooth muscle cell attachment, and proliferation (Bilsel & Abci, 2012)

continued on following page

Table 4. Continued

Materials	Application	Advantages
Collagen (type I, II, and III), cellulose	Tissue engineering scaffold (Alamein et al., 2013; M. Li et al., 2005), urinary bladder regeneration (Han & Gouma, 2006), mesh for pelvic floor reconstruction in abdominal wall rat model (Horan et al., 2009)	Providing microstructure and cell adhesion, cell attach and proliferation (Alamein et al., 2013; M. Li et al., 2005), cell penetration through implant, proliferation, matrix deposition (Han & Gouma, 2006)
BM/SM	Urinary bladder regeneration (Han & Gouma, 2006), mesh for pelvic floor reconstruction in abdominal wall rat model (Horan et al., 2009)	Natural ECM analogues to promote cell-ECM interaction (Crapo et al., 2011; H. Lu et al., 2012)

(X. Wu et al., 2020)

On the other hand, natural biomaterials are also employed in tissue engineering, and they are broadly categorized into two classes: protein-based biomaterials (including collagen, silk fibroin, gelatin, fibronectin, and keratin), typically derived from human or animal sources and contain bioactive molecules that mimic the native ECM, and polysaccharide-based biomaterials (such as hyaluronic acid, cellulose, glucose, alginate, chondroitin, chitin, and its derivative chitosan), originate primarily from algae (e.g., agar and alginate) or microbial sources (e.g., dextran and its derivatives) (Chow et al., 2008; Cui et al., 2007).

The ECM has two structurally and functionally components, the basement membrane (BM) and the stromal matrix (SM). The BM includes macromolecules such as type IV collagen, laminin, glycoproteins, nidogens, and perlecan, which together organize a supportive membranous network for peripheral cells, facilitate epithelial integrity, and maintain cell polarity (Badylak et al., 2009; Crapo et al., 2011). In contrast, the SM is composed of larger fibrous proteins that provide tensile strength and structural support.

Although acellular, native ECM materials are rich in cytokines and bioactive molecules. When applied to compromised pelvic floor tissues, these components elicit critical regulatory functions, guiding cellular processes including adhesion, proliferation, differentiation, and matrix remodeling (Uebersax et al., 2009). As a preformed, immune-compatible scaffold, ECM-based biomaterials are increasingly recognized for their therapeutic potential in regenerative medicine and pelvic floor reconstruction.

SUI IN MALES

Taking into account that the SUI is common among men, particularly following radical prostatectomy, stem cell therapy can be a feasible procedure for the treatment of this condition, with short-term safety and promising results reported in the literature. However, it is important to remember the possible complications associated with the surgery, like strictures, scars, and fibrotic areas in the membranous urethra, prior injection of bulking agents or internal urethrotomy as well as radiation therapy negatively, and the bad influenced the success rates. (Gallo et al., 2019)

Authors like Young et al. used ADSC in six men with persistent urinary incontinence after the surgery injected transurethrally into the submucosal space of the membranous urethra. They reported an improvement in terms of urine leakage volume in 4 patients in the 24-h pad test and temporal deterioration in 2 patients. The improvement regarding the subjective symptoms and quality of life were similar. MRI showed an increase in functional urethral length (from 6.1 to 8.3 mm) between the lower rim of the pubic bone and the bladder neck. They did not observe any adverse events, such as pelvic pain, inflammation, or de novo urgency during follow-up. (Choi et al., 2016a)

Based on his previous data in women, the first clinical trial for male SUI was performed by Mitterberger et al. using the same technique of ultrasound-guided injection of MDSCs in 63 male patients with SUI after radical prostatectomy. They reported 41 patients (65%) with the recovery of the continence and 17 (27%) showed improvement after the injection, while 5 (8%) did not show any progress. (Mitterberger, Marksteiner, et al., 2008)

In 2012, Gerullis et al. reported on the largest clinical trial therein on 222 males receiving five injections of autologous MDSCs around the rhabdosphincter, reporting 26 (12%) with the recovery of the continence, and 94 (42%) with improvement after 1 year of follow-up (Gerullis et al., 2012b). In 2012, Yamamoto et al also used a periurethral injection of ADSCs in 3 patients, finding a progressive improvement of the sphincter function after six months in 2, and SUI improved in terms of leakage volume measured by a 24-h pad test in all cases and disappeared in the third patient. No significant adverse events were assessed. (Yamamoto et al., 2012)

One study in 11 men with SUI following radical prostatectomy who received ADSC injections into the rhabdosphincter reported a 40.9% improvement in mean pad weight ($P = 0.01$), MUCP, and incontinence episodes at 12 months72. ADSCs were isolated from 250 ml of adipose tissue aspirated from the abdominal wall, which resulted in minimal donor site morbidity. The group used rapid cell isolation and harvesting technology, which enabled them to perform cell harvesting and injection in the same operative setting.

In 2014, Gotoh et al. reported their experience with 11 patients after periurethral injection of ADSCs into the external sphincters of patients who had undergone radical prostatectomy or transurethral resection of the prostate with Holmium laser, and therapeutic effects and safety were observed at both one-year and long-term follow-up, describing a progressive improvement in sphincter function, proved by an increase in the maximum urethral closing pressure and functional profile length and no significant adverse events were observed. They hypothesize about the role of the bulking effect produced by the injected adipose tissue fraction mixed with ADSCs, and also the potential of the paracrine effect of the ADSCs which leads to differentiation into mature adipose tissue and possibly into contractile cells. (Gotoh et al., 2014)

In 2016, Choi et al informed similar efficacy and safety with the use of a similar technique in a group of 6 patients with SUI after prostate surgery. (Choi et al., 2016a; Shin et al., 2020a)

Next to the lack of good data, the numbers of clinical trials and of patients in current trials are small and the amount of evidence is limited, a health-economic analysis will be needed to establish the potential role of this therapy in the future. Dosage of cells injected, the number of injections, a consensus on the most suitable source and injection route of SC, and the method of delivery are other factors that need to be addressed in a future study.

URGENCY URINARY INCONTINENCE

The Urgency Urinary Incontinence defined by ICS as:" Observation of involuntary leakage from the urethral orifice associated with the individual reporting the sensation of a sudden, compelling desire to void." (D'Ancona et al., 2019)

Urinary urgency, Overactive Bladder (OAB) Syndrome refers to a constellation of symptoms including increased daytime frequency and/or nocturia, with urinary incontinence (OAB-wet) or without (OAB-dry), in the absence of urinary tract infection or another detectable disease. The uninhibited detrusor contractions are the cause of OAB symptoms. Bladder outlet obstruction (BOO) is a well-known cause of bladder dysfunction, it could be induced by benign prostatic hyperplasia, pelvic floor contracture or urethral stricture. Other causes for bladder dysfunction include chronic ischemia, hyperlipidemia, and diabetes. (Kim et al., 2014)

The Overflow Urinary Incontinence (OUI) is defined by ICS as:" Complaint of urinary incontinence in the symptomatic presence of an excessively (over-) full bladder (no cause identified)." (D'Ancona et al., 2019) On the other hand, Detrusor underactivity (DUA) is a urodynamic diagnosis defined by the International Continence Society as "a contraction of reduced strength and/or duration, resulting in prolonged bladder emptying and/or failure to achieve complete bladder emptying within a normal time span."

Different lines of treatments can be found to offer to the patients. As first-line, conservative therapies, including behavioral therapy, e.g., bladder training, moderate the consumption of liquids, meditation, etc. Next level of treatment includes medications, such as anticholinergic and Beta 3 agonist, with the secondary side effect related to these agents, adding the fact that, while anticholinergic medications improve symptoms, the number of dry patients is relatively low, making the discontinuation rates high. Formulas such as transdermal delivery systems and sustained-release oral agents, have improved the tolerability of these drugs.

In the United States, approved antimuscarinic agents include oxybutynin, tolterodine, trospium chloride, solifenacin, and darifenacin. (Elser, 2012; Saks & Arya, 2009) As the third line, it can be offer modalities such injection of medication into the detrusor muscle like onabotulinumtoxinA (Botox®) and neuromodulation (Percutaneous Tibial Nerve Stimulation (PTNS), implantable sacral neuromodulation (InterStim®; Medronic, Inc.)). The patient needs to keep in mind that even though she is receiving additional treatment, behavioral therapy is the key to improvement. When the treatments fail, bladder dysfunction is associated with an increase in the thickness of the detrusor wall, and secondarily, with an increase in retrograde pressure towards the kidney. It is caused by collagen deposition in the bladder as a result of various pathological processes, resulting in bladder fibrosis, reduced smooth muscle, bladder instability, and induces a flaccid, noncompliant bladder. (Elbadawi et al., 1993; Steers & De Groat, 1988) Compensated bladder dysfunction with an OAB often occurs after 6 weeks. (Elbadawi et al., 1993)

In order to protect the upper urinary tract, the only remaining treatment option is surgical reconstruction of the bladder. Unfortunately, the potential complications of metabolic imbalance from the use of a portion of the gastrointestinal tract to perform augmentation cystoplasty are well characterized. (Stanasel et al., 2010)

Bladder dysfunction is an excellent candidate for SC therapy due to its chronic nature and high prevalence, which may help overcome the limitations of current treatment. Most studies that have dealt with models of decompensated bladder dysfunction carry a disease period of more than 6 weeks. SC therapy in BOO is described in Table 4, which demonstrates the detailed characteristics of each study.

Nitta et al., transplanted skeletal muscle-derived SC (SkMSCs) (CD34-/45- and CD34-/CD45-) from green fluorescent protein (GFT) transgenic mouse muscles into a nude rat model for the regeneration of bladder dysfunction.(Nitta et al., 2010) After SkMSCs transplantation, significantly higher functional recovery was noted. Yokoyama et al. reported a differentiation from SkMSCs into bladder smooth muscle cells (SMCs) (Yokoyama et al., 2001) and demonstrate contractility after seeding. (S. H. Lu et al., 2003) The main advantage of SkMSCs is that they are a source of autologous transplantation. Skeletal muscle is the largest organ in the body, and SkMSCs can be obtained relatively easily and safely during surgery. (Kim et al., 2014)

Lee et al. reported that transplantation of human MSCs labeled with nanoparticles containing superparamagnetic iron oxide into the bladder inhibited bladder fibrosis and induced improvement of bladder dysfunction, demonstrated by serial T2-weighted MR images, showing a clear hypointense signal induced by the labeled

MSCs. While the expression of collagen, Hepatocyte growth factor (HGF), c-met, and Transforming growth factor-beta (TGF-β) proteins increased after BOO, their expression returned to the original levels after MSC transplantation. Also, the maximal voiding pressure and residual urine volume were recovered after the treatment. (H. J. Lee et al., 2012)

Table 5. Stem cell therapy in preclinical models of OAB. Jung Hyun Shin1

Author	Method	Aninal/ Sex	Cell type	Route of injection	Model of action
Liang et al (2018)(Liang et al., 2018)	Bilateral iliac artery ischemia	female rat	hAFSCs	Intravenous tail injection	TNFα
Liang et al (2017)(Liang et al., 2017)	Cerebral ischemia	female rat	hAFSCs	Bladder injection	NGF, M2, M3, P2X1
Liang et al (2016)(Liang et al., 2016)	Cerebral ischemia	female rat	hUCB-derived CD34+	Intravenous tail injection	NGF, M2, M3
Chen et al (2012)(Chen et al., 2012)	Bilateral iliac artery ischemia	female rat	BM-MSCs	iliac artery	regenerates bladder tissue
Song et al (2014)(Song et al., 2014)	Bladder outflow obstruction (BOO)	female rat	hAD-MSCs	Bladder injection	SDF-1, HGF, paracrine effects
Kanno et al (2014)(Kanno et al., 2014)	Partial bladder outlet obstruction (pBOO) obstruction (BOO)	female rat	Rat BM-MSCs	Bladder injection	morphological changes
Zhang et al (2012)(H. Zhang et al., 2012a)	Diabetes	male rat	ADSCs	Intravenous tail injection	Apoptosis vascular integrity

Recently, the contribution of BMSC to the regeneration of bladder detrusor muscle and anal sphincter with improved contractility has been reported. (Kanematsu et al., 2005; Lorenzi et al., 2008)

Nishijima et al. using cells labeled with GFP, demonstrated the differentiation into smooth muscle tissue, suggesting that transplanted BMSCs may improve bladder contractility by differentiating into smooth muscle-like cells. (Nishijima et al., 2007)

The problem is that the isolation techniques used were not specific for BMSCs, so other SC could have differentiated into smooth muscle tissue in vivo instead. Woo et al. performed the transplantation by systemic injection and observed evidence of histological and functional improvement, reporting that engrafted MSCs had improved compliance compared to those without engraftment, as well as the benefits of SC transplantation for bladder remodeling and the significance of paracrine mechanisms by demonstrating a positive functional outcome after systemic injection of a few homing transplanted SCs. (Kim et al., 2014; Woo et al., 2011)

It is also known that hyperlipidemia can induce chronic ischemia, with a secondary increase of TGFβ-1 in the bladder, which leads to fibrosis. Huang et al. showed that bladder injection or systemic intravenous transplantation of ADSCs improved functional outcomes in their hyperlipidemia-induced overactive bladder model, with better results in those who received a local injection. They demonstrated that SC transplantation could enhance angiogenesis and cholinergic innervation. (Huang et al., 2010)

Zhang et al. reported that improved voiding function was noted in ADSC-treated rats with an underactive DBD model. Moreover, the paracrine pathway appears to play a major role in this process, resulting in reduced apoptosis and preservation of the "suburothelial capillary network". (Kim et al., 2014; H. Zhang et al., 2012b)

Based on the pathophysiology of bladder ischemia induced by bilateral iliac arterial partial occlusion BPAO and secondary oxidative stress, Chiang et al developed an arterial deocclusive treatment in animal model with ADSC and ADSC derivatives, with the administration of ADSC and ADSC-derived microvesicles (MV) through femoral artery injection to improve vecical microcirculation and reduce inflammatory cell infiltration, oxidative stress, among others. (Chiang et al., 2021) Follow-up was done with Doppler flowmetry and real-time laser speckle contrast imaging to quantify blood flow in the common iliac arteries and bladder microcirculation at 24 hours and 2 weeks after treatment. Bladder histology, immunostaining, and lipid peroxidation tests were performed, as well as measurement of expressions of P2X2, P2X3, M2, and M3 receptors and nerve growth factor (NGF). This group reported an improvement in the adverse effects of BPAO, including bladder hyperactivity, bladder ischemia, and oxidative stress, and hypothesized that muscarinic inflammation, purinergic signaling, and NGF could be involved in the therapeutic mechanism.

Liang et al. induced detrusor hyperactivity with significant decreases in evacuated volume and interval between contractions by occluding the cerebral artery in female rats. Subsequently, they injected stem cells from human amniotic fluid (hAFSCs) improving this bladder dysfunction with increased expression of nerve growth factor (NGF), M2 and M3 muscarinic receptors, and P2X1 purinergic receptors (Liang et al., 2016, 2017) Liang's group induced detrusor hyperactivity in female rats by 2Fr Fogarty balloon injury of the common iliac artery and increased serum cholesterol, finding that hAFSC injected into the tail vein improved histological abnormalities and voiding parameters by downregulation of inducers of oxidative stress and TNF-α expression. (Liang et al., 2018)

In 2015, Levanovich et al. reported a pilot study on intradetrusor injection of adult muscle-derived cells (AMDC) in a human 79-year-old DUA patient, with a history of 2 transurethral resections of the prostate and suffered from recurrent urinary retention despite the administration of alpha-blockers and bethanechol. The patient was on clean intermittent catheterization 4 to 6 times per day and suffered from repeated episodes of gross hematuria and febrile urinary tract infections. Approximately 150mg of quadriceps femoris muscle was collected, and a final concentration of 250 million AMDCs was harvested. Under local anesthesia, with a flexible cystoscope, 15 mL of AMDC was injected at a 2 mm depth. The patient's subjective symptoms improved, and the ability to void small volumes at 3 months' post-treatment was observed, but the patient still required intermittent catheterization over the one-year follow-up period. (Levanovich et al., 2015; Shin et al., 2020b)

CONCLUSION

SUI is a relatively benign condition with a strong negative impact on the quality of life. Despite the broad spectrum of minimally invasive surgical procedures, the incidence of long-term recurrence remains high, as well as the complications derived from these therapies. The hope is in emerging therapies that aim to replace or restore the functionality of tissues and organs, and that can ensure long-term efficacy. New insights into growth factors, chemokines, and cell-free or cell-free secretomes may play a key role in treating patients with incontinence of any type.

Despite these findings, however, stem cell therapy is not yet available for clinical use. The biggest obstacle in current preclinical studies of bladder dysfunction is that there are no intermediary examinations, such as urodynamics or bladder biopsy at a midpoint in the experiment, researchers cannot judge whether SCs may serve a more preventative or ameliorative role. It is hard to know whether the current BOO model could be later reversed or not.

Neovascularization and reinnervation are critical hurdles to be overcome for future application to bladder reconstruction, complete regeneration with functional integration is much more challenging. Although preclinical data and clinical case reports have shown positive effects of CS treatment, including CS transplantation and tissue engineering for regeneration, more research is needed to elucidate the mechanisms of the local and systemic effects of CS. before they can be fully integrated into clinical practice for bladder dysfunction.

REFERENCES

Alamein, M. A., Stephens, S., Liu, Q., Skabo, S., & Warnke, P. H. (2013). Mass production of nanofibrous extracellular matrix with controlled 3D morphology for large-scale soft tissue regeneration. *Tissue Engineering. Part C, Methods*, *19*(6), 458–472. DOI: 10.1089/ten.tec.2012.0417 PMID: 23102268

Arjmand, B., Safavi, M., Heidari, R., Aghayan, H., Bazargani, S. T., Dehghani, S., & Pourmand, G. (2017). Concomitant transurethral and transvaginal-periurethral injection of autologous adipose derived stem cells for treatment of female stress urinary incontinence: A phase one clinical trial. *Acta Medica Iranica*, •••, 368–374.

Atala, A., Bauer, S. B., Soker, S., Yoo, J. J., & Retik, A. B. (2006). Tissue-engineered autologous bladders for patients needing cystoplasty. *Lancet*, *367*(9518), 1241–1246. DOI: 10.1016/S0140-6736(06)68438-9 PMID: 16631879

Badra, S., Andersson, K. E., Dean, A., Mourad, S., & Williams, J. K. (2013). Long-term structural and functional effects of autologous muscle precursor cell therapy in a nonhuman primate model of urinary sphincter deficiency. *The Journal of Urology*, *190*(5), 1938–1945. DOI: 10.1016/j.juro.2013.04.052 PMID: 23618586

Badylak, S. F., Freytes, D. O., & Gilbert, T. W. (2009). Extracellular matrix as a biological scaffold material: Structure and function. *Acta Biomaterialia*, *5*(1), 1–13. DOI: 10.1016/j.actbio.2008.09.013 PMID: 18938117

Barakat, B., Franke, K., Schakaki, S., Hijazi, S., Hasselhof, V., & Vögeli, T. A. (2020a). Stem cell applications in regenerative medicine for stress urinary incontinence: A review of effectiveness based on clinical trials. *Arab Journal of Urology*, *18*(3), 194–205. DOI: 10.1080/2090598X.2020.1750864 PMID: 33029431

Barakat, B., Franke, K., Schakaki, S., Hijazi, S., Hasselhof, V., & Vögeli, T. A. (2020b). Stem cell applications in regenerative medicine for stress urinary incontinence: A review of effectiveness based on clinical trials. En *Arab Journal of Urology* (Vol. 18, Número 3, pp. 194-205). Taylor and Francis Ltd. DOI: 10.1080/2090598X.2020.1750864

Bilsel, Y., & Abci, I. (2012). The search for ideal hernia repair; mesh materials and types. *International Journal of Surgery*, *10*(6), 317–321. DOI: 10.1016/j.ijsu.2012.05.002 PMID: 22588090

Blaganje, M., & Lukanović, A. (2012). Intrasphincteric autologous myoblast injections with electrical stimulation for stress urinary incontinence. *International Journal of Gynaecology and Obstetrics: the Official Organ of the International Federation of Gynaecology and Obstetrics, 117*(2), 164–167. DOI: 10.1016/j.ijgo.2011.11.029 PMID: 22342057

Blaivas, J. G., & Olsson, C. A. (1988). Stress incontinence: Classification and surgical approach. *Journal of Urology, 139*(4 I), 727-731. DOI: 10.1016/S0022-5347(17)42611-5

Boffito, M., Sirianni, P., Di Rienzo, A. M., & Chiono, V. (2015). Thermosensitive block copolymer hydrogels based on poly(ε-caprolactone) and polyethylene glycol for biomedical applications: State of the art and future perspectives. *Journal of Biomedical Materials Research. Part A, 103*(3), 1276–1290. DOI: 10.1002/jbm.a.35253 PMID: 24912941

Burkhard, F. C., Bosch, J. L. H. R., Cruz, F., Lemack, G. E., Nambiar, A. K., Thiruchelvam, N., & Tubaro, A. (2018). EAU Guidelines on Urinary Incontinence in Adults. *European Association of Urology*, 99.

Cannon, T. W., Lee, J. Y., Somogyi, G., Pruchnic, R., Smith, C. P., Huard, J., & Chancellor, M. B. (2003). Improved sphincter contractility after allogenic muscle-derived progenitor cell injection into the denervated rat urethra. *Urology, 62*(5), 958–963. DOI: 10.1016/S0090-4295(03)00679-4 PMID: 14624934

Cao, B., Wang, Y., Wen, D., Liu, W., Wang, J., Fan, G., Ruan, L., Song, B., Cai, Y., Wei, M., Li, X., Xia, J., Chen, N., Xiang, J., Yu, T., Bai, T., Xie, X., Zhang, L., Li, C., & Wang, C. (2020). A trial of lopinavir-ritonavir in adults hospitalized with severe covid-19. *The New England Journal of Medicine, 382*(19), 1787–1799. DOI: 10.1056/NEJMoa2001282 PMID: 32187464

Carr, L. K., Robert, M., Kultgen, P. L., Herschorn, S., Birch, C., Murphy, M., & Chancellor, M. B. (2013). Autologous muscle derived cell therapy for stress urinary incontinence: A prospective, dose ranging study. *The Journal of Urology, 189*(2), 595–601. DOI: 10.1016/j.juro.2012.09.028 PMID: 23260547

Carr, L. K., Steele, D., Steele, S., Wagner, D., Pruchnic, R., Jankowski, R., Erickson, J., Huard, J., & Chancellor, M. B. (2008). 1-Year Follow-Up of Autologous Muscle-Derived Stem Cell Injection Pilot Study To Treat Stress Urinary Incontinence. *International Urogynecology Journal, 19*(6), 881–883. DOI: 10.1007/s00192-007-0553-z PMID: 18204978

Chancellor, M. B., Yokoyama, T., Tirney, S., Mattes, C. E., Ozawa, H., Yoshimura, N., De Groat, W. C., & Huard, J. (2000). Preliminary results of myoblast injection into the urethra and bladder wall: A possible method for the treatment of stress urinary incontinence and impaired detrusor contractility. *Neurourology and Urodynamics, 19*(3), 279–287. DOI: 10.1002/(SICI)1520-6777(2000)19:3<279::AID-NAU9>3.0.CO;2-M PMID: 10797585

Chen, S., Zhang, H. Y., Zhang, N., Li, W. H., Shan, H., Liu, K., & Yang, Y. (2012). Treatment for chronic ischaemia-induced bladder detrusor dysfunction using bone marrow mesenchymal stem cells: An experimental study. *International Journal of Molecular Medicine, 29*(3), 416–422. DOI: 10.3892/ijmm.2011.846 PMID: 22109789

Chiang, B. J., Liao, C. H., Mao, S. H., & Chien, C. T. (2021). Adipose-derived stem cells and their derived microvesicles ameliorate detrusor overactivity secondary to bilateral partial iliac arterial occlusion-induced bladder ischemia. *International Journal of Molecular Sciences, 22*(13), 7000. Advance online publication. DOI: 10.3390/ijms22137000 PMID: 34210091

Choi, J. Y., Kim, T. H., Yang, J. D., Suh, J. S., & Kwon, T. G. (2016a). Adipose-derived regenerative cell injection therapy for postprostatectomy incontinence: A phase i clinical study. *Yonsei Medical Journal, 57*(5), 1152–1158. DOI: 10.3349/ymj.2016.57.5.1152 PMID: 27401646

Choi, J. Y., Kim, T. H., Yang, J. D., Suh, J. S., & Kwon, T. G. (2016b). Adipose-derived regenerative cell injection therapy for postprostatectomy incontinence: A phase i clinical study. *Yonsei Medical Journal, 57*(5), 1152–1158. DOI: 10.3349/ymj.2016.57.5.1152 PMID: 27401646

Chow, D., Nunalee, M. L., Lim, D. W., Simnick, A. J., & Chilkoti, A. (2008). Peptide-based Biopolymers in Biomedicine and Biotechnology. *Materials Science & Engineering. R, Reports: A Review Journal, 62*(4), 125-155. DOI: 10.1016/j.mser.2008.04.004

Coburn, J. M., Gibson, M., Monagle, S., Patterson, Z., & Elisseeff, J. H. (2012). Bioinspired nanofibers support chondrogenesis for articular cartilage repair. *Proceedings of the National Academy of Sciences of the United States of America, 109*(25), 10012–10017. DOI: 10.1073/pnas.1121605109 PMID: 22665791

Crapo, P. M., Gilbert, T. W., & Badylak, S. F. (2011). An overview of tissue and whole organ decellularization processes. *Biomaterials, 32*(12), 3233–3243. DOI: 10.1016/j.biomaterials.2011.01.057 PMID: 21296410

Cui, F.-Z., Li, Y., & Ge, J. (2007). Self-assembly of mineralized collagen composites. *Materials Science and Engineering R Reports*, *57*(1), 1–27. DOI: 10.1016/j.mser.2007.04.001

D'Ancona, C., Haylen, B., Oelke, M., Abranches-Monteiro, L., Arnold, E., Goldman, H., Hamid, R., Homma, Y., Marcelissen, T., Rademakers, K., Schizas, A., Singla, A., Soto, I., Tse, V., de Wachter, S., & Herschorn, S. (2019). The International Continence Society (ICS) report on the terminology for adult male lower urinary tract and pelvic floor symptoms and dysfunction. *Neurourology and Urodynamics*, *38*(2), 433–477. DOI: 10.1002/nau.23897 PMID: 30681183

Dissaranan, C., Cruz, M. A., Kiedrowski, M. J., Balog, B. M., Gill, B. C., Penn, M. S., Goldman, H. B., & Damaser, M. S. (2014). Rat mesenchymal stem cell secretome promotes elastogenesis and facilitates recovery from simulated childbirth injury. *Cell Transplantation*, *23*(11), 1395–1406. DOI: 10.3727/096368913X670921 PMID: 23866688

Eberli, D., Andersson, K. E., Yoo, J. J., & Atala, A. (2009). A canine model of irreversible urethral sphincter insufficiency. *BJU International*, *103*(2), 248–253. DOI: 10.1111/j.1464-410X.2008.08001.x PMID: 18782310

Elbadawi, A., Yalla, S. V., & Resnick, N. M. (1993). Structural basis of geriatric voiding dysfunction. III. Detrusor overactivity. *Journal of Urology, 150*(5 II), 1668-1680. DOI: 10.1016/S0022-5347(17)35868-8

Elser, D. M. (2012). Stress urinary incontinence and overactive bladder syndrome: Current options and new targets for management. *Postgraduate Medicine*, *124*(3), 42–49. DOI: 10.3810/pgm.2012.05.2547 PMID: 22691898

Furuta, A., Jankowski, R. J., Honda, M., Pruchnic, R., Yoshimura, N., & Chancellor, M. B. (2007). State of the art of where we are at using stem cells for stress urinary incontinence. *Neurourology and Urodynamics*, *26*(7), 966–971. DOI: 10.1002/nau.20448 PMID: 17580339

Gallo, F., Ninotta, G., Schenone, M., Cortese, P., & Giberti, C. (2019). Advances in stem cell therapy for male stress urinary incontinence. En *Expert Opinion on Biological Therapy* (Vol. 19, Número 4, pp. 293-300). Taylor and Francis Ltd. DOI: 10.1080/14712598.2019.1578343

Garcia-Arranz, M., Alonso-Gregorio, S., Fontana-Portella, P., Bravo, E., Diez Sebastian, J., Fernandez-Santos, M. E., & Garcia-Olmo, D. (2020). Two phase I/II clinical trials for the treatment of urinary incontinence with autologous mesenchymal stem cells. *Stem Cells Translational Medicine*, *9*(12), 1500–1508. DOI: 10.1002/sctm.19-0431 PMID: 32864818

Gerullis, H., Eimer, C., Georgas, E., Homburger, M., El-Baz, A. G., Wishahi, M., Borós, M., Ecke, T. H., & Otto, T. (2012a). Muscle-derived cells for treatment of iatrogenic sphincter damage and urinary incontinence in men. *TheScientificWorld-Journal, 2012*, 1–6. Advance online publication. DOI: 10.1100/2012/898535 PMID: 22919359

Gerullis, H., Eimer, C., Georgas, E., Homburger, M., El-Baz, A. G., Wishahi, M., Borós, M., Ecke, T. H., & Otto, T. (2012b). Muscle-derived cells for treatment of iatrogenic sphincter damage and urinary incontinence in men. *TheScientificWorld-Journal, 2012*, 1–6. Advance online publication. DOI: 10.1100/2012/898535 PMID: 22919359

Gluckman, E., Broxmeyer, H. E., Auerbach, A. D., Friedman, H. S., Douglas, G. W., Devergie, A., Esperou, H., Thierry, D., Socie, G., Lehn, P., Cooper, S., English, D., Kurtzberg, J., Bard, J., & Boyse, E. A. (1989). Hematopoietic Reconstitution in a Patient with Fanconi's Anemia by Means of Umbilical-Cord Blood from an HLA-Identical Sibling. *The New England Journal of Medicine, 321*(17), 1174–1178. DOI: 10.1056/NEJM198910263211707 PMID: 2571931

Gluckman, E., Rocha, V., Boyer-Chammard, A., Locatelli, F., Arcese, W., Pasquini, R., Ortega, J., Souillet, G., Ferreira, E., Laporte, J.-P., Fernandez, M., & Chastang, C. (1997). Outcome of Cord-Blood Transplantation from Related and Unrelated Donors. *The New England Journal of Medicine, 337*(6), 373–381. DOI: 10.1056/NEJM199708073370602 PMID: 9241126

Goldman, H. B., Sievert, K. D., & Damaser, M. S. (2012). Will we ever use stem cells for the treatment of SUI? ICI-RS 2011. *Neurourology and Urodynamics, 31*(3), 386–389. DOI: 10.1002/nau.22217 PMID: 22431263

Gotoh, M., Yamamoto, T., Kato, M., Majima, T., Toriyama, K., Kamei, Y., Matsukawa, Y., Hirakawa, A., & Funahashi, Y. (2014). Regenerative treatment of male stress urinary incontinence by periurethral injection of autologous adipose-derived regenerative cells: 1-year outcomes in 11 patients. *International Journal of Urology, 21*(3), 294–300. DOI: 10.1111/iju.12266 PMID: 24033774

Gul, M., Hildorf, S., Dong, L., Thorup, J., Hoffmann, E. R., Jensen, C. F. S., Sønksen, J., Cortes, D., Fedder, J., Andersen, C. Y., & Goossens, E. (2020). Review of injection techniques for spermatogonial stem cell transplantation. En *Human Reproduction Update* (Vol. 26, Número 3, pp. 368-391). Oxford University Press. DOI: 10.1093/humupd/dmaa003

Gunetti, M., Tomasi, S., Giammò, A., Boido, M., Rustichelli, D., Mareschi, K., Errichiello, E., Parola, M., Ferrero, I., Fagioli, F., Vercelli, A., & Carone, R. (2012). Myogenic Potential of Whole Bone Marrow Mesenchymal Stem Cells In Vitro and In Vivo for Usage in Urinary Incontinence. *PLoS One*, *7*(9), e45538. Advance online publication. DOI: 10.1371/journal.pone.0045538 PMID: 23029081

Hakim, L., De Ridder, D., & Van der Aa, F. (2015). Slings for urinary incontinence and the application of cell-based therapy. In *Advanced Drug Delivery Reviews* (Vol. 82, pp. 22–30). Elsevier B.V., DOI: 10.1016/j.addr.2014.11.006

Han, D., & Gouma, P.-I. (2006). Electrospun bioscaffolds that mimic the topology of extracellular matrix. *Nanomedicine; Nanotechnology, Biology, and Medicine*, *2*(1), 37–41. DOI: 10.1016/j.nano.2006.01.002 PMID: 17292114

Hillary, C. J., Roman, S., MacNeil, S., Aicher, W. K., Stenzl, A., & Chapple, C. R. (2020). Regenerative medicine and injection therapies in stress urinary incontinence. En *Nature Reviews Urology* (Vol. 17, Número 3, pp. 151-161). Nature Research. DOI: 10.1038/s41585-019-0273-4

Horan, R. L., Bramono, D. S., Stanley, J. R. L., Simmons, Q., Chen, J., Boepple, H. E., & Altman, G. H. (2009). Biological and biomechanical assessment of a long-term bioresorbable silk-derived surgical mesh in an abdominal body wall defect model. *Hernia*, *13*(2), 189–199. DOI: 10.1007/s10029-008-0459-9 PMID: 19198755

Huang, Y. C., Shindel, A. W., Ning, H., Lin, G., Harraz, A. M., Wang, G., Garcia, M., Lue, T. F., & Lin, C. S. (2010). Adipose Derived Stem Cells Ameliorate Hyperlipidemia Associated Detrusor Overactivity in a Rat Model. *The Journal of Urology*, *183*(3), 1232–1240. DOI: 10.1016/j.juro.2009.11.012 PMID: 20096880

Ishikawa, F., Drake, C. J., Yang, S., Fleming, P. A., Minamiguchi, H., Visconti, R. P., Crosby, C. V., Argraves, W. S., Harada, M., Key, L. L.Jr, Livingston, A. G., Wingard, J. R., & Ogawa, M. (2003). Transplanted human cord blood cells give rise to hepatocytes in engrafted mice. *Annals of the New York Academy of Sciences*, *996*(1), 174–185. DOI: 10.1111/j.1749-6632.2003.tb03245.x PMID: 12799295

Kanematsu, A., Yamamoto, S., Iwai-Kanai, E., Kanatani, I., Imamura, M., Adam, R. M., Tabata, Y., & Ogawa, O. (2005). Induction of smooth muscle cell-like phenotype in marrow-derived cells among regenerating urinary bladder smooth muscle cells. *American Journal of Pathology*, *166*(2), 565–573. DOI: 10.1016/S0002-9440(10)62278-X PMID: 15681839

Kanno, Y., Mitsui, T., Sano, H., Kitta, T., Moriya, K., & Nonomura, K. (2014). Contribution of bone marrow-derived mesenchymal stem cells to the morphological changes in the bladder after partial outlet obstruction: A preliminary study. *International journal of urology : official journal of the Japanese Urological Association, 21*(7), 714-718. DOI: 10.1111/iju.12406

Kaufman, M. R. (2020). Contemporary application of autologous muscle-derived cells for urinary sphincter regeneration. *World Journal of Urology, 38*(9), 2095–2099. DOI: 10.1007/s00345-019-03018-9 PMID: 31834471

Kim, J. H., Lee, H. J., & Song, Y. S. (2014). Treatment of bladder dysfunction using stem cell or tissue engineering technique. En *Korean Journal of Urology* (Vol. 55, Número 4, pp. 228-238). Korean Urological Association. DOI: 10.4111/kju.2014.55.4.228

Kinebuchi, Y., Aizawa, N., Imamura, T., Ishizuka, O., Igawa, Y., & Nishizawa, O. (2010). Autologous bone-marrow-derived mesenchymal stem cell transplantation into injured rat urethral sphincter. *International Journal of Urology, 17*(4), 359–368. DOI: 10.1111/j.1442-2042.2010.02471.x PMID: 20202003

Knoll, J., Amend, B., Harland, N., Isser, S., Bézière, N., Kraushaar, U., Stenzl, A., & Aicher, W. K. (2024). Cell Therapy by Mesenchymal Stromal Cells Versus Myoblasts in a Pig Model of Urinary Incontinence. *Tissue Engineering. Part A, 30*(1-2), 14–30. DOI: 10.1089/ten.tea.2023.0103 PMID: 37933911

Kobashi, K. C., Albo, M. E., Dmochowski, R. R., Ginsberg, D. A., Goldman, H. B., Gomelsky, A., Kraus, S. R., Sandhu, J. S., Shepler, T., Treadwell, J. R., Vasavada, S., & Lemack, G. E. (2017). Surgical Treatment of Female Stress Urinary Incontinence: AUA/SUFU Guideline. *The Journal of Urology, 198*(4), 875–883. DOI: 10.1016/j.juro.2017.06.061 PMID: 28625508

Kong, K. Y., Ren, J., Kraus, M., Finklestein, S. P., & Brown, R. H.Jr. (2004). Human Umbilical Cord Blood Cells Differentiate into Muscle in sjl Muscular Dystrophy Mice. *Stem Cells (Dayton, Ohio), 22*(6), 981–993. DOI: 10.1634/stemcells.22-6-981 PMID: 15536189

Kuismanen, K., Sartoneva, R., Haimi, S., Mannerström, B., Tomás, E., Miettinen, S., & Nieminen, K. (2014). Autologous Adipose Stem Cells in Treatment of Female Stress Urinary Incontinence: Results of a Pilot Study. *Stem Cells Translational Medicine, 3*(8), 936–941. DOI: 10.5966/sctm.2013-0197 PMID: 24985079

Lee, C. N., Jang, J. B., Kim, J. Y., Koh, C., Baek, J. Y., & Lee, K. J. (2010). Human cord blood stem cell therapy for treatment of stress urinary incontinence. *Journal of Korean Medical Science*, *25*(6), 813–816. DOI: 10.3346/jkms.2010.25.6.813 PMID: 20514298

Lee, H. J., Won, J. H., Doo, S. H., Kim, J. H., Song, K. Y., Lee, S. J., Lim, I., Chang, K. T., Song, Y. S., & Kim, S. U. (2012). Inhibition of collagen deposit in obstructed rat bladder outlet by transplantation of superparamagnetic iron oxide-labeled human mesenchymal stem cells as monitored by molecular magnetic resonance imaging (MRI). *Cell Transplantation*, *21*(5), 959–970. DOI: 10.3727/096368911X627516 PMID: 22449414

Lee, J. Y., Qu-Petersen, Z., Cao, B., Kimura, S., Jankowski, R., Cummins, J., Usas, A., Gates, C., Robbins, P., Wernig, A., & Huard, J. (2000). Clonal isolation of muscle-derived cells capable of enhancing muscle regeneration and bone healing. *The Journal of Cell Biology*, *150*(5), 1085–1099. DOI: 10.1083/jcb.150.5.1085 PMID: 10973997

Levanovich, P. E., Diokno, A., Hasenau, D. L., Lajiness, M., Pruchnic, R., & Chancellor, M. B. (2015). Intradetrusor injection of adult muscle-derived cells for the treatment of underactive bladder: Pilot study. *International Urology and Nephrology*, *47*(3), 465–467. DOI: 10.1007/s11255-015-0924-1 PMID: 25663053

Li, G., Yu, C., Yu, P., Peng, Q., Wang, Q., Ren, S., Li, H., Li, M., Li, P., & He, R. (2022). Periurethral and Intravenous Injections of Adipose-Derived Stem Cells to Promote Local Tissue Recovery in a Rat Model of Stress Urinary Incontinence. *Urology*, *167*, 82–89. DOI: 10.1016/j.urology.2022.05.018 PMID: 35654272

Li, M., Mondrinos, M. J., Gandhi, M. R., Ko, F. K., Weiss, A. S., & Lelkes, P. I. (2005). Electrospun protein fibers as matrices for tissue engineering. *Biomaterials*, *26*(30), 5999–6008. DOI: 10.1016/j.biomaterials.2005.03.030 PMID: 15894371

Li, X., Xie, J., Yuan, X., & Xia, Y. (2008). Coating electrospun poly(epsilon-caprolactone) fibers with gelatin and calcium phosphate and their use as biomimetic scaffolds for bone tissue engineering. *Langmuir*, *24*(24), 14145–14150. DOI: 10.1021/la802984a PMID: 19053657

Li, Y., Pan, H., & Huard, J. (2010). Isolating stem cells from soft musculoskeletal tissues. *Journal of Visualized Experiments*, *41*(41). Advance online publication. DOI: 10.3791/2011 PMID: 20644509

Li, Z., & Tan, B. H. (2014). Towards the development of polycaprolactone based amphiphilic block copolymers: Molecular design, self-assembly and biomedical applications. *Materials Science and Engineering C*, *45*, 620–634. DOI: 10.1016/j.msec.2014.06.003 PMID: 25491872

Liang, C. C., Lee, T. H., & Chang, S. D. (2016). Effect of umbilical cord blood stem cells transplantation on bladder dysfunction induced by cerebral ischemia in rats. *Taiwanese Journal of Obstetrics & Gynecology*, *55*(5), 672–679. DOI: 10.1016/j.tjog.2015.11.002 PMID: 27751414

Liang, C. C., Shaw, S. W. S., Lin, Y. H., & Lee, T. H. (2018). Amniotic fluid stem cells ameliorate bladder dysfunction induced by chronic bladder ischemia in rat. *Neurourology and Urodynamics*, *37*(1), 123–131. DOI: 10.1002/nau.23316 PMID: 28605059

Liang, C. C., Steven Shaw, S. W., Huang, Y. H., Lin, Y. H., & Lee, T. H. (2017). Bladder Transplantation of Amniotic Fluid Stem Cell may Ameliorate Bladder Dysfunction After Focal Cerebral Ischemia in Rat. *Stem Cells Translational Medicine*, *6*(4), 1227–1236. DOI: 10.1002/sctm.16-0212 PMID: 28186672

Lin, A. S., Carrier, S., Morgan, D. M., & Lue, T. F. (1998). Effect of simulated birth trauma on the urinary continence mechanism in the rat. *Urology*, *52*(1), 143–151. DOI: 10.1016/S0090-4295(98)00136-8 PMID: 9671889

Lin, H.-K., Madihally, S. V., Palmer, B., Frimberger, D., Fung, K.-M., & Kropp, B. P. (2015). Biomatrices for bladder reconstruction. *Advanced Drug Delivery Reviews*, *82-83*, 47–63. DOI: 10.1016/j.addr.2014.11.020 PMID: 25477305

Lin, M., Lu, Y., & Chen, J. (2022). Tissue-engineered repair material for pelvic floor dysfunction. *Frontiers in Bioengineering and Biotechnology*, *10*(September), 1–12. DOI: 10.3389/fbioe.2022.968482 PMID: 36147522

Liu, M., Lu, Y., Sun, F., Li, Y., Wu, J., & Zou, Q. (2025). The Nerve-Induced Adipose Stem Cells Promote Nerve Repair in Stress Urinary Incontinence by Regulating Schwann Cell Repair Phenotype Conversion Through Activation of the Notch Pathway. *Molecular Neurobiology*, *62*(6), 7330–7344. DOI: 10.1007/s12035-025-04704-z PMID: 39881114

Liu, X., Holzwarth, J. M., & Ma, P. X. (2012). Functionalized synthetic biodegradable polymer scaffolds for tissue engineering. *Macromolecular Bioscience*, *12*(7), 911–919. DOI: 10.1002/mabi.201100466 PMID: 22396193

Lorenzi, B., Pessina, F., Lorenzoni, P., Urbani, S., Vernillo, R., Sgaragli, G., Gerli, R., Mazzanti, B., Bosi, A., Saccardi, R., & Lorenzi, M. (2008). Treatment of experimental injury of anal sphincters with primary surgical repair and injection of bone marrow-derived mesenchymal stem cells. *Diseases of the Colon and Rectum*, *51*(4), 411–420. DOI: 10.1007/s10350-007-9153-8 PMID: 18224375

Lu, H., Hoshiba, T., Kawazoe, N., & Chen, G. (2012). Comparison of decellularization techniques for preparation of extracellular matrix scaffolds derived from three-dimensional cell culture. *Journal of Biomedical Materials Research. Part A*, *100*(9), 2507–2516. DOI: 10.1002/jbm.a.34150 PMID: 22623317

Lu, L., Peter, S. J., Lyman, M. D., Lai, H. L., Leite, S. M., Tamada, J. A., Uyama, S., Vacanti, J. P., Langer, R., & Mikos, A. G. (2000). In vitro and in vivo degradation of porous poly(DL-lactic-co-glycolic acid) foams. *Biomaterials*, *21*(18), 1837–1845. DOI: 10.1016/S0142-9612(00)00047-8 PMID: 10919687

Lu, S. H., Cannon, T. W., Chermanski, C., Pruchnic, R., Somogyi, G., Sacks, M., De Groat, W. C., Huard, J., & Chancellor, M. B. (2003). Muscle-derived stem cells seeded into acellular scaffolds develop calcium-dependent contractile activity that is modulated by nicotinic receptors. *Urology*, *61*(6), 1285–1291. DOI: 10.1016/S0090-4295(03)00020-7 PMID: 12809930

Mangera, A., Bullock, A. J., Roman, S., Chapple, C. R., & MacNeil, S. (2013). Comparison of candidate scaffolds for tissue engineering for stress urinary incontinence and pelvic organ prolapse repair. *BJU International*, *112*(5), 674–685. DOI: 10.1111/bju.12186 PMID: 23773418

Mangır, N., Bullock, A. J., Roman, S., Osman, N., Chapple, C., & MacNeil, S. (2016). Production of ascorbic acid releasing biomaterials for pelvic floor repair. *Acta Biomaterialia*, *29*, 188–197. DOI: 10.1016/j.actbio.2015.10.019 PMID: 26478470

Mitterberger, M., Marksteiner, R., Margreiter, E., Pinggera, G. M., Frauscher, F., Ulmer, H., Fussenegger, M., Bartsch, G., & Strasser, H. (2008). Myoblast and Fibroblast Therapy for Post-Prostatectomy Urinary Incontinence: 1-Year Followup of 63 Patients. *The Journal of Urology*, *179*(1), 226–231. DOI: 10.1016/j.juro.2007.08.154 PMID: 18001790

Mitterberger, M., Pinggera, G. M., Marksteiner, R., Margreiter, E., Fussenegger, M., Frauscher, F., Ulmer, H., Hering, S., Bartsch, G., & Strasser, H. (2008). Adult Stem Cell Therapy of Female Stress Urinary Incontinence. *European Urology*, *53*(1), 169–175. DOI: 10.1016/j.eururo.2007.07.026 PMID: 17683852

Nguyen, M. K., Jeon, O., Krebs, M. D., Schapira, D., & Alsberg, E. (2014). Sustained localized presentation of RNA interfering molecules from in situ forming hydrogels to guide stem cell osteogenic differentiation. *Biomaterials, 35*(24), 6278–6286. DOI: 10.1016/j.biomaterials.2014.04.048 PMID: 24831973

Ning, H., Lin, G., Lue, T. F., & Lin, C. S. (2006). Neuron-like differentiation of adipose tissue-derived stromal cells and vascular smooth muscle cells. *Differentiation, 74*(9-10), 510–518. DOI: 10.1111/j.1432-0436.2006.00081.x PMID: 17177848

Nishijima, S., Sugaya, K., Miyazato, M., Kadekawa, K., Oshiro, Y., Uchida, A., Hokama, S., & Ogawa, Y. (2007). Restoration of bladder contraction by bone marrow transplantation in rats with underactive bladder. *Biomedical Research, 28*(5), 275–280. DOI: 10.2220/biomedres.28.275 PMID: 18000341

Nitta, M., Tamaki, T., Tono, K., Okada, Y., Masuda, M., Akatsuka, A., Hoshi, A., Usui, Y., & Terachi, T. (2010). Reconstitution of experimental neurogenic bladder dysfunction using skeletal muscle-derived multipotent stem cells. *Transplantation, 89*(9), 1043–1049. DOI: 10.1097/TP.0b013e3181d45a7f PMID: 20150836

Okamoto, M., & John, B. (2013). Synthetic biopolymer nanocomposites for tissue engineering scaffolds. *Progress in Polymer Science, 38*(10), 1487–1503. DOI: 10.1016/j.progpolymsci.2013.06.001

Pan, Z., & Ding, J. (2012). Poly(lactide-co-glycolide) porous scaffolds for tissue engineering and regenerative medicine. *Interface Focus, 2*(3), 366–377. DOI: 10.1098/rsfs.2011.0123 PMID: 23741612

Pesce, M., Orlandi, A., Iachininoto, M. G., Straino, S., Torella, A. R., Rizzuti, V., Pompilio, G., Bonanno, G., Scambia, G., & Capogrossi, M. C. (2003). Myoendothelial differentiation of human umbilical cord blood-derived stem cells in ischemic limb tissues. *Circulation Research, 93*(5). Advance online publication. DOI: 10.1161/01.RES.0000090624.04507.45 PMID: 12919944

Peters, K. M., Dmochowski, R. R., Carr, L. K., Robert, M., Kaufman, M. R., Sirls, L. T., Herschorn, S., Birch, C., Kultgen, P. L., & Chancellor, M. B. (2014). Autologous muscle derived cells for treatment of stress urinary incontinence in women. *The Journal of Urology, 192*(2), 469–476. DOI: 10.1016/j.juro.2014.02.047 PMID: 24582537

Pokrywczynska, M., Adamowicz, J., Czapiewska, M., Balcerczyk, D., Jundzill, A., Nowacki, M., Petros, P., & Drewa, T. (2016). Targeted therapy for stress urinary incontinence: A systematic review based on clinical trials. *Expert Opinion on Biological Therapy, 16*(2), 233–242. DOI: 10.1517/14712598.2016.1118459 PMID: 26560683

Qu-Petersen, Z., Deasy, B., Jankowski, R., Ikezawa, M., Cummins, J., Pruchnic, R., Mytinger, J., Cao, B., Gates, C., Wernig, A., & Huard, J. (2002). Identification of a novel population of muscle stem cells in mice: Potential for muscle regeneration. *The Journal of Cell Biology*, *157*(5), 851–864. DOI: 10.1083/jcb.200108150 PMID: 12021255

Raya-Rivera, A., Esquiliano, D. R., Yoo, J. J., Lopez-Bayghen, E., Soker, S., & Atala, A. (2011). Tissue-engineered autologous urethras for patients who need reconstruction: An observational study. *Lancet*, *377*(9772), 1175–1182. DOI: 10.1016/S0140-6736(10)62354-9 PMID: 21388673

Ribeiro-Filho, L. A., & Sievert, K.-D. (2015). Acellular matrix in urethral reconstruction. *Advanced Drug Delivery Reviews*, *82-83*, 38–46. DOI: 10.1016/j.addr.2014.11.019 PMID: 25477304

Rice, J. J., Martino, M. M., De Laporte, L., Tortelli, F., Briquez, P. S., & Hubbell, J. A. (2013). Engineering the regenerative microenvironment with biomaterials. *Advanced Healthcare Materials*, *2*(1), 57–71. DOI: 10.1002/adhm.201200197 PMID: 23184739

Rocha, V., Wagner, J. E.Jr, Sobocinski, K. A., Klein, J. P., Zhang, M.-J., Horowitz, M. M., & Gluckman, E. (2000). Graft-Versus-Host Disease in Children Who Have Received a Cord-Blood or Bone Marrow Transplant from an HLA-Identical Sibling. *The New England Journal of Medicine*, *342*(25), 1846–1854. DOI: 10.1056/NEJM200006223422501 PMID: 10861319

Rodríguez, L. V., Alfonso, Z., Zhang, R., Leung, J., Wu, B., & Ignarro, L. J. (2006). Clonogenic multipotent stem cells in human adipose tissue differentiate into functional smooth muscle cells. *Proceedings of the National Academy of Sciences of the United States of America*, *103*(32), 12167–12172. DOI: 10.1073/pnas.0604850103 PMID: 16880387

Roman, S., Mangir, N., Bissoli, J., Chapple, C. R., & MacNeil, S. (2016). Biodegradable scaffolds designed to mimic fascia-like properties for the treatment of pelvic organ prolapse and stress urinary incontinence. *Journal of Biomaterials Applications*, *30*(10), 1578–1588. DOI: 10.1177/0885328216633373 PMID: 26896234

Roobrouck, V. D., Ulloa-Montoya, F., & Verfaillie, C. M. (2008). Self-renewal and differentiation capacity of young and aged stem cells. *Experimental Cell Research*, *314*(9), 1937–1944. DOI: 10.1016/j.yexcr.2008.03.006 PMID: 18439579

Rortveit, G., Hannestad, Y. S., Kjersti Daltveit, A., & Hunskaar, S. (2001). Age- and type-dependent effects of parity on urinary incontinence: The Norwegian EPINCONT study. *Obstetrics and Gynecology*, *98*(6), 1004–1010. DOI: 10.1097/00006250-200112000-00004 PMID: 11755545

Rossi, F., & van Griensven, M. (2014). Polymer functionalization as a powerful tool to improve scaffold performances. *Tissue Engineering. Part A*, *20*(15-16), 2043–2051. DOI: 10.1089/ten.tea.2013.0367 PMID: 24206079

Saks, E. K., & Arya, L. A. (2009). Pharmacologic Management of Urinary Incontinence, Voiding Dysfunction, and Overactive Bladder. En *Obstetrics and Gynecology Clinics of North America* (Vol. 36, Número 3, pp. 493-507). Obstet Gynecol Clin North Am. DOI: 10.1016/j.ogc.2009.08.001

Salem, H. K., & Thiemermann, C. (2010). Mesenchymal stromal cells: Current understanding and clinical status. *Stem Cells (Dayton, Ohio)*, *28*(3), 585–596. DOI: 10.1002/stem.269 PMID: 19967788

Sanchez-Ramos, J. R., Song, S., Kamath, S. G., Zigova, T., Willing, A., Cardozo-Pelaez, F., Stedeford, T., Chopp, M., & Sanberg, P. R. (2001). Expression of neural markers in human umbilical cord blood. *Experimental Neurology*, *171*(1), 109–115. DOI: 10.1006/exnr.2001.7748 PMID: 11520125

Sèbe, P., Doucet, C., Cornu, J. N., Ciofu, C., Costa, P., De Medina, S. G. D., Pinset, C., & Haab, F. (2011). Intrasphincteric injections of autologous muscular cells in women with refractory stress urinary incontinence: A prospective study. *International Urogynecology Journal*, *22*(2), 183–189. DOI: 10.1007/s00192-010-1255-5 PMID: 20821309

Seval, M. M., & Koyuncu, K. (2022). Current status of stem cell treatments and innovative approaches for stress urinary incontinence. *Frontiers in Medicine, 9*(December 2009). DOI: 10.3389/fmed.2022.1073758

Sharifiaghdas, F., Tajalli, F., Taheri, M., Naji, M., Moghadasali, R., Aghdami, N., Baharvand, H., Azimian, V., & Jaroughi, N. (2016). Effect of autologous muscle-derived cells in the treatment of urinary incontinence in female patients with intrinsic sphincter deficiency and epispadias: A prospective study. *International journal of urology : official journal of the Japanese Urological Association, 23*(7), 581-586. DOI: 10.1111/iju.13097

Shin, J. H., Ryu, C. M., Yu, H. Y., Shin, D. M., & Choo, M. S. (2020a). Current and Future Directions of Stem Cell Therapy for Bladder Dysfunction. En *Stem Cell Reviews and Reports* (Vol. 16, Número 1, pp. 82-93). Springer. DOI: 10.1007/s12015-019-09922-2

Shin, J. H., Ryu, C. M., Yu, H. Y., Shin, D. M., & Choo, M. S. (2020b). Current and Future Directions of Stem Cell Therapy for Bladder Dysfunction. En *Stem Cell Reviews and Reports* (Vol. 16, Número 1, pp. 82-93). Springer. DOI: 10.1007/s12015-019-09922-2

Shirvan, M. K., Alamdari, D. H., Mahboub, M. D., Ghanadi, A., Rahimi, H. R., & Seifalian, A. M. (2013). A novel cell therapy for stress urinary incontinence, short-term outcome. *Neurourology and Urodynamics*, *32*(4), 377–382. DOI: 10.1002/nau.22301 PMID: 22972395

Song, M., Heo, J., Chun, J. Y., Bae, H. S., Kang, J. W., Kang, H., Cho, Y. M., Kim, S. W., Shin, D. M., & Choo, M. S. (2014). The paracrine effects of mesenchymal stem cells stimulate the regeneration capacity of endogenous stem cells in the repair of a bladder-outlet-obstruction-induced overactive bladder. *Stem Cells and Development*, *23*(6), 654–663. DOI: 10.1089/scd.2013.0277 PMID: 24192209

Stanasel, I., Mirzazadeh, M., & Smith, J. J. (2010). Bladder tissue engineering. En *Urologic Clinics of North America* (Vol. 37, Número 4, pp. 593-599). Urol Clin North Am. DOI: 10.1016/j.ucl.2010.06.008

Stangel-Wojcikiewicz, K., Jarocha, D., Piwowar, M., Jach, R., Uhl, T., Basta, A., & Majka, M. (2014). Autologous muscle-derived cells for the treatment of female stress urinary incontinence: A 2-year follow-up of a polish investigation. *Neurourology and Urodynamics*, *33*(3), 324–330. DOI: 10.1002/nau.22404 PMID: 23606303

Steers, W. D., & De Groat, W. C. (1988). Effect of bladder outlet obstruction on micturition reflex pathways in the rat. *The Journal of Urology*, *140*(4), 864–871. DOI: 10.1016/S0022-5347(17)41846-5 PMID: 3418824

Strasser, H., Marksteiner, R., Margreiter, E., Mitterberger, M., Pinggera, G. M., Frauscher, F., Fussenegger, M., Kofler, K., & Bartsch, G. (2007). Transurethral ultrasonography-guided injection of adult autologous stem cells versus transurethral endoscopic injection of collagen in treatment of urinary incontinence. *World Journal of Urology*, *25*(4), 385–392. DOI: 10.1007/s00345-007-0190-7 PMID: 17701044

Uebersax, L., Merkle, H. P., & Meinel, L. (2009). Biopolymer-based growth factor delivery for tissue repair: From natural concepts to engineered systems. *Tissue Engineering. Part B, Reviews*, *15*(3), 263–289. DOI: 10.1089/ten.teb.2008.0668 PMID: 19416020

Uhlén, M., Fagerberg, L., Hallström, B. M., Lindskog, C., Oksvold, P., Mardinoglu, A., Sivertsson, Å., Kampf, C., Sjöstedt, E., Asplund, A., Olsson, I. M., Edlund, K., Lundberg, E., Navani, S., Szigyarto, C. A. K., Odeberg, J., Djureinovic, D., Takanen, J. O., Hober, S., & Pontén, F. (2015). Tissue-based map of the human proteome. *Science*, *347*(6220), 1260419. Advance online publication. DOI: 10.1126/science.1260419 PMID: 25613900

Vacanti, N. M., Cheng, H., Hill, P. S., Guerreiro, J. D. T., Dang, T. T., Ma, M., Watson, S., Hwang, N. S., Langer, R., & Anderson, D. G. (2012). Localized delivery of dexamethasone from electrospun fibers reduces the foreign body response. *Biomacromolecules*, *13*(10), 3031–3038. DOI: 10.1021/bm300520u PMID: 22920794

Vaegler, M., Lenis, A. T., Daum, L., Amend, B., Stenzl, A., Toomey, P., Renninger, M., Damaser, M. S., & Sievert, K. D. (2012). Stem cell therapy for voiding and erectile dysfunction. En *Nature Reviews Urology* (Vol. 9, Número 8, pp. 435-447). DOI: 10.1038/nrurol.2012.111

Vashaghian, M., Ruiz-Zapata, A. M., Kerkhof, M. H., Zandieh-Doulabi, B., Werner, A., Roovers, J. P., & Smit, T. H. (2017). Toward a new generation of pelvic floor implants with electrospun nanofibrous matrices: A feasibility study. *Neurourology and Urodynamics*, *36*(3), 565–573. DOI: 10.1002/nau.22969 PMID: 26840206

Wang, B., Zhou, J., Banie, L., Reed-Maldonado, A. B., Ning, H., Lu, Z., Ruan, Y., Zhou, T., Wang, H. S., Oh, B. S., Wang, G., Qi, S. L., Lin, G., & Lue, T. F. (2018). Low-intensity extracorporeal shock wave therapy promotes myogenesis through PERK/ATF4 pathway. *Neurourology and Urodynamics*, *37*(2), 699–707. DOI: 10.1002/nau.23380 PMID: 28763567

Wang, G., Lin, G., Zhang, H., Qiu, X., Ning, H., Banie, L., Fandel, T., Albersen, M., Lue, T. F., & Lin, C. S. (2011). Effects of prolonged vaginal distension and β-aminopropionitrile on urinary continence and urethral structure. *Urology*, *78*(4), 968.e13–968.e19. DOI: 10.1016/j.urology.2011.07.1381 PMID: 21982017

Wang, H. J., Chuang, Y. C., & Chancellor, M. B. (2011). Development of cellular therapy for the treatment of stress urinary incontinence. *International Urogynecology Journal*, *22*(9), 1075–1083. DOI: 10.1007/s00192-011-1432-1 PMID: 21505907

Woo, L. L., Tanaka, S. T., Anumanthan, G., Pope, J. C.IV, Thomas, J. C., Adams, M. C., Brock, J. W., & Bhowmick, N. A. (2011). Mesenchymal stem cell recruitment and improved bladder function after bladder outlet obstruction: Preliminary data. *The Journal of Urology*, *185*(3), 1132–1138. DOI: 10.1016/j.juro.2010.10.033 PMID: 21255803

Wu, J. M., Hundley, A. F., Fulton, R. G., & Myers, E. R. (2009). Forecasting the prevalence of pelvic floor disorders in U.S. Women: 2010 to 2050. *Obstetrics and Gynecology*, *114*(6), 1278–1283. DOI: 10.1097/AOG.0b013e3181c2ce96 PMID: 19935030

Wu, X., Jia, Y., Sun, X., & Wang, J. (2020). Tissue engineering in female pelvic floor reconstruction. *Engineering in Life Sciences*, *20*(7), 275–286. DOI: 10.1002/elsc.202000003 PMID: 32647506

Xie, M., Song, L., Wang, J., Fan, S., Zhang, Y., & Xu, Y. (2013). Evaluation of stretched electrospun silk fibroin matrices seeded with urothelial cells for urethra reconstruction. *The Journal of Surgical Research*, *184*(2), 774–781. DOI: 10.1016/j.jss.2013.04.016 PMID: 23706393

Yamamoto, T., Gotoh, M., Kato, M., Majima, T., Toriyama, K., Kamei, Y., Iwaguro, H., Matsukawa, Y., & Funahashi, Y. (2012). Periurethral injection of autologous adipose-derived regenerative cells for the treatment of male stress urinary incontinence: Report of three initial cases. *International Journal of Urology*, *19*(7), 652–659. DOI: 10.1111/j.1442-2042.2012.02999.x PMID: 22435469

Yokoyama, T., Huard, J., Pruchnic, R., Yoshimura, N., Qu, Z., Cao, B., de Groat, W. C., Kumon, H., & Chancellor, M. B. (2001). Muscle-derived cell transplantation and differentiation into lower urinary tract smooth muscle. *Urology*, *57*(4), 826–831. DOI: 10.1016/S0090-4295(00)01083-9 PMID: 11306423

Zambon, J. P., Williams, K. J., Bennington, J., & Badlani, G. H. (2019). Applicability of regenerative medicine and tissue engineering for the treatment of stress urinary incontinence in female patients. *Neurourology and Urodynamics*, *38*(S4), S76–S83. DOI: 10.1002/nau.24033 PMID: 31099087

Zhang, H., Qiu, X., Shindel, A. W., Ning, H., Ferretti, L., Jin, X., Lin, G., Lin, C. S., & Lue, T. F. (2012a). Adipose tissue-derived stem cells ameliorate diabetic bladder dysfunction in a type II diabetic rat model. *Stem Cells and Development*, *21*(9), 1391–1400. DOI: 10.1089/scd.2011.0244 PMID: 22008016

Zhang, H., Qiu, X., Shindel, A. W., Ning, H., Ferretti, L., Jin, X., Lin, G., Lin, C. S., & Lue, T. F. (2012b). Adipose tissue-derived stem cells ameliorate diabetic bladder dysfunction in a type II diabetic rat model. *Stem Cells and Development*, *21*(9), 1391–1400. DOI: 10.1089/scd.2011.0244 PMID: 22008016

Zhang, Y., McNeill, E., Tian, H., Soker, S., Andersson, K. E., Yoo, J. J., & Atala, A. (2008). Urine derived cells are a potential source for urological tissue reconstruction. *The Journal of Urology*, *180*(5), 2226–2233. DOI: 10.1016/j.juro.2008.07.023 PMID: 18804817

Zhou, S., Zhang, K., Atala, A., Khoury, O., Murphy, S. V., Zhao, W., & Fu, Q. (2016). Stem Cell Therapy for Treatment of Stress Urinary Incontinence: The Current Status and Challenges. *Stem Cells International*, *2016*(1), 7060975. Advance online publication. DOI: 10.1155/2016/7060975 PMID: 26880983

Zhu, J. (2010). Bioactive modification of poly(ethylene glycol) hydrogels for tissue engineering. *Biomaterials*, *31*(17), 4639–4656. DOI: 10.1016/j.biomaterials.2010.02.044 PMID: 20303169

Zuk, P. A., Zhu, M., Mizuno, H., Huang, J., Futrell, J. W., Katz, A. J., Benhaim, P., Lorenz, H. P., & Hedrick, M. H. (2001). Multilineage cells from human adipose tissue: Implications for cell-based therapies. *Tissue Engineering*, *7*(2), 211–228. DOI: 10.1089/107632701300062859 PMID: 11304456

Chapter 4
Other Bladder Conditions

ABSTRACT

There are many conditions that affect the urothelium, like the underactive bladder or detrusor underactivity, Interstitial Cystitis and Bladder Pain Syndrome. These affections can have a high impact in the quality of life of the patients, and increase the requirements of hospitalizations, absence from work, psychological impact, among others. The detrusor underactivity characterized by the impaired contractility of the detrusor muscle, is a debilitating condition often accompanied by a complex constellation of pelvic pain symptoms. This chapter delves into the intricate relationship between these two conditions, exploring their etiologies, clinical manifestations, diagnostic challenges, and therapeutic approaches. By understanding the underlying mechanisms and the impact of detrusor atony on pelvic pain, clinicians can develop comprehensive and effective management strategies to improve the quality of life for affected patients.

BACKGROUND

SC is an emerging field with promising potential for treating overactive bladder (OAB). OAB is characterized by symptoms such as urinary urgency, frequency, and incontinence, which significantly impact the quality of life. But we face the opposite problem in the real life. By introducing SC into the bladder wall, it is conceivable that they could differentiate into smooth muscle cells, thereby regenerating damaged detrusor muscle tissue and improving bladder contractility. Detrusor atony, characterized by impaired detrusor muscle contractility, is a condition that significantly impacts patients' quality of life.

SC, particularly MSCs, have shown potential in preclinical models for treating bladder dysfunctions, including OAB and underactivity. These cells can differentiate into various cell types and promote tissue regeneration by releasing chemokines and progenitor cells. MSCs are favored due to their enhanced proliferation and

DOI: 10.4018/979-8-3693-8829-7.ch004

differentiation capabilities, as well as their lower risk of tumorigenicity compared to other stem cells (Shin et al., 2020a).

The multifaceted nature of detrusor atony and pelvic pain necessitates a comprehensive understanding of their pathophysiology, clinical manifestations, and management strategies. Research has demonstrated that MSCs can be derived from the patient's own bone marrow and have phenotypic and physiological similarities with bladder smooth muscle cells. This similarity suggests that MSCs could serve as an alternative cell source for repairing damaged bladder tissues (Bohne & Urwiller, 1956).

Challenges remain, such as ensuring the survival and proper integration of SC in the bladder tissue. Additionally, more clinical trials are needed to establish the efficacy and safety of stem cell therapy for OAB (Shin et al., 2020a).

Overall, while SC therapy for OAB and other bladder conditions is still in the experimental stage, it holds great promise for providing a more effective and long-lasting treatment option in the future. Frequently intertwined with a spectrum of pelvic pain symptoms, it presents a complex diagnostic and therapeutic challenge. The prevalence of this combined condition is substantial, with estimates suggesting.

Despite advancements in urological and pelvic medicine, significant knowledge gaps persist regarding the underlying mechanisms and optimal treatment approaches. A multidisciplinary approach, encompassing urology, gynecology, pain management, and psychology, is essential to address the multifaceted needs of affected patients.

Some of the mechanisms that the regenerative therapy can intervene would be by modulation of inflammation. Pelvic pain often involves chronic inflammation. SC possess anti-inflammatory properties that could potentially reduce pain and improve overall pelvic health. Regarding the neuromuscular re-education, the SC therapy might aid in the restoration of normal neuromuscular communication within the bladder and pelvic floor, contributing to improved bladder function and reduced pain.

Whit the tissue engineering, the SC could be used to create engineered bladder tissue, providing a potential replacement for severely damaged bladders. While these possibilities are exciting, it's essential to note that SC research is still in its early stages, and significant challenges remain before widespread clinical application. Rigorous research, ethical considerations, and regulatory approval are crucial steps towards realizing the full potential of stem cell therapy in treating detrusor atony and pelvic pain.

DETRUSOR UNDERACTIVITY (DUA)

Bladder underactivity, or detrusor underactivity, is a condition where the bladder muscle (detrusor) cannot contract effectively, leading to incomplete bladder emptying. Detrusor underactivity (DUA) is a urodynamic diagnosis defined by the International Continence Society as "a contraction of reduced strength and/or duration, resulting in prolonged bladder emptying and/or inability to achieve emptying of the bladder within a normal period of time. DUA is estimated to affect between 12 and 45% of women and between 9 and 28% of men under 50 years of age, but increases to approximately 48% in men over 70 years of age. The pathophysiology remains unknown, but there are clear triggering risk factors including aging, neurological deficits, diabetes mellitus, and persistent BOO (Shin et al., 2020a).

The mechanism may be idiopathic, changes in afferent or efferent nerve control, myogenic through detrusor degeneration, changes in ion exchange and transport. There is no definitive treatment for DUA, and drugs such as cholinergics to increase contractility have limited efficacy, and currently treatment is based on bladder emptying by intermittent or permanent catheterization, or suprapubic cystostomy. (Aldamanhori & Chapple, 2017; Drake et al., 2014; Shin et al., 2020a; Yoshida & Yamaguchi, 2014)

Among the different models of DUA, diabetic mice with intraperitoneal injection of streptozotocin (STZ) with or without a diet high in cholesterol, induction of chronic ischemia (vessel ligation or hyperlipidemia), pelvic nerve or spinal cord injury (direct trauma or cold injury), cerebral ischemia, diabetes, and persistent BOO. Considering that these same techniques are used to induce OAB in animal models, it is essential to perform functional evaluations. (J. H. Kim et al., 2013) The different animal models for OAB or DUA and results of preclinical stem cell therapy in these models are summarized in Table 1 (Shin et al., 2020a).

Zhang et al. used autologous AD-MSC in a group of 8-week-old female rats with induced type 2 diabetes with low-dose intraperitoneal STZ and a high-fat diet to achieve bladder dysfunction. (Zhang et al., 2012) Experimental AD-MSCs were treated with 10μM 5-ethynyl-2′-deoxyuridine (EdU) fluorescent staining for their monitoring after being injected into the tail vein or directly into the bladder. After injection into the bladder, EdU-labeled AD-MSCs were visualized predominantly in the submucosal layer, with some EdU nuclei in smooth muscle with actin expression. Intravenous injection of AD-MSC differentiated directly into detrusor smooth muscle cells, with improvement in voiding dysfunction, associated with paracrine effects with inhibition of apoptosis and promotion of angiogenesis.

However, improvement in micturition was only demonstrated in 40 to 60% of diabetic animals. Subsequently, the same study group then added defocused low-energy shock waves (DLSW) to the protocol to maximize the therapeutic effect of

AD-MSC. (Y. Jin et al., 2017; Zhang et al., 2019) Zhang et al compared the effects of autologous AD-MSCs versus DLSW-pretreated AD-MSCs in diabetic rat models. AD-MSC pretreated with DLSW showed higher efficacy and more secretion of VEGF and NGF.

Liang et al. also used an STZ-induced diabetic rat model to assess the effects of hAFSCs. (Liang et al., 2018) Liang's group measured the immunoreactivities and mRNAs of NGF and M2 and M3 muscarinic receptors in rats with STZ-induced bladder dysfunction, subsequently treated with hAFSCs, demonstrating recovery of NGF and muscarinic receptors, suggesting this as a potential mechanism for the recovery of function.(Shin et al., 2020a)

Table 1. Stem cell therapy in preclinical models of DUA

Author	Method	Aninal/Sex	Cell type	Route of injection	Model of action
Tremp et al (2014)(Tremp et al., 2014)	Bilateral iliac artery ischemia	female rat	ADSCs	Bladder injection	TNFá
Nitta et al (2010)(Nitta et al., 2010)	Cerebral ischemia	female rat	Sk-MSC	Damaged nerve lesion	NGF, M2, M3, P2X1
Kwon et al (2005)(Kwon et al., 2005)	Cerebral ischemia	female rat	MDCs	Damaged nerve lesion	NGF, M2, M3
Huard et al (2002)(Huard et al., 2002)	Bilateral iliac artery ischemia	female rat	MDCs	Bladder injection	regenerates bladder tissue
Sakuma et al (2009) (Sakuma et al., 2009)	Bladder outflow obstruction (BOO)	female rat	ADSCs	Bladder injection	SMC TGFâ paracrine effects
Liang et al (2018)(Liang et al., 2018)	Partial bladder outlet obstruction (pBOO) obstruction (BOO)	female rat	hAFSCs	Bladder injection	NGF M2, M3 receptors changes

In 2015, Levanovich et al. conducted a study with intradetrusor injection of adult muscle-derived cells (AMDC) in a human DUA patient, (Levanovich et al., 2015; Shin et al., 2020a) a 79-year-old man who, after transurethral resection of the prostate, suffered from recurrent urinary retention despite the administration of alpha-blockers and bethanechol. The patient required intermittent catheterization 4 to 6 times per day, but suffered from episodes of gross hematuria and febrile urinary tract infections. A final concentration of 250 million CDMA was obtained from 150 mg of quadriceps femoris muscle under local anesthesia. CDMAs were injected to a depth of 2 mm by direct visualization with a flexible cystoscope. Each injection volume was 0.5 ml, with a total volume of 15 mL. Subjective symptoms of the patient improved, and the ability to urinate small volumes at 3 months was observed post-treatment, but the patient still required intermittent catheterization during the one-year follow-up period.

NEUROGENIC BLADDER

Neurogenic bladder dysfunction is a common complication of neurological disorders, affecting patients with varying degrees of severity, which significantly impacts quality of life, making it a critical focus of preclinical research. Individuals with nerve damage often experience urinary frequency, incontinence, urgency, and urinary tract infections. The efficacy of SC therapy in promoting bladder functional recovery following spinal cord injury (SCI), multiple sclerosis (MS), Parkinson's disease (PaD), and stroke has a potential role. (Salehi-Pourmehr et al., 2020)

In the United States, a substantial proportion of patients with MS (40-90%), PaD (37-72%), and stroke (15%) develop neurogenic bladder. Furthermore, between 50-90% of individuals with MS exhibit hyperreflexia, while the remainder present with areflexia. Following SCI, a significant majority (70-84%) experience some form of bladder dysfunction. (S. J. Kim et al., 2020; Wyndaele et al., 2001)

A variety of approaches have been developed to manage neurogenic bladder, including conservative therapies (lifestyle modifications, bladder retraining, pelvic floor exercises), pharmacologic treatments (anticholinergics, beta-adrenergic agonists), non-invasive interventions (electrical stimulation, clean intermittent catheterization, indwelling catheters), and surgical procedures (augmentation cystoplasty). However, achieving complete control of voiding dysfunction remains challenging, and existing treatments often have associated side effects. (de Sèze et al., 2007; Jahn et al., 2012; Sauerwein, 1990)

Laboratory animals, primarily mice and rats, are extensively employed in preclinical research due to their biological similarities to humans. Consequently, these models are commonly used to investigate neurogenic bladder dysfunction. Micturition is a complex process regulated by intricate neural pathways connecting the brain, spinal cord, and peripheral nervous system. (Bugos et al., 2009)

Damage to the cerebral cortex, brainstem, basal ganglia, or spinal cord, as occurs in various neurological disorders, can disrupt this neural control, leading to NGB. Typically, upper motor neuron lesions (above the sacral cord) result in neurogenic detrusor overactivity, whereas lower motor neuron injuries, such as those affecting the lumbar cord, conus medullaris, cauda equina, or peripheral nerves, often manifest as underactive or acontractile detrusor. (Apodaca et al., 2003; Craggs et al., 2006; Vizzard, 1999)

Patients with neurogenic bladder or idiopathic detrusor overactivity necessitate bladder tissue engineering to circumvent the complications associated with augmentation cystoplasty. Researchers have explored both natural and synthetic biomaterials as potential bladder substitutes, often incorporating functional molecules, nanotechnology, or exogenous cells to enhance regenerative capabilities. While preclinical studies demonstrate promising outcomes, translation to large animal models and human clinical trials has yielded inconsistent results. Standardizing biomaterial fabrication, bioactive agent integration, nanotechnology implementation, and stem cell application is essential for achieving functional bladder regeneration. Advanced technologies such as next-generation sequencing and magnetic resonance imaging can provide valuable insights into underlying mechanisms and facilitate non-invasive monitoring of regenerative processes. (Salehi-Pourmehr et al., 2020)

While SC therapies hold promise, concerns remain regarding cellular rejection by the host immune system. A significant proportion of transplanted cells perish within days of administration. (Marra et al., 2016) Animal models, particularly rodents, offer valuable platforms to investigate the efficacy and safety of novel therapeutic interventions like stem cell therapy. These models effectively replicate many disease characteristics, enabling the exploration of underlying mechanisms and potential adverse effects.

Salehi-Pourmehr et al published a systematic review, by identified 1,189 relevant studies published up to June 2019. Of these, 20 met our inclusion criteria for evaluation of SC therapy in rodent models of neurogenic bladder dysfunction. Fifteen studies focused on SCI, while two addressed PaD, two stroke, and one MS. A meta-analysis was conducted on SCI experiments, and detailed urodynamic data were reported for all neurological conditions (Table 2).

Common SC sources included neural progenitor cells, BMSC, hAFSCs, and UCSCs. SC transplantation significantly increased micturition pressure in both contusion and transection SCI models at four and eight weeks' post-intervention. Improvements in residual urine volume, micturition volume, and bladder capacity were observed at 28 days post-transplantation exclusively in the transection SCI model. Recovery of non-voiding contractions occurred at 56 days post-transplantation in the contusion model. They concluded SC therapy has shown promise in partially restoring bladder function in spinal cord injury models. However, due to the limited number of studies in other neurological diseases, further research is required to elucidate the precise mechanisms underlying bladder recovery in these conditions

Table 2. Characteristics of included studies in systematic review and meta-analysis

Species	Disease	Oucome	Mechanism (s)	cell type	cell density
Female rats (Campeau et al., 2014)	PaD	Improvement at 42 days	Tyrosine hydroxylase positive neurons in the treated substantia nigra pars compacta (juxtacrine effect).	rBMSCs	1×10^5
female athymic nude rats (Soler et al., 2012)	PaD	Temporarily amelioration of bladder dysfunction	Juxtacrine or paracrine effects.	hAFSCs and Right MFB hBMSCs	-
Female Sprague-Dawley rats (Liang et al., 2017)	Stroke	Improvement of bladder dysfunction	Expressions of bladder NGF, muscarinic and P2X1 receptors.	hAFSCs	1×10^6
Female Sprague-Dawley rats (Liang et al., 2016)	Cerebral ischemia	Restoration of bladder dysfunction	Expressions of NGF, M2, and M3 in the bladder.	hUCMSCs	1×10^6
Male Sprague-Dawley (Cho et al., 2014)	SCI	Improvement of bladder dysfunction	Inhibiting apoptosis and enhancing cell proliferation.	Oral mucosa stem cells of rats	-
Immune-deficient B6.CB17 mice (Fandel et al., 2016)	SCI	Improvement of NVC, RV and voiding efficiency	Differentiate into GABAergic neuron subtypes and receive synaptic inputs functional integration into host spinal cord.	hESC-MGEs	3×10^5 (low dose) or 6×10^5 and 8×10^5 (high dose)
Male Sprague-Dawley rats (Hu et al., 2012)	SCI	Recovery of lower urinary tract function	Inhibition unmediated C-fiber bladder afferent sprouting and decrease a C-fiber bladder-to- bladder spinal micturition reflex.	BMSCs	3×10^6
Female Sprague-Dawley rats (H. J. Lee et al., 2015)	SCI	Recovery of bladder function	Differentiation into smooth muscle cells	BM3.B10 (B10)	1×10^6
Female athymic rats (Y. Jin et al., 2011)	SCI	Attenuation of hyperactive bladder reflexes	Differentiation into glia and astrocytes.	hGRP and hGDA	1×10^5
Female Wistar rats (Mitsui et al., 2003)	SCI	Recovery of voiding function		Immortalized neural stem cells (EG6)	1×10^6

continued on following page

Table 2. Continued

Species	Disease	Oucome	Mechanism (s)	cell type	cell density
Female Sprague-Dawley rats (Mitsui et al., 2005)	SCI	Improvement of bladder function	Neuroprotection, axonal growth	NRPs and GRPs	10 µL
Female Sprague-Dawley rats (Mitsui et al., 2011)	SCI	Recovery of micturition function	Neuroprotection	Neuronal precursors	1× 10^5 cells/ ìL
Female Sprague Dawley rats (B et al., 2008)	SCI	Improvement of bladder function in 2 models	A mixed population of lineage restricted neural precursor cells	NPCs	1×106 indirect injection 2×106 NPCs in LP delivery
Female Sprague Dawley rats (Park et al., 2010)	SCI	Not recovery in bladder dysfunction	-	hMSCs	30×105/5 ìL
Female Fischer 344 rats (Sandner et al., 2016)	SCI	No changes in the recovery of bladder function	-	BMSCs and fibroblasts	BMSCs (105 cells/ ìL) or fibroblast (105 cells/ ìL)
Sprague-Dawley rats (Temeltas et al., 2009)	SCI	Improvement some lower urinary system function	Neuronal differentiation by E-NCAM and A2B5	NRPs and GRPs or nBMSCs	1×106
Female rats (Y. Jin et al., 2016)	SCI	Only bladder recovery after the combined treatment in the N/C/ G group	5-HT expression	NPCs	1×105
Female rats (Salehi-Pourmehr et al., 2019)	SCI	Improvement some lower urinary system function	-	BM-MSCs	1×106

continued on following page

Table 2. Continued

Species	Disease	Oucome	Mechanism (s)	cell type	cell density
Sprague-Dawley rats (DePaul et al., 2015)	SCI	Void volume and frequency 8 and 10 weeks postinjury	Enhance tissue sparing and decrease astroglial reactivity	human MAPCs	4×106
C57BL/6 J mice (Z. Jin et al., 2017)	EAE	Improvement of bladder dysfunction	c-Kit expression amelioration of pannexin 1 and connexin43	Interstitial cells of Cajal	

(Salehi-Pourmehr et al., 2020)

PaD, Parkinson disease; SCI, spinal cord injury; EAE, experimental autoimmune encephalomyelitis; rBMSC, rat bone marrow mesenchymal stem cells; hBMSC, human bone marrow mesenchymal stem cells; hAFSC, human amniotic fluid stem cells; NPCs, neural progenitor cells; GRPs, glial-restricted progenitors; hUCMSC, human umbilical cord blood stem cells; MSC, mesenchymal stromal cells; hESC-MGEs, human embryonic stem cells- medial ganglionic eminence; hGRP, human glial-restricted progenitors; hGDA, astrocytes derived from hGRP; MFB, medial forebrain bundle; LP, lumbar puncture; MCAO, middle cerebral artery occlusion; N/C/G, NPC+ lentivirus vectors expressing chondroitinase (Chase/LV) + Growth factors; MPA, multipotent adult progenitor cell; NGF, nerve growth factor; NRPs, neuronal restrict

Previous meta-analyses have evaluated bladder recovery following SC therapy in SCI-induced bladder dysfunction by controlling for SCI model heterogeneity. (J. H. Kim et al., 2015) In the Salehi-Pourmehr study expanded on this by considering additional factors influencing heterogeneity, such as timing of urodynamic assessment (28 vs. 8 weeks post-transplantation).

While cell type, route, and timing of transplantation could potentially impact outcomes, these factors were relatively consistent across included studies. The majority of SCs were administered intrathecally, and the timing of transplantation was predominantly 9 days post-injury. Due to limited heterogeneity in cell type and transplantation parameters, subgroup analyses were not performed for these variables. However, we acknowledge the potential influence of these factors and recommend further investigation in future studies.

By focusing on intrathecal SC administration and controlling for SCI model and urodynamic assessment timing, our study provides a more refined analysis of bladder recovery following SC therapy. Importantly, this review extends beyond SCI to include other neurological conditions associated with neurogenic bladder, offering a broader perspective on the potential of SC therapy in this patient population.

They concluded that the systematic review and meta-analysis demonstrate that SC therapy significantly improves micturition pressure in both contusion and transection SCI models at 4 and 8 weeks' post-transplantation. While residual urine volume, micturition volume, and bladder capacity improved in the transection model at 28 days, non-voiding contraction recovery was observed only at 56 days in the contusion model.

To advance our understanding of bladder recovery, high-quality SCI studies with rigorous methodology are essential to minimize bias. Moreover, given the limited research on neurogenic bladder dysfunction (NGB) in Parkinson's disease, stroke, and multiple sclerosis, additional studies with robust designs are needed to elucidate underlying mechanisms and develop effective treatment strategies.

INTERSTITIAL CYSTITIS/BLADDER PAIN SYNDROME (IC/BPS)

Pelvic pain syndrome, including conditions like interstitial cystitis/bladder pain syndrome (IC/BPS), is characterized by chronic pelvic pain, urinary urgency, and frequency. SC therapy is being investigated as a treatment option due to its anti-inflammatory and regenerative properties2. MSCs can help reduce inflammation and promote healing of the bladder lining, potentially alleviating pain and improving bladder function2. Early studies in animal models have shown promising results, but more research is needed to establish the safety and efficacy of stem cell therapy for this condition in humans (L & Ke, 2018; Shin et al., 2024).

Interstitial cystitis/bladder pain syndrome (IC / BPS) presents as chronic pelvic pain usually accompanied by frequency, nocturia, voiding dysfunction or urgency, in the absence of identifiable pathology. (D'Ancona et al., 2019) IC/BPS is overall uncommon, with a female predominance (M:F, 1:9). The worldwide prevalence varies among studies, but recent evidence suggests that IC/BPS prevalence might be 2% or more in the female population. Also can be associated with sexual dysfunction, sleep disturbance, anxiety, depression, and chronic stress, leading to significantly decreased quality of life.

The pathophysiology of the syndrome remains unclear, but hypotheses include defects in the glycosaminoglycan (GAG) layer, impaired urothelial permeability, denudation of the epithelium, autoimmunity, infection, and increased inflammatory responses, with mast cell infiltration and neurofilament production. The treatment modalities used for other types of bladder dysfunction are the current therapy for this disease, such as behavior modification, drug therapy, intravesical instillation with hyaluronic acid, and cystoscopic surgery, but they only can provide temporary control of symptoms and carry a high risk of recurrence and side effects. (Shin et al., 2020b)

Lines of research have been conducted to find the pathophysiology and a treatment strategy for IC / BPS in a variety of animal models, including the bladder-centric model, induced by the addition of toxic substances to the urine, and models with more complex mechanisms, such as modulation of bladder function through alterations to the central nervous system and psychological and physical stress induction. (Birder & Andersson, 2018; A. Kim, Han, et al., 2017)

Currently, no ongoing clinical trials on human IC / BPS SC therapy have been registered on Clinicaltrial.gov. Eleven preclinical trials evaluating the therapeutic effects of SC treatments in an IC / BPS rat model have been published from South Korea, China, Japan, and Taiwan (Table 3), all using chemical instillation bladder toxicity to reproduce the clinical features of IC / BPS, which are urothelial denudation, inflammation, and bladder fibrosis with frequent urination and small voiding volumes. Chen et al. demonstrated that a combination of melatonin and AD-MSC were superior to either alone in a rat model with cyclophosphamide-induced cystitis (CYP-IC) due to synergistic antioxidant effects. (Chen et al., 2014a; Shin et al., 2020b)

Song et al. evaluated the therapeutic effects of HUCB in the hydrochloric acid-induced cystitis (HCL-IC) rat model. The toxic effect of the HCL-IC is related to the decreased growth factors, and increased in those receiving transplanted HUCB, with an amelioration on the voiding interval, and activated signaling cascade needed for the growth factors production (WNT signaling pathway). This is the first report that provides an experimental evidence of the therapeutic effects and molecular mechanisms of MSC therapy to IC using an orthodox rat animal model. (Song et al., 2015)

Table 3. Stem cell therapy in preclinical models of IC/BPS

Author	Aninal/Sex	Cell type	Route of injection	Model of action
Ryu et al 74	female rat	hESC-MSC	Bladder injection	IGF
Lee et al (2018)(S. W. Lee et al., 2018)	female rat	hESC-MSC	Bladder injection	Anti-Fibrosis
Xie et al (2018)(Xie et al., 2018)	female rat	hUCB-MSCs	Intravenous tail injection	AKT mTOR
Furuta et al (2018)(Furuta et al., 2018)	female rat	AD-MSCs	Bladder injection	TNFá TGFâ VEGF
Kim et al (2017)(A. Kim, Yu, et al., 2017a)	female rat	hESC-MSC	Bladder injection	Wnt

continued on following page

Table 3. Continued

Author	Aninal/Sex	Cell type	Route of injection	Model of action
Li et al (2017)(Li et al., 2017)	female rat	hUSCs	Bladder instillation injection	Anti-inflammatory
Xiao et al (2017)(*TGF-â/MAPK signaling mediates the effects of bone marrow mesenchymal stem cells on urinary control and interstitial cystitis after urinary bladder transplantation - PubMed*, s. f.)	female rat	BM-MSCs	Bladder transplan-tation, intraperitoneal injection	TGF-â/MAPK
Kim et al (2016)(A. Kim et al., 2016)	female/male rat	hUCB-MSC	Bladder injection	Anti-fibrotic
Hirose et al (2016)(Hirose et al., 2016)	female rat	DP-SC (Dental pulp)	Bladder injection	Anti-inflammatory
Song et al (2015)(Song et al., 2015)	female rat	hUCB-MSC	Bladder injection	Wnt
Chen et al (2014)(Chen et al., 2014b)	male rat	AD-MSCs	Intravenous tail injection	Anti-inflammatory

The WNT signaling pathway was also activated by multipotent stem cells (M-MSCs), and these stem cells directly differentiated into urothelium or pericytes of vascular endothelial tissue. Kim et al. studied the injection hydrochloric-acid instillation-induced IC/BPS animal model, then after treated with a single local injection of M-MSCs, reporting an ameliorated bladder symptoms of IC/BPS, as well on the bladder architecture. The infused M-MSCs differentiated into multiple cell types and gradually integrated into vascular-like structures. Micro-PET-MRI was performed to evaluate tumorigenicity immediately post-injection and 3, 6, 9, and 12 months afterward, and rats were sacrificed for necropsy after their last image workup. These necropsies showed that the MMSCs did not migrate to other organs, and no signs of tumorigenesis were observed. (A. Kim, Yu, et al., 2017b; Shin et al., 2020b)

The same group subsequently presented several papers studying a rat model with ketamine-induced cystitis (KC) replicating the clinical symptoms of IC described previously. (A. Kim et al., 2016; S. W. Lee et al., 2018; Song et al., 2016) This group reported a direct relationship in decreased bladder capacity and increased ketamine dose, with increased fibrosis and apoptosis of the submucosal ketamine dose-dependent, similar to the symptoms of human patients with IC / BPS. M-MSCs were injected directly into the bladder once, and histologic and functional abnormalities were shown to relieve one week later. (S. W. Lee et al., 2018) An anti-fibrotic mechanism of M-MSCs was proposed in the KC model. (A. Kim et al., 2016; S. W. Lee et al., 2018) A limitation of the HCL-IC model is the acute nature of the urothelial damage caused, whereas in humans, IC / BPS is a chronic disease. (Shin et al., 2020b)

To generate a model of IC that can mimic the chronic symptoms, Choo and his colleagues used an LPS-IC model by instillation with protamine sulfate (PS) and lipopolysaccharide (LPS) once a week for five weeks in 8 weeks old female rats. (Ryu et al., 2018, 2019) The distribution, perivascular integration, and therapeutic effects of M-MSCs were monitored by in vivo confocal microscopic and endoscopic imaging, together with immunofluorescence analysis of bladder tissues, demonstrated that the transplanted M-MSCs were grafted after differentiation into multiple cell types, and gradually integrated into the adjacent perivascular structure up to 30 days after transplantation. Injection of M-MSC also significantly relieved functional and histological damage in this LPS-IC model. (Ryu et al., 2018)

Treatment with M-MSC was superior to BM-MSC in the LPS-IC model, and transplantation and M-MSC differentiation were enhanced by growth factors activated by the SHH-WNT signaling pathway, including EGF, IGF, and FGF. These reports from the Choo group provide the first evidence for therapeutic efficacy, with long-term safety and in vivo distribution and cellular properties of hESC derivatives in preclinical IC / BPS models. (A. Kim, Yu, et al., 2017b; Ryu et al., 2018; Shin et al., 2020b)

CONCLUSIONS

DUA and pelvic pain represent a complex clinical challenge with significant implications for patients' quality of life. While advancements in diagnostic and therapeutic modalities have improved management, the interplay between these conditions necessitates a comprehensive understanding of their pathophysiology and the development of innovative treatment approaches.

The potential of stem cell therapy to revolutionize the treatment of detrusor atony and pelvic pain is particularly promising. By harnessing the regenerative capabilities of stem cells, researchers and clinicians may be able to restore damaged detrusor muscle tissue, reduce inflammation, and improve bladder function. However, it is essential to approach this emerging field with caution and rigorous scientific inquiry.

Multidisciplinary collaboration, patient education, and continued research are fundamental to optimizing patient care and advancing our understanding of these conditions. Through concerted efforts, we can improve the lives of individuals affected by detrusor atony and pelvic pain.

SC therapy holds significant promise as a potential treatment for neurogenic bladder dysfunction. Preclinical studies, primarily in rodent models, have demonstrated encouraging results in improving bladder function, particularly in terms of micturition pressure and, to a lesser extent, bladder capacity and residual urine volume.

Overall, while stem cell therapy for these conditions is still in the experimental stage, it holds promise for providing new treatment options in the future. However, it's essential to acknowledge that research is still in its early stages. While promising, the long-term efficacy and safety of stem cell therapy for neurogenic bladder dysfunction in humans remain to be established through rigorous clinical trials.

REFERENCES

Aldamanhori, R., & Chapple, C. R. (2017). Underactive bladder, detrusor underactivity, definition, symptoms, epidemiology, etiopathogenesis, and risk factors. *Current Opinion in Urology*, *27*(3), 293–299. DOI: 10.1097/MOU.0000000000000381 PMID: 28221218

Apodaca, G., Kiss, S., Ruiz, W., Meyers, S., Zeidel, M., & Birder, L. (2003). Disruption of bladder epithelium barrier function after spinal cord injury. *American Journal of Physiology. Renal Physiology*, *284*(5), F966–F976. DOI: 10.1152/ajprenal.00359.2002 PMID: 12527557

Birder, L., & Andersson, K. E. (2018). Animal Modelling of Interstitial Cystitis/ Bladder Pain Syndrome. *International Neurourology Journal*, *22*(Suppl 1), S3–S9. DOI: 10.5213/inj.1835062.531 PMID: 29385788

Bohne, A. W., & Urwiller, K. L. (1956). Experience with urinary bladder regeneration. *Postgraduate Seminar. American Urological Association. North Central Section*, 24-27.

Bugos, O., Bhide, M., & Zilka, N. (2009). Beyond the rat models of human neurodegenerative disorders. *Cellular and Molecular Neurobiology*, *29*(6-7), 859–869. DOI: 10.1007/s10571-009-9367-5 PMID: 19263215

Campeau, L., Soler, R., Sittadjody, S., Pareta, R., Nomiya, M., Zarifpour, M., Opara, E. C., Yoo, J. J., & Andersson, K.-E. (2014). Effects of allogeneic bone marrow derived mesenchymal stromal cell therapy on voiding function in a rat model of Parkinson disease. *The Journal of Urology*, *191*(3), 850–859. DOI: 10.1016/j.juro.2013.08.026 PMID: 23973520

Chen, Y. T., Chiang, H. J., Chen, C. H., Sung, P. H., Lee, F. Y., Tsai, T. H., Lo Chang, C., Chen, H. H., Sun, C. K., Leu, S., Chang, H. W., Yang, C. C., & Yip, H. K. (2014a). Melatonin treatment further improves adipose-derived mesenchymal stem cell therapy for acute interstitial cystitis in rat. *Journal of Pineal Research*, *57*(3), 248–261. DOI: 10.1111/jpi.12164 PMID: 25132326

Chen, Y. T., Chiang, H. J., Chen, C. H., Sung, P. H., Lee, F. Y., Tsai, T. H., Lo Chang, C., Chen, H. H., Sun, C. K., Leu, S., Chang, H. W., Yang, C. C., & Yip, H. K. (2014b). Melatonin treatment further improves adipose-derived mesenchymal stem cell therapy for acute interstitial cystitis in rat. *Journal of Pineal Research*, *57*(3), 248–261. DOI: 10.1111/jpi.12164 PMID: 25132326

Cho, Y.-S., Ko, I.-G., Kim, S.-E., Lee, S.-M., Shin, M.-S., Kim, C.-J., Kim, S.-H., Jin, J.-J., & Kim, K.-H. (2014). Oral mucosa stem cells alleviates spinal cord injury-induced neurogenic bladder symptoms in rats. *Journal of Biomedical Science*, *21*(1), 43. DOI: 10.1186/1423-0127-21-43 PMID: 24884998

Craggs, M. D., Balasubramaniam, A. V., Chung, E. A. L., & Emmanuel, A. V. (2006). Aberrant reflexes and function of the pelvic organs following spinal cord injury in man. *Autonomic Neuroscience*, *126-127*, 355–370. DOI: 10.1016/j.autneu.2006.03.010 PMID: 16731049

D'Ancona, C., Haylen, B., Oelke, M., Abranches-Monteiro, L., Arnold, E., Goldman, H., Hamid, R., Homma, Y., Marcelissen, T., Rademakers, K., Schizas, A., Singla, A., Soto, I., Tse, V., de Wachter, S., & Herschorn, S. (2019). The International Continence Society (ICS) report on the terminology for adult male lower urinary tract and pelvic floor symptoms and dysfunction. *Neurourology and Urodynamics*, *38*(2), 433–477. DOI: 10.1002/nau.23897 PMID: 30681183

de Sèze, M., Ruffion, A., Denys, P., Joseph, P.-A., & Perrouin-Verbe, B. (2007). The neurogenic bladder in multiple sclerosis: Review of the literature and proposal of management guidelines. *Multiple Sclerosis*, *13*(7), 915–928. DOI: 10.1177/1352458506075651 PMID: 17881401

DePaul, M. A., Palmer, M., Lang, B. T., Cutrone, R., Tran, A. P., Madalena, K. M., Bogaerts, A., Hamilton, J. A., Deans, R. J., Mays, R. W., Busch, S. A., & Silver, J. (2015). Intravenous multipotent adult progenitor cell treatment decreases inflammation leading to functional recovery following spinal cord injury. *Scientific Reports*, *5*(1), 16795. DOI: 10.1038/srep16795 PMID: 26582249

Drake, M. J., Williams, J., & Bijos, D. A. (2014). Voiding dysfunction due to detrusor underactivity: An overview. *Nature Reviews. Urology*, *11*(8), 454–464. DOI: 10.1038/nrurol.2014.156 PMID: 25002201

Fandel, T. M., Trivedi, A., Nicholas, C. R., Zhang, H., Chen, J., Martinez, A. F., Noble-Haeusslein, L. J., & Kriegstein, A. R. (2016). Transplanted Human Stem Cell-Derived Interneuron Precursors Mitigate Mouse Bladder Dysfunction and Central Neuropathic Pain after Spinal Cord Injury. *Cell Stem Cell*, *19*(4), 544–557. DOI: 10.1016/j.stem.2016.08.020 PMID: 27666009

Furuta, A., Yamamoto, T., Igarashi, T., Suzuki, Y., Egawa, S., & Yoshimura, N. (2018). Bladder wall injection of mesenchymal stem cells ameliorates bladder inflammation, overactivity, and nociception in a chemically induced interstitial cystitis-like rat model. *International Urogynecology Journal*, *29*(11), 1615–1622. DOI: 10.1007/s00192-018-3592-8 PMID: 29508044

Hirose, Y., Yamamoto, T., Nakashima, M., Funahashi, Y., Matsukawa, Y., Yamaguchi, M., Kawabata, S., & Gotoh, M. (2016). Injection of Dental Pulp Stem Cells Promotes Healing of Damaged Bladder Tissue in a Rat Model of Chemically Induced Cystitis. *Cell Transplantation*, *25*(3), 425–436. DOI: 10.3727/096368915X689523 PMID: 26395427

Hu, Y., Liao, L. M., Ju, Y. H., Fu, G., Zhang, H. Y., & Wu, H. X. (2012). Intravenously transplanted bone marrow stromal cells promote recovery of lower urinary tract function in rats with complete spinal cord injury. *Spinal Cord*, *50*(3), 202–207. DOI: 10.1038/sc.2011.128 PMID: 22124344

Huard, J., Yokoyama, T., Pruchnic, R., Qu, Z., Li, Y., Lee, J. Y., Somogyi, G. T., de Groat, W. C., & Chancellor, M. B. (2002). Muscle-derived cell-mediated ex vivo gene therapy for urological dysfunction. *Gene Therapy*, *9*(23), 1617–1626. DOI: 10.1038/sj.gt.3301816 PMID: 12424614

Jahn, P., Beutner, K., & Langer, G. (2012). Types of indwelling urinary catheters for long-term bladder drainage in adults. *Cochrane Database of Systematic Reviews*, *10*(10), CD004997. Advance online publication. DOI: 10.1002/14651858.CD004997.pub3 PMID: 23076911

Jin, Y., Bouyer, J., Shumsky, J. S., Haas, C., & Fischer, I. (2016). Transplantation of neural progenitor cells in chronic spinal cord injury. *Neuroscience*, *320*, 69–82. DOI: 10.1016/j.neuroscience.2016.01.066 PMID: 26852702

Jin, Y., Neuhuber, B., Singh, A., Bouyer, J., Lepore, A., Bonner, J., Himes, T., Campanelli, J. T., & Fischer, I. (2011). Transplantation of human glial restricted progenitors and derived astrocytes into a contusion model of spinal cord injury. *Journal of Neurotrauma*, *28*(4), 579–594. DOI: 10.1089/neu.2010.1626 PMID: 21222572

Jin, Y., Xu, L., Zhao, Y., Wang, M., Jin, X., & Zhang, H. (2017). Endogenous Stem Cells Were Recruited by Defocused Low-Energy Shock Wave in Treating Diabetic Bladder Dysfunction. *Stem Cell Reviews and Reports*, *13*(2), 287–298. DOI: 10.1007/s12015-016-9705-1 PMID: 27921202

Jin, Z., Ding, Y., Xue, R., Jia, Z., Huang, Z., Ding, Y., Gu, C., & Yang, J. (2017). Involvement of interstitial cells of Cajal in bladder dysfunction in mice with experimental autoimmune encephalomyelitis. *International Urology and Nephrology*, *49*(8), 1353–1359. DOI: 10.1007/s11255-017-1597-8 PMID: 28425078

Kim, A., Han, J.-Y., Ryu, C.-M., Yu, H. Y., Lee, S., Kim, Y., Jeong, S. U., Cho, Y. M., Shin, D.-M., & Choo, M.-S. (2017). Histopathological characteristics of interstitial cystitis/bladder pain syndrome without Hunner lesion. *Histopathology*, *71*(3), 415–424. DOI: 10.1111/his.13235 PMID: 28394416

Kim, A., Yu, H. Y., Heo, J., Song, M., Shin, J. H., Lim, J., Yoon, S. J., Kim, Y., Lee, S., Kim, S. W., Oh, W., Choi, S. J., Shin, D. M., & Choo, M. S. (2016). Mesenchymal stem cells protect against the tissue fibrosis of ketamine-induced cystitis in rat bladder. *Scientific Reports*, *6*(1), 30881. Advance online publication. DOI: 10.1038/srep30881 PMID: 27481042

Kim, A., Yu, H. Y., Lim, J., Ryu, C. M., Kim, Y. H., Heo, J., Han, J. Y., Lee, S., Bae, Y. S., Kim, J. Y., Bae, D. J., Kim, S. Y., Noh, B. J., Hong, K. S., Han, J. Y., Lee, S. W., Song, M., Chung, H. M., Kim, J. K., & Choo, M. S. (2017a). Improved efficacy and in vivo cellular properties of human embryonic stem cell derivative in a preclinical model of bladder pain syndrome. *Scientific Reports*, *7*(1), 8872. Advance online publication. DOI: 10.1038/s41598-017-09330-x PMID: 28827631

Kim, A., Yu, H. Y., Lim, J., Ryu, C. M., Kim, Y. H., Heo, J., Han, J. Y., Lee, S., Bae, Y. S., Kim, J. Y., Bae, D. J., Kim, S. Y., Noh, B. J., Hong, K. S., Han, J. Y., Lee, S. W., Song, M., Chung, H. M., Kim, J. K., & Choo, M. S. (2017b). Improved efficacy and in vivo cellular properties of human embryonic stem cell derivative in a preclinical model of bladder pain syndrome. *Scientific Reports*, *7*(1), 8872. Advance online publication. DOI: 10.1038/s41598-017-09330-x PMID: 28827631

Kim, J. H., Lee, S. R., Song, Y. S., & Lee, H. J. (2013). Stem cell therapy in bladder dysfunction: Where are we? And where do we have to go? En *BioMed Research International* (Vol. 2013). DOI: 10.1155/2013/930713

Kim, J. H., Shim, S. R., Doo, S. W., Yang, W. J., Yoo, B. W., Kim, J. M., Ko, Y. M., Song, E. S., Lim, I. S., Lee, H. J., & Song, Y. S. (2015). Bladder recovery by stem cell based cell therapy in the bladder dysfunction induced by spinal cord injury: Systematic review and meta-analysis. *PLoS One*, *10*(3), e0113491. DOI: 10.1371/journal.pone.0113491 PMID: 25781610

Kim, S. J., Cho, Y. S., Park, J. M., Na, Y. G., & Kim, K. H. (2020). Stem Cell Therapy for Neurogenic Bladder After Spinal Cord Injury: Clinically Possible? *International Neurourology Journal*, *24*(Suppl 1), S3–S10. DOI: 10.5213/inj.2040150.075 PMID: 32482052

Kwon, D., Minnery, B., Kim, Y., Jang, H. K., De Miguel, F., Yoshimura, N., & Chancellor, M. B. (2005). Neurologic recovery and improved detrusor contractility using muscle-derived cells in rat model of unilateral pelvic nerve transection. *Urology*, *65*(6), 1249–1253. DOI: 10.1016/j.urology.2005.01.037 PMID: 15922415

Lee, H. J., An, J., Doo, S. W., Kim, J. H., Choi, S. S., Lee, S.-R., Park, S. W., Song, Y. S., & Kim, S. U. (2015). Improvement in Spinal Cord Injury-Induced Bladder Fibrosis Using Mesenchymal Stem Cell Transplantation Into the Bladder Wall. *Cell Transplantation*, *24*(7), 1253–1263. DOI: 10.3727/096368914X682125 PMID: 24912020

Lee, S. W., Ryu, C. M., Shin, J. H., Choi, D., Kim, A., Yu, H. Y., Han, J. Y., Lee, H. Y., Lim, J., Kim, Y. H., Heo, J., Lee, S., Ju, H., Kim, S., Hong, K. S., Han, J. Y., Song, M., Chung, H. M., Kim, J. K., & Choo, M. S. (2018). The therapeutic effect of human embryonic stem cell-derived multipotent mesenchymal stem cells on chemical-induced cystitis in rats. *International Neurourology Journal*, *22*(Suppl 1), S34–S45. DOI: 10.5213/inj.1836014.007 PMID: 29385783

Levanovich, P. E., Diokno, A., Hasenau, D. L., Lajiness, M., Pruchnic, R., & Chancellor, M. B. (2015). Intradetrusor injection of adult muscle-derived cells for the treatment of underactive bladder: Pilot study. *International Urology and Nephrology*, *47*(3), 465–467. DOI: 10.1007/s11255-015-0924-1 PMID: 25663053

Li, J., Luo, H., Dong, X., Liu, Q., Wu, C., Zhang, T., Hu, X., Zhang, Y., Song, B., & Li, L. (2017). Therapeutic effect of urine-derived stem cells for protamine/lipopolysaccharide-induced interstitial cystitis in a rat model. *Stem Cell Research & Therapy*, *8*(1), 107. Advance online publication. DOI: 10.1186/s13287-017-0547-9 PMID: 28482861

Liang, C. C., Lee, T. H., & Chang, S. D. (2016). Effect of umbilical cord blood stem cells transplantation on bladder dysfunction induced by cerebral ischemia in rats. *Taiwanese Journal of Obstetrics & Gynecology*, *55*(5), 672–679. DOI: 10.1016/j.tjog.2015.11.002 PMID: 27751414

Liang, C. C., Shaw, S. W. S., Huang, Y. H., Lin, Y. H., & Lee, T. H. (2018). Improvement in bladder dysfunction after bladder transplantation of amniotic fluid stem cells in diabetic rats. *Scientific Reports*, *8*(1), 2105. Advance online publication. DOI: 10.1038/s41598-018-20512-z PMID: 29391467

Liang, C. C., Steven Shaw, S. W., Huang, Y. H., Lin, Y. H., & Lee, T. H. (2017). Bladder Transplantation of Amniotic Fluid Stem Cell may Ameliorate Bladder Dysfunction After Focal Cerebral Ischemia in Rat. *Stem Cells Translational Medicine*, *6*(4), 1227–1236. DOI: 10.1002/sctm.16-0212 PMID: 28186672

Marra, G., Sturch, P., Oderda, M., Tabatabaei, S., Muir, G., & Gontero, P. (2016). Systematic review of lower urinary tract symptoms/benign prostatic hyperplasia surgical treatments on men's ejaculatory function: Time for a bespoke approach? *International journal of urology : official journal of the Japanese Urological Association, 23*(1), 22-35. DOI: 10.1111/iju.12866

Mitsui, T., Fischer, I., Shumsky, J. S., & Murray, M. (2005). Transplants of fibroblasts expressing BDNF and NT-3 promote recovery of bladder and hindlimb function following spinal contusion injury in rats. *Experimental Neurology, 194*(2), 410–431. DOI: 10.1016/j.expneurol.2005.02.022 PMID: 16022868

Mitsui, T., Kakizaki, H., Tanaka, H., Shibata, T., Matsuoka, I., & Koyanagi, T. (2003). Immortalized neural stem cells transplanted into the injured spinal cord promote recovery of voiding function in the rat. *The Journal of Urology, 170*(4 Pt 1), 1421–1425. DOI: 10.1097/01.ju.0000075501.05758.33 PMID: 14501781

Mitsui, T., Neuhuber, B., & Fischer, I. (2011). Acute administration of AMPA/Kainate blocker combined with delayed transplantation of neural precursors improves lower urinary tract function in spinal injured rats. *Brain Research, 1418*, 23–31. DOI: 10.1016/j.brainres.2011.08.023 PMID: 21937028

Nitta, M., Tamaki, T., Tono, K., Okada, Y., Masuda, M., Akatsuka, A., Hoshi, A., Usui, Y., & Terachi, T. (2010). Reconstitution of experimental neurogenic bladder dysfunction using skeletal muscle-derived multipotent stem cells. *Transplantation, 89*(9), 1043–1049. DOI: 10.1097/TP.0b013e3181d45a7f PMID: 20150836

Park, W. B., Kim, S. Y., Lee, S. H., Kim, H.-W., Park, J.-S., & Hyun, J. K. (2010). The effect of mesenchymal stem cell transplantation on the recovery of bladder and hindlimb function after spinal cord contusion in rats. *BMC Neuroscience, 11*(1), 119. DOI: 10.1186/1471-2202-11-119 PMID: 20846445

Ryu, C. M., Shin, J. H., Yu, H. Y., Ju, H., Kim, S., Lim, J., Heo, J., Lee, S., Shin, D. M., & Choo, M. S. (2019). N-acetylcysteine prevents bladder tissue fibrosis in a lipopolysaccharide-induced cystitis rat model. *Scientific Reports*, *9*(1), 8134. Advance online publication. DOI: 10.1038/s41598-019-44631-3 PMID: 31148586

Ryu, C. M., Yu, H. Y., Lee, H. Y., Shin, J. H., Lee, S., Ju, H., Paulson, B., Lee, S., Kim, S., Lim, J., Heo, J., Hong, K. S., Chung, H. M., Kim, J. K., Shin, D. M., & Choo, M. S. (2018). Longitudinal intravital imaging of transplanted mesenchymal stem cells elucidates their functional integration and therapeutic potency in an animal model of interstitial cystitis/bladder pain syndrome. *Theranostics*, *8*(20), 5610–5624. DOI: 10.7150/thno.27559 PMID: 30555567

Sakuma, T., Matsumoto, T., Kano, K., Fukuda, N., Obinata, D., Yamaguchi, K., Yoshida, T., Takahashi, S., & Mugishima, H. (2009). Mature, adipocyte derived, dedifferentiated fat cells can differentiate into smooth muscle-like cells and contribute to bladder tissue regeneration. *The Journal of Urology*, *182*(1), 355–365. DOI: 10.1016/j.juro.2009.02.103 PMID: 19457498

Salehi-Pourmehr, H., Hajebrahimi, S., Rahbarghazi, R., Pashazadeh, F., Mahmoudi, J., Maasoumi, N., & Sadigh-Eteghad, S. (2020). Stem Cell Therapy for Neurogenic Bladder Dysfunction in Rodent Models: A Systematic Review. *International Neurourology Journal*, *24*(3), 241–257. DOI: 10.5213/inj.2040058.029 PMID: 33017895

Salehi-Pourmehr, H., Rahbarghazi, R., Mahmoudi, J., Roshangar, L., Chapple, C. R., Hajebrahimi, S., Abolhasanpour, N., & Azghani, M.-R. (2019). Intra-bladder wall transplantation of bone marrow mesenchymal stem cells improved urinary bladder dysfunction following spinal cord injury. *Life Sciences*, *221*, 20–28. DOI: 10.1016/j.lfs.2019.02.011 PMID: 30735734

Sandner, B., Ciatipis, M., Motsch, M., Soljanik, I., Weidner, N., & Blesch, A. (2016). Limited Functional Effects of Subacute Syngeneic Bone Marrow Stromal Cell Transplantation After Rat Spinal Cord Contusion Injury. *Cell Transplantation*, *25*(1), 125–139. DOI: 10.3727/096368915X687679 PMID: 25812176

Sauerwein, D. (1990). [Surgical treatment of spastic bladder paralysis in paraplegic patients. Sacral deafferentation with implantation of a sacral anterior root stimulator]. *Der Urologe. Ausg. A*, *29*(4), 196–203. PMID: 2399626

Shin, J. H., Park, J. H., Ryu, C.-M., Shin, D.-M., & Choo, M.-S. (2024). Stem cell therapy for interstitial cystitis/bladder pain syndrome. *Lower Urinary Tract Symptoms*, *16*(4), e12527. DOI: 10.1111/luts.12527 PMID: 38867432

Shin, J. H., Ryu, C. M., Yu, H. Y., Shin, D. M., & Choo, M. S. (2020a). Current and Future Directions of Stem Cell Therapy for Bladder Dysfunction. En *Stem Cell Reviews and Reports* (Vol. 16, Número 1, pp. 82-93). Springer. DOI: 10.1007/s12015-019-09922-2

Shin, J. H., Ryu, C. M., Yu, H. Y., Shin, D. M., & Choo, M. S. (2020b). Current and Future Directions of Stem Cell Therapy for Bladder Dysfunction. En *Stem Cell Reviews and Reports* (Vol. 16, Número 1, pp. 82-93). Springer. DOI: 10.1007/s12015-019-09922-2

Soler, R., Füllhase, C., Hanson, A., Campeau, L., Santos, C., & Andersson, K.-E. (2012). Stem cell therapy ameliorates bladder dysfunction in an animal model of Parkinson disease. *The Journal of Urology*, *187*(4), 1491–1497. DOI: 10.1016/j.juro.2011.11.079 PMID: 22341818

Song, M., Lim, J., Yu, H. Y., Park, J., Chun, J. Y., Jeong, J., Heo, J., Kang, H., Kim, Y., Cho, Y. M., Kim, S. W., Oh, W., Choi, S. J., Jang, S. W., Park, S., Shin, D. M., & Choo, M. S. (2015). Mesenchymal Stem Cell Therapy Alleviates Interstitial Cystitis by Activating Wnt Signaling Pathway. *Stem Cells and Development*, *24*(14), 1648–1657. DOI: 10.1089/scd.2014.0459 PMID: 25745847

Song, M., Yu, H. Y., Chun, J. Y., Shin, D. M., Song, S. H., Choo, M. S., & Song, Y. S. (2016). The fibrosis of ketamine, a noncompetitive N-methyl-D-aspartic acid receptor antagonist dose-dependent change in a ketamine-induced cystitis rat model. *Drug and Chemical Toxicology*, *39*(2), 206–212. DOI: 10.3109/01480545.2015.1079916 PMID: 26324997

Temeltas, G., Dagci, T., Evren, V., & Lekili, M. (2009). Effects of neuronal and glial restricted precursor cells transplantation on erectile function after experimentally induced spinal cord injury. *Journal of Sexual Medicine*, *6*(12), 3265–3273. DOI: 10.1111/j.1743-6109.2009.01376.x PMID: 19570038

Tremp, M., Salemi, S., Largo, R., Andersson, K. E. A., Plock, J., Aboushwareb, T., Sulser, T., & Eberli, D. (2014). Adipose-derived stem cells (ADSCs) and muscle precursor cells (MPCs) for the treatment of bladder voiding dysfunction. *World Journal of Urology*, *32*(5), 1241–1248. DOI: 10.1007/s00345-013-1200-6 PMID: 24217741

Vizzard, M. A. (1999). Alterations in growth-associated protein (GAP-43) expression in lower urinary tract pathways following chronic spinal cord injury. *Somatosensory & Motor Research*, *16*(4), 369–381. DOI: 10.1080/08990229970429 PMID: 10632033

Wyndaele, J. J., Madersbacher, H., & Kovindha, A. (2001). Conservative treatment of the neuropathic bladder in spinal cord injured patients. *Spinal Cord*, *39*(6), 294–300. DOI: 10.1038/sj.sc.3101160 PMID: 11438850

Xiao, Y., Song, Y. J., Song, B., Huang, C. B., Ling, Q., & Yu, X. (2017). TGF-β/MAPK signaling mediates the effects of bone marrow mesenchymal stem cells on urinary control and interstitial cystitis after urinary bladder transplantation. *American Journal of Translational Research*, *9*(3), 1193.

Xie, J., Liu, B., Chen, J., Xu, Y., Zhan, H., Yang, F., Li, W., & Zhou, X. (2018). Umbilical cord-derived mesenchymal stem cells alleviated inflammation and inhibited apoptosis in interstitial cystitis via AKT/mTOR signaling pathway. *Biochemical and Biophysical Research Communications*, *495*(1), 546–552. DOI: 10.1016/j.bbrc.2017.11.072 PMID: 29137981

Yoshida, M., & Yamaguchi, O. (2014). Detrusor Underactivity: The Current Concept of the Pathophysiology. *Lower Urinary Tract Symptoms*, *6*(3), 131–137. DOI: 10.1111/luts.12070 PMID: 26663593

Zhang, H., Qiu, X., Shindel, A. W., Ning, H., Ferretti, L., Jin, X., Lin, G., Lin, C. S., & Lue, T. F. (2012). Adipose tissue-derived stem cells ameliorate diabetic bladder dysfunction in a type II diabetic rat model. *Stem Cells and Development*, *21*(9), 1391–1400. DOI: 10.1089/scd.2011.0244 PMID: 22008016

Zhang, H., Zhao, Y., Wang, M., Song, W., Sun, P., & Jin, X. (2019). A promising therapeutic option for diabetic bladder dysfunction: Adipose tissue-derived stem cells pretreated by defocused low-energy shock wave. *Journal of Tissue Engineering and Regenerative Medicine*, *13*(6), 986–996. DOI: 10.1002/term.2844 PMID: 30811857

Chapter 5
Vaginal and Urethral Conditions

ABSTRACT

This chapter explores the potential of stem cell therapy to address a spectrum of vaginal conditions, including menopause-related atrophy, fistulas, prolapses, recurrent urinary tract infections, and sexually related issues. We delve into the underlying pathophysiology of these conditions and the limitations of conventional treatments. The focus is on the emerging role of SC therapy as a potential treatment modality and their application, such as mesenchymal stem cells, endometrial SC, and induced pluripotent SC, for vaginal tissue regeneration and functional restoration. This chapter examines the mechanisms of action, preclinical and clinical trial data, and challenges associated with stem cell-based therapies. By highlighting the promising advancements in this field, we aim to provide insights into the future of stem cell therapy as a transformative approach for women's health and quality of life, and contribute to the growing body of knowledge in this field and inform healthcare providers and patients about this promising therapeutic approach.

BACKGROUND

SC therapy is being explored as a promising treatment for various vaginal problems, including pelvic organ prolapse (POP), vaginal atrophy, and vaginal reconstruction.

POP occurs when the muscles and tissues supporting the pelvic organs weaken, causing the organs to drop into or out of the vagina. MSCs have shown potential in repairing weak vaginal tissues by promoting extracellular matrix ingrowth, neovascularization, and smooth muscle formation1. This can improve the biomechanical properties of the vagina, offering a new treatment perspective for POP (Zhang, Ma, et al., 2021).

DOI: 10.4018/979-8-3693-8829-7.ch005

Vaginal atrophy, often associated with menopause, involves thinning, drying, and inflammation of the vaginal walls. SC therapy can help regenerate the vaginal tissue, improving its elasticity and moisture levels2. This can alleviate symptoms such as dryness, itching, and discomfort during intercourse.

For women with congenital anomalies or those who have undergone surgery, stem cell therapy combined with scaffold strategies is being investigated for vaginal reconstruction (Zhang, Ma, et al., 2021). This approach can provide mechanical support and promote tissue repair, offering an alternative to traditional surgical methods.

Overall, while SC therapy for vaginal problems is still in the experimental stage, it holds significant promise for providing effective and long-lasting treatment options.

MENOPAUSE AND VAGINAL ATROPHY

Menopause is a stage in a woman's life, full of physical and psychological changes, which will have an impact on her quality of life and her interpersonal relationships. Hormonal decrease causes decreased vascular flow, vaginal atrophy, depletion of normal flora, mainly lactic acid-producing Lactobacillus, which increases the risk of suffering from urinary infections due to growth of coliform bacteria causing inflammation, antibiotic abuse, local discomfort, dyspaurenia, psychological problems, etc. (Bachmann & Nevadunsky, 2000; G. A. Casarotti et al., 2018; Heinemann & Reid, 2005; Semmens & Wagner, 1982).

Genitourinary Syndrome of Menopause (GSM) is being defined by vaginal dryness, dyspareunia, and other genitourinary symptoms, significantly impacts the quality of life for postmenopausal women. Prevalence studies report GSM rates of 57% to 85% among menopausal women, underscoring the condition's widespread occurrence. Despite the availability of treatments, none have consistently provided long-term relief. (Erceg Ivkošić et al., 2023)

The urogenital tract is sensitive to the effect of oestrogen and progesterone throughout adult life. The inflamed epithelium causes overactive bladder symptoms, such as dysuria, urgency, frequency, pelvic pain and incontinence. The local collagen decreases affect the soft tissues and muscles, and can causes also vaginal prolapses. Due to the hypoestrogenism of the menopause, the urethra suffers a decrease in superficial squamous and intermediate cells, as well as the muscles cells of the distal genitourinary tract, due to this degenerative process. The reduction in the submucosal vascular plexus and the decrease in intramural connective tissue tension decreases pressure, promoting urine incontinence. (Bergman et al., 1990; Semmelink et al., 1990, p. 1).

Since hypoestrogenism is the main cause of all these changes, current therapy is based on replacement with topical or systematic supplements, although the role of oestrogen replacement therapy remains controversial, because requires a continuous therapy, with side effects such as risks of thrombosis, hemangiomas, impossibility of offering it to patients with hormone-dependent cancer such as breast cancer. Another problems is that it not always easy to find this supplements in the market, especially depending on the local culture of some countries, and most of the times, they are not covered by the insurances.

Because of this inconveniencies, it is estimated that only 25% of women opt for these therapies, while the rest of women do not use any treatment, or use lubricating gels or moisturizers, which do not stop the degenerative process. (Jackson et al., 2002; Nappi & Nijland, 2008; Robinson & Cardozo, 2011)

Taking into account the causes of this degenerative damage, and, as we have explored in previous chapters, with regenerative therapies we can improve vascular flow, transform adipocytes into myocytes to improve tissue strength and trophism, replenish collagen, reduce the inflammatory effect, anti-scarring, anti-apoptotic, immunomodulatory, and anti-microbial actions with the apocrine effect of the SC, growth factors and cytokines, being an ideal and easily accessible tissue to receive these therapies, and may have promising results.

Genitourinary Syndrome of Menopause (GSM) and Adipose-Derived Mesenchymal Stem Cells (AD-MSCs)

Some research has focused on non-hormonal alternatives, with AD-MSCs emerging as a promising therapeutic option. Lipogems®, a technology developed by G. Casarotti and C. Tremolada, involves the processing of adipose tissue into a product containing AD-MSCs. This minimally manipulated, FDA-cleared device has demonstrated efficacy in treating GSM symptoms. (G. A. Casarotti et al., 2018). The preliminare study presented by Casarotti et al, was with three patients with postmenopausal urogenital atrophy were treated with injections of autologous, micro-fragmented adipose tissue (Lipogems®). The lower or the lateral abdomen was chosen as donor site for adipose tissue harvesting. The patients reported a significant improvement in symptoms within six months, with complete resolution by nine months. This subjective improvement was corroborated by clinical evaluation and persisted without recurrence for at least 36 months.

Immunohistochemical analysis at 36 months demonstrated full restoration of vaginal tissue, including glycogen production, increased vascularization, and regeneration of epithelium and subcutaneous tissue. Additionally, vaginal discharge analysis revealed a return to acidic pH and lactobacillus colonization. No postoperative complications or adverse events were reported. (G. A. Casarotti et al., 2018)

A case series of 35 postmenopausal women with severe GSM showed remarkable improvement following a single local application of Lipogems®. Symptoms resolved or significantly reduced within six months and persisted for at least two years. Histological analysis revealed complete restoration of vaginal mucosa to a premenopausal state. (G. Casarotti & Tremolada, 2020, 2020)

These findings suggest that AD-MSCs offer a safe and effective treatment for GSM, addressing a significant unmet clinical need. Further research is warranted to elucidate the underlying mechanisms of action and explore the long-term benefits of this innovative approach.

ENDOMETRIUM

It important to remark that the human endometrium exhibits remarkable regenerative capacity, recovering fully after menstruation, childbirth, extensive curettage, and even hormone replacement therapy in postmenopause. This regeneration, involving the restoration of endometrial glands and stroma, originates from basal layer remnants under the influence of estrogen. While the concept of endometrial stem cells driving this regeneration was proposed earlier, concrete evidence was lacking. The main problem is the retrieval of SC in the endometrial tissue, because of the challenge of directly identifying adult SC within tissues. Gargett et al hypothesized the existence of a small population of epithelial and stromal stem cells in the endometrial basal layer capable of forming colonies in vitro. (Gargett, 2004)

By culturing purified single cell suspensions from hysterectomy tissue at low densities, they confirmed this hypothesis: $0.22 \pm 0.07\%$ of epithelial cells and $1.25 \pm 0.18\%$ of stromal cells formed colonies. Two distinct colony types emerged for both cell types during culture: large colonies comprised densely packed, highly proliferative small cells, potentially initiated by stem or progenitor cells; and smaller colonies containing loosely arranged, less proliferative large cells, likely representing transit-amplifying (TA) cells. While 37% of epithelial colonies exhibited the large phenotype, only 1.6% of stromal colonies did.

These colonies were able to expressed cytokeratin, epithelial cell adhesion molecule, and α6-integrin, some others express CD90-positive fibroblasts, and some exhibited myofibroblast differentiation with α-smooth muscle actin expression. They identify a rare clonogenic epithelial and stromal cells with high proliferative potential, representing putative stem/progenitor cells. More common clonogenic cells with lower proliferative potential likely represent TA cells. These cell types exhibit distinct growth factor dependencies and differentiation capacities.

To definitively characterize these cells as SC, we will compare proliferative potential, assess self-renewal through subcloning, and conduct gene expression profiling to identify specific markers. Ultimately, in vivo reconstitution experiments will be essential to establish their SC nature. Reconstitution of endometrial glands and stromal tissue, respectively, following xenografting into immunocompromised mice would provide compelling evidence for the SC nature of large colony-initiating cells within the human endometrium.

This group is working in identify the precise location of the SC within the endometrial basal layer by immunophenotyping, using established stem cell markers, in order to explore for their potential involvement in the pathogenesis of endometriosis, adenomyosis, endometrial hyperplasia, and endometrial cancer, which are characterized by abnormal endometrial proliferation. Additionally, the impact of hormone replacement therapy on the postmenopausal endometrium can be explored. This research holds promise for developing novel therapeutic approaches that target endometrial growth and regression, offering alternatives to traditional hormone-based treatments for these gynecological diseases. (Gargett, 2004)

Asherman syndrome is a condition where scar tissue (adhesions) forms inside the uterus. These adhesions can cause the walls of the uterus to stick together, reducing the available space within the uterine cavity. It can be caused by previous uterine surgeries, such as a dilation and curettage, severe pelvic infections, tuberculosis or schistosomiasis. The symptoms include irregular or absent periods, pelvic pain, infertility and miscarriage.

Zhang et al. (Zhang, Shi, et al., 2021) conducted a study on 17 women aged 20-40 with Asherman syndrome, previous failed hysteroscopic adhesiolysis, and inadequate endometrial response to hormone therapy. Patients with congenital uterine abnormalities or other implantation barriers were excluded. The study aimed to evaluate the efficacy of collagen scaffold/umbilical cord mesenchymal SC (CS/UC-MSC) transplantation in regenerating the endometrium and improving pregnancy outcomes.

They reported an endometrial thickness increased significantly in 12 patients, with six achieving a thickness of 6 mm or more. Of the 17 patients, 15 underwent 22 frozen-thawed embryo transfers (FET), resulting in three pregnancies: two live births and one miscarriage. Notably, one patient achieved a spontaneous pregnancy without FET. Endometrial microvascular density and expression of proliferation and differentiation markers (Ki67, ERα, PR) were elevated, indicating increased angiogenesis and endometrial remodeling.

A meta-analysis by Zhao Y. et al. (Zhao et al., 2020) involving 77 patients across seven studies demonstrated improved menstrual cycles and pregnancy rates with stem cell therapy compared with conventional treatments. Endometrial thickness increased in 101 out of 105 patients assessed across eight studies.

Jichun Tan I et al. (Tan et al., 2016) investigated the use of menstrual blood-derived MSCs (Men-MSCs) in treating Asherman syndrome. Seven infertile women with severe, recurrent Asherman syndrome underwent autologous Men-MSC transplantation. Endometrial thickness increased significantly in all patients, with five achieving a thickness of 7 mm. One spontaneous pregnancy and two successful embryo transfers were reported. These studies collectively suggest that stem cell therapy holds promise for treating Asherman syndrome by promoting endometrial regeneration and improving fertility outcomes.

Hailan Ma et al. (Ma et al., 2020) investigated the impact of intrauterine transplantation of menstrual blood-derived mesenchymal stem cells (Men-MSCs) on endometrial thickness and pregnancy outcomes in 12 infertile women aged 20-40 with refractory Asherman syndrome. These women had failed to respond to conventional treatments. Autologous Men-MSCs were successfully isolated and expanded from menstrual blood for transplantation.

LICHEN SCLEROSUS (LS)

Lichen sclerosus (LS) is a chronic inflammatory skin condition with potential autoimmune origins. The disease often manifests in the anogenital region, where it can lead to scarring and increased cancer risk if left untreated. While prevalent in postmenopausal women, LS can affect individuals of all ages and genders, because it can affect the vagina and penis. (Erceg Ivkošić et al., 2023)

The autoimmune nature of vulvar and foreskin of the penis LS is supported by T-cell infiltration, elevated Th1 cytokines, increased BIC/miR-155 expression, and autoantibodies against specific proteins. Genetic predisposition is evident with increased HLA-DQ and HLA-DR allele frequencies in LS patients. Clinically, LS presents with a spectrum of symptoms ranging from asymptomatic to severe itching, burning, pain, and dysuria, and sometimes, could be the cause of recurrent urinary tract infections.

Characteristic skin changes include ivory-white plaques, atrophy, and scarring. Progressive disease can lead to phimosis, urinary complications, vulvar stenosis, severe sexual dysfunction, and dyspareunia, and an increased risk of developing penile cancer, including its precursor, penile intraepithelial neoplasia, as well as elevated risk of vulvar squamous cell carcinoma, emphasizing the importance of early diagnosis and treatment. Recent research has explored the therapeutic potential of ADSC and PRP in managing LS. (Erceg Ivkošić et al., 2023; Kravvas et al., 2022)

These treatments have shown promise in alleviating symptoms and promoting tissue regeneration. ADSC exert their effects through anti-inflammatory, immunomodulatory, and regenerative properties, while PRP stimulates healing and tissue repair. Clinical studies have reported significant symptom improvement and quality of life enhancement following ADSC and PRP treatments. These findings suggest a promising therapeutic approach for LS, although further research is needed to fully elucidate the mechanisms of action and long-term outcomes.

Eshtiaghi et al. (Eshtiaghi & Sadownik, 2019) conducted a review encompassing multiple studies on the treatment of LS with ADSC and PRP. Results from cohorts of 15, 36, and 8 patients demonstrated significant symptom improvement and enhanced quality of life following treatment.

ADSC, known for their regenerative and immunomodulatory properties, are believed to restore damaged tissue through differentiation and cytokine secretion. The efficacy of PRP is attributed to its high concentration of growth factors, including PDGF, TGF-β, and EGF, which stimulate mesenchymal cell proliferation and extracellular matrix synthesis. (Krapf et al., 2020)

The therapeutic potential of ADSC is attributed to their abundance within the adipose-derived stromal vascular fraction (AD-SVF), which includes various cell types and bioactive factors. These cells exhibit anti-inflammatory effects by interacting with immune cells and modulating T-cell responses. Additionally, PRP contributes to tissue healing by releasing growth factors. These findings collectively suggest that ADSC and PRP hold promise as effective treatments for LS, offering potential benefits in terms of symptom relief and tissue regeneration.

Tedesco et al. (Tedesco et al., 2020) conducted a randomized controlled trial involving 40 patients with LS. Participants were divided into two groups: AD-SVF and AD-SVF plus PRP. Both groups demonstrated improvements in symptoms, skin elasticity, hydration, and atrophy. Notably, 13 patients experienced complete symptom resolution, while 23 reported significant symptom alleviation. Interestingly, two patients observed a reduction in white lesions. However, no statistically significant difference was found between the two treatment groups in terms of overall clinical improvement (AD-SVF: mean score 2.0 ± 0.8; AD-SVF plus PRP: mean score 1.8 ± 1.1, $p = 0.42$).

PELVIC FLOOR DYSFUNCTION (PFD)

Pelvic floor dysfunction (PFD) is a group of clinical conditions including SUI and pelvic organ prolapse (POP), usualy commonly associated with aging, affects a substantial number of women and is demonstrating an increasing prevalence over time. It occurs when the muscles and tissues supporting the pelvic organs weaken and stretch, allowing the organs to slip down into or bulge out of the vagina. This condition is more common in women who have had multiple vaginal deliveries or have undergone menopause. POP affects people's quality of life due to high treatment costs, frequent postoperative complications, debilitating symptoms, and limited success with conservative management.

Current treatment options, including surgery, pessaries, and pelvic floor rehabilitation, often yield suboptimal results. Surgical interventions are associated with high recurrence rates and mesh-related complications, while pessaries Pelvic floor rehabilitation has demonstrated limited efficacy. Pessaries are commonly recommended as a first-line approach for managing pelvic floor conditions; however, anatomical variations among women may limit the effectiveness of commercially available devices due to insufficient customization, and can cause discomfort and infections, and prevent for having a typical sexual life. Consequently, there is a critical need for novel, safe, and effective therapeutic approaches.

There is a roll of tissue engineering for this condition as well, but, despite the prevalence of POP, fundamental research into its pathophysiology remains limited. Studies investigating cellular behavior, histological characteristics, protein expression alterations, and potential cell-based therapies are relatively scarce. SC-based therapies, including those utilizing induced pluripotent SC, reprogrammed fibroblasts, and endometrial stem progenitor cells, hold significant promise for treating complex diseases. (Trounson & DeWitt, 2016; Wen et al., 2013)

Tseng et al published a study using a 3D printing vaginal pessary to treat pelvic organ prolapse, suggesting that customized pessaries may offer a more effective alternative for women with POP, particularly when standard commercial devices prove inadequate. By integrating transvaginal ultrasound imaging with 3D printing technology, individualized pessaries were designed to better conform to each patient's anatomical profile. A Gellhorn-type pessary was initially inserted into the vaginal canal to reposition prolapsed pelvic structures. Under transvaginal ultrasound (US) guidance, residual anatomical gaps not covered by the device were identified. To address these, Ototoform—an impression-grade silicone—was molded into a spherical shape and affixed to the Gellhorn pessary, serving as a template for customization. This combined prototype was then scanned and utilized in a 3D printing workflow to fabricate a personalized silicone vaginal pessary, tailored to the patient's unique anatomical requirements, and having the potential to significantly enhance comfort, retention, and overall patient satisfaction (Tseng et al., s. f.).

While stem cell applications in clinical settings remain limited, they offer a safer and more adaptable approach compared to traditional gene regulation methods. The research targeting pathogenic gene regulation within SC, particularly Mfn2 modulation for POP treatment, remains scarce.

BMSCs have emerged as a promising candidate due to their regenerative potential, they are well-characterized stem cell sources with robust differentiation potential and a capacity to secrete bioactive factors that promote tissue repair. (Al Demour et al., 2018) Jin et al. use BMSCs, genetically modified to express elastin, to restore normal collagen and elastin distribution within connective tissues. (Jin et al., 2016) Subsequently, these elastin-expressing BMSCs were differentiated into fibroblasts using FGF to stimulate collagen and elastin production. For sustained FGF delivery, they encapsulated the growth factor within poly (lactic-co-glycolic acid) (PLGA) nanoparticles. In vitro, a seven-day cell culture system demonstrated that a single administration of PLGA-loaded bFGF nanoparticles induced sustained collagen and elastin production by elastin-expressing BMSCs without additional FGF supplementation.

In vivo, co-injection of these nanoparticles with elastin-expressing BMSCs into PFD rats significantly enhanced urodynamic outcomes. These findings collectively establish an effective model for connective tissue engineering utilizing BMSCs and injectable PLGA-loaded growth factors.

Wang et al studied the senescence-related protein mitofusin 2 (Mfn2) and the association with POP through rat model, constructed by ovariectomy and traction. The study demonstrated that downregulating Mfn2 expression using stem cells decreased Mfn2 levels while increasing procollagen 1A1/1A2/3A1 expression within the uterosacral ligaments of POP rats. These effects were notably pronounced between 14 and 21 days' post-injection. These findings suggest that Mfn2 could serve as a potential therapeutic target for POP management.

The ECM is essential for the healing and tissue restoration process, providing structural and growth factors that allow the tissue to regenerate in an organized way to close the wound. Acellular dermal matrix (ADM) products have been developed to mimic the benefits of host ECM, allowing for better results in different pathologies, derived from different sources, including human (HADM), porcine (PADM), and bovine (BADM), with multiple processing protocols depending on the origin of ADM. (Liang et al., 2015)

A distinct subclass of natural biomaterials includes tissue-derived scaffolds, such as acellular matrix meshes—novel allogeneic constructs that retain ECM integrity while eliminating immunogenic cellular components. Cellular adhesion within these matrices is mediated by integrin-ligand interactions, making basement membrane and dermal ECM structures ideal candidates for scaffolding purposes (Murphy et al., 2013). These matrices promote adhesion and proliferation yet often suffer from limited mechanical strength and suboptimal degradation rates, which may contribute to recurrence of pelvic floor prolapse (Barnes et al., 2007).

To address these limitations, bioengineered scaffolds have been developed to emulate native ECM ligands. Such constructs are commonly used in vitro to replicate ECM microenvironments and enhance cell–matrix interactions (Liu et al., 2009; Ravindran et al., 2010). Extensive research has focused on optimizing decellularization protocols, ensuring complete removal of cellular debris while preserving bioactive cues and ECM architecture (Crapo et al., 2011; Lu et al., 2012).

A prior study, conducted by Rui Liang, and Pamela Maolli, utilized a biologic, non-crosslinked acellular matrix in primate sacrocolpopexy. These investigators demonstrated that the acellular matrix mitigated the adverse effects of polypropylene mesh on vaginal flora. (Liang, Knight, Barone, et al., 2017; Liang, Knight, Easley, et al., 2017)

Karon et al, published a study in 211 patients with laparoscopic sacrocolpopexy for POP using ADM between January 6, 2012, and December 31, 2017. (Karon & Chatterjee, 2019) Each patient was classified according to the POP-Q [Quantification] system chart using the tool interactive test provided by the American Urogynecologic Society prior to the procedure. The patients filled out the pelvic floor symptoms survey (PFDI-20) and the Pelvic Floor Impact Questionnaire (PFIQ-7), to analyze for patient satisfaction and the results after 4 weeks of surgery and during annual follow-up. Biopsies were taken from the sacrocolpopexy area in those patients who required another gynecological procedure unrelated to the prolapse or at the time of repeat sacrocolpopexy for POP, which were analyzed after paraffin fixation by the McGowan Institute for Tissue Regeneration. 105 responded to the survey. Charts were completed for 106. Most of the patients referred being "a little better" or "much better" (77/88; 87.5%), the third quartile the PFDI-20 score was 93 with a median of 60 and the PFIQ-7 score was 43 with a median of 29. 5 patients underwent reoperation (4.76%).

The main postoperative complaint was overactive bladder, followed by vaginal discharge. Histology showed a lack of regenerative scar tissue in the site of the prolapse in those patients who required re-intervention, or neovascularization and presence of connective and ligamentous tissue around the matrix in those in whom the surgery was successful. No severe fibrosis or neoplastic formation was reported.

Following this results, ADM patch can be a good alternative to synthetic polypropylene mesh in patients undergoing sacrocolpopexy, either laparoscopically or vaginally. More studies are required to have an alternative to synthetic meshes and to be able to offer an adequate treatment with the minimum possible complications.

RECTOVAGINAL OR VESICOVAGINAL FISTULAS AND STEM CELL THERAPY

The abnormal connections between the rectum or bladder with the vagina are known as Rectovaginal or Vesicovaginal, often result from childbirth, infections, malignancies, pelvic surgeries, or inflammatory bowel diseases. Treatment challenges and high recurrence rates highlight the need for innovative therapies.

The World Health Organization estimates that approximately 3 million women worldwide are currently living with obstetric fistula, with 50,000 to 130,000 new cases occurring annually. (Diallo et al., 2015) In the United States, the incidence of fistula is relatively low, with rates below 0.5% following simple hysterectomy and around 10% after radical hysterectomy. In contrast, Sub-Saharan Africa reports a significantly higher incidence of vesicovaginal fistula, with an estimated 33,000 new cases each year and approximately 10,000 fistula repairs performed annually. (*Urogenital tract fistulas in females - UpToDate*, s. f.)

Cell-based tissue engineering may offer an alternative way for genotourinary tract reconstruction, especially for urothelial cells (UCs). ADSCs have been explored for urinary tract tissue engineering applications. Jack et al. (Jack et al., 2005) demonstrated the viability of human ADSCs in bladder and urethral smooth muscle for up to 12 weeks post-delivery, with evidence of morphological and phenotypic integration into the surrounding tissue. In subsequent work, the same group differentiated HADSCs into bladder muscle cells using smooth muscle inductive media and implanted them on PLGA scaffolds for bladder augmentation in an athymic rat model (Jack et al., 2009). Zhu et al. (Zhu et al., 2010) investigated the use of undifferentiated ADSCs seeded on bladder acellular matrix grafts for bladder reconstruction in a rabbit model, observing regeneration of smooth muscle and nervous tissue.

As we mentioned in previous chapters, studies have demonstrated the potential of various factors in inducing SC differentiation. Borowiak et al. (Borowiak et al., 2009) showed that small molecules can direct mouse SC towards the endoderm lineage. Li et al. (Li et al., 2009) highlighted the role of embryonic urogenital sinus mesenchyme and TGF-beta signaling in prostate development from SC. Additionally, FGF-2 has been implicated in the proliferation and differentiation of ADSC via the Wnt pathway. (Rider et al., 2008)

Tian et al. create a study using this model reported increased VEGF and PDGF-BB levels in conditioned medium. (Tian et al., 2010) However, initial VEGF and PDGF-BB levels in the conditioned medium were comparable to controls, suggesting that the observed elevations during the differentiation period were induced by SC activation rather than direct medium components. The precise mechanisms underlying this transdifferentiation process remain to be elucidated. Future studies should focus on identifying specific factors within the conditioned medium that initiate the transdifferentiation process.

Conditioned media derived from organ-specific cells have been successfully employed to induce stem cell differentiation into functional cell types. Shi et al. (Shi et al., 2012) cultured ADSCs in urothelial-conditioned medium for 14 days, resulting in a morphological transformation of approximately 25-50% of cells into a polygonal epithelium-like shape. These cells exhibited upregulated expression of urothelial markers (uroplakin 2 and cytokeratin-18) and downregulated expression of the mesenchymal marker vimentin at both RNA and protein levels, confirming their transdifferentiation into UC-like cells. To elucidate the underlying molecular mechanisms, they analyzed PDGF-BB and VEGF expression levels in the conditioned medium, that were significantly elevated ($p < 0.05$) compared to controls, suggesting their pivotal role in the induction process, proposing that culturing HADSCs in a urothelial-conditioned medium promotes their transdifferentiation into UC-like cells, as evidenced by morphological changes and altered marker expression. These results highlight the potential of ADSCs as a cell source for urinary tract tissue engineering.

A potential mechanism for this cellular transformation involves paracrine signaling, whereby cells release cytokines that influence the fate of neighboring cells. Effective SC differentiation protocols often leverage knowledge of extracellular signals and gene regulatory factors that orchestrate embryonic development. Given that many signal transduction pathways are initiated by extracellular protein ligands, such as growth factors, protein-based materials have emerged as popular tools for inducing stem cell differentiation. (Zaret, 2009)

A study by Zambon et al. (Zambon et al., 2010) reported 23 cases of complex urogenital fistulas between 2004 and 2007, highlighting the challenges associated with these conditions, particularly in cases of failed abdominal repair. One of the problems is wound healing in the bladder area takes longer than that for skin, and postoperative recurrence rates remain high.

MSCs, particularly those derived from adipose tissue, have emerged as promising candidates due to their anti-inflammatory and regenerative properties. While traditionally obtained through culture expansion, recent studies have explored the use of minimally processed adipose tissue, such as SVF. (Erceg Ivkošić et al., 2023)

Clinical studies have demonstrated the potential of MSC-based therapies in treating rectovaginal fistulas. A meta-analysis by Cao Y. et al. (Cao et al., 2021) reported significantly higher healing rates in patients receiving stem cell treatment compared to placebo.

Additionally, a study by García-Arranz M et al. (García-Arranz et al., 2016) showed promising results with allogeneic adipose-derived MSCs in treating fistulas associated with Crohn's disease. These findings suggest that stem cell therapy may represent a valuable treatment option for patients with refractory rectovaginal fistulas, offering hope for improved outcomes and quality of life.

Microfragmented adipose tissue (MFAT) has demonstrated efficacy in treating perianal fistulas, particularly in Crohn's disease. A single MFAT injection led to complete clinical and instrumental remission in 67% of patients within six weeks. (G. Casarotti & Tremolada, 2020; Image Regenerative Clinic & Tremolada, 2019)

De Weerd et al. explored the use of fat grafting on six patients with recalcitrant rectovaginal fistulas (RVFs), four due to obstetric injury and two associated with Crohn’s disease. (de Weerd et al., 2015) A single or two-step procedure involving transperineal fat injection and fistula transection was performed, with a fistula tract transversely transection. Most of the patients required a second treatment after 6 weeks. While most patients experienced healing, recurrence occurred in two patients with Crohn's disease-related RVFs after 23 and 25 months.

These findings highlight the potential of adipose-derived therapies in managing complex fistulas. However, further research is needed to optimize treatment protocols and identify patient subgroups most likely to benefit from these approaches.

To investigate the potential of human freeze-dried amnion as a MSCs scaffold for vesicovaginal fistula repair, Kurniawati et al. conducted a two-stage study, using a New Zealand rabbit model. (Kurniawati et al., 2022) The first stage involved creating a vesicovaginal fistula, followed by surgical closure using either human freeze-dried amnion or a control treatment in the second stage. Wound healing was assessed by evaluating the expression of PDGF, VEGF, FGF, occludin, and claudin-4. The expression of these levels were significantly higher in vesicovaginal fistula models treated with human freeze-dried amnion compared to the control group. The authors suggest that human freeze-dried amnion plays a crucial role in enhancing wound healing processes during vesicovaginal fistula repair.

URETHRA

Urethral stricture is a narrowing of the anterior urethra caused by fibrosis and scarring of the urethral mucosa and surrounding spongy tissue (spongiofibrosis). In contrast, the term “stenosis” is typically used for the posterior urethra, which lacks spongy tissue. Meatal stenosis specifically refers to a short narrowing at the urethral opening without involving the fossa navicularis. (Latini et al., 2014; Mundy & Andrich, 2011)

A standardized definition for female urethral stricture (FUS) remains elusive. Most literature characterizes FUS as a fixed anatomical narrowing resulting in reduced urethral caliber. While the normal urethral caliber is generally accepted as 18-30 Fr, the specific threshold for diagnosing FUS varies widely, ranging from

<10 Fr to <20 Fr, with <14 Fr being the most common criterion. Notably, the concept of urethral stricture has also been extended to transgender patients, despite the absence of surrounding spongy tissue in the reconstructed urethra.

The incidence of urethral stricture in males is estimated at 229-627 cases per 100,000, with the anterior urethra, particularly the bulbar region, being most commonly affected (92.2%, 46.9%, respectively). In women, urethral stricture is less prevalent, occurring in 0.08-5.4% of patients with refractory lower urinary tract symptoms (LUTS). The condition is more common in women over 64. (Montorsi et al., 2002; Singh et al., 2013)

Several factors can contribute to this condition, such trauma injuries (Straddle injuries, pelvic fractures, or direct trauma to the urethra), surgical procedures (Prostate surgery, urethral catheterization, or other surgical interventions in the area), infections (urinary tract infections, sexually transmitted infections (STIs), urethritis), lichen sclerosus, cancer (Tumors in the urethra or prostate), radiation therapy for prostate or bladder cancer. It is important to note that multiple factors can contribute to the development of urethral stricture, and often, a combination of causes is involved.

Urethral strictures can also occur in specific populations. Hypospadias repair is associated with a 1.3-20% risk of meatal stenosis or urethral stricture, with higher rates observed in developed countries due to increased surgical intervention. Circumcision-related meatal stenosis is rare, occurring in less than 0.2% of cases. Transgender individuals face unique challenges. Urethral strictures affect 2-56% of female-to-male (FtM) transgender patients, primarily at the neomeatus. In male-to-female (MtF) transgender patients, strictures occur in 4-40% of cases, almost exclusively at the neomeatus of the reconstructed urethra. (*EAU Guidelines on Urethral Strictures - DEFINITION EPIDEMIOLOGY AETIOLOGY AND PREVENTION - Uroweb*, s. f.)

There is a possibility of using regenerative therapies also for urethral strictures. Choosing an appropriate biomaterial for scaffolding is an important step in tissue engineering. Ideal scaffolds must provide a suitable microenvironment to promote cell growth, differentiation, and the formation of functional tissue substitutes. Additionally, seamless integration into host tissue upon implantation is crucial. Biological scaffolds, derived from natural sources, are particularly attractive due to their inherent incorporation of bioactive molecules, which can significantly enhance cellular processes.

Acellular grafts, derived from cadaveric or animal sources such as porcine small intestinal submucosa (SIS), acellular bladder matrix, or acellular dermal matrix, serve as cell-free scaffolds for host cell ingrowth. Their primary advantage is off-the-shelf availability (el-Kassaby et al., 2008; Fiala et al., 2007; Palminteri et al., 2012). A small randomized controlled trial (RCT) involving 30 patients compared acellular bladder matrix to buccal mucosal grafts (BMGs), reporting patency rates of 66.6%

and 100%, respectively. The inferior performance of acellular bladder matrix was particularly evident in cases of poor urethral bed health. (Fiala et al., 2007)

Several retrospective case series examining primarily porcine SIS have reported variable patency rates ranging from 20% to 110%. These studies are often limited by heterogeneity in patient populations, treatment approaches, and outcome definitions (Table 1). (el-Kassaby et al., 2008; Tang et al., 2020, 2020; Xu et al., 2013, 2013) Factors such as extensive spongiofibrosis, poor graft vascularization, previous treatments, and longer stricture lengths have been associated with poorer outcomes.

Fue et at. has previously demonstrated the feasibility of using a biodegradable urethral stent seeded with autologous urethral epithelial cells for treating post-traumatic urethral strictures in a rabbit model (Fu et al., 2009).

Table 1. Studies for the use of SC for urethral repair

Study	N	Follow up in months	Type of graft	Patency Rate (%)
Fiala et al., 2007	50	31.2	porcine small intestinal submucosa	80
el-Kassaby et al. 2008	15	25	cadaveric acellular bladder matrix	33-88
Palminteri et al. 2012	30	71	porcine small intestinal submucosa	76
Xu et al. 2013	28	24.8	porcine small intestinal submucosa	92
Tang et al. 2020	49	15	allogeneic acellular dermal matrix	85.7

Building upon this, future investigations will explore the potential of induced adipose-derived stem cells for similar applications in the ureter, bladder, and urethra.

CONCLUSIONS

SC therapy emerges as a promising avenue for addressing a spectrum of gynecological conditions, including vaginal atrophy, prolapse, fistulas, lichen sclerosus, urethral strictures, and sacrocolpopexy. The inherent regenerative capabilities of SC, coupled with their immunomodulatory and anti-inflammatory properties, offer potential solutions to these complex and often debilitating conditions.

While significant advancements have been made, further research is imperative to fully understand the underlying mechanisms and optimize therapeutic strategies. Overcoming challenges related to cell sourcing, differentiation, and delivery will be crucial for translating preclinical findings into safe and effective clinical applications. Ultimately, stem cell-based therapies hold the potential to revolutionize the management of gynecological disorders, improving quality of life for countless women.

REFERENCES

Al Demour, S., Jafar, H., Adwan, S., AlSharif, A., Alhawari, H., Alrabadi, A., Zayed, A., Jaradat, A., & Awidi, A. (2018). Safety and Potential Therapeutic Effect of Two Intracavernous Autologous Bone Marrow Derived Mesenchymal Stem Cells injections in Diabetic Patients with Erectile Dysfunction: An Open Label Phase I Clinical Trial. *Urologia Internationalis*, *101*(3), 358–365. DOI: 10.1159/000492120 PMID: 30173210

Bachmann, G. A., & Nevadunsky, N. S. (2000). Diagnosis and treatment of atrophic vaginitis. *American Family Physician*, *61*(10), 3090–3096. PMID: 10839558

Barnes, C. P., Pemble, C. W.IV, Brand, D. D., Simpson, D. G., & Bowlin, G. L. (2007). Cross-linking electrospun type II collagen tissue engineering scaffolds with carbodiimide in ethanol. *Tissue Engineering*, *13*(7), 1593–1605. DOI: 10.1089/ten.2006.0292 PMID: 17523878

Bergman, A., Karram, M. M., & Bhatia, N. N. (1990). Changes in urethral cytology following estrogen administration. *Gynecologic and Obstetric Investigation*, *29*(3), 211–213. DOI: 10.1159/000293384 PMID: 2358196

Borowiak, M., Maehr, R., Chen, S., Chen, A. E., Tang, W., Fox, J. L., Schreiber, S. L., & Melton, D. A. (2009). Small molecules efficiently direct endodermal differentiation of mouse and human embryonic stem cells. *Cell Stem Cell*, *4*(4), 348–358. DOI: 10.1016/j.stem.2009.01.014 PMID: 19341624

Cao, Y., Su, Q., Zhang, B., Shen, F., & Li, S. (2021). Efficacy of stem cells therapy for Crohn's fistula: A meta-analysis and systematic review. *Stem Cell Research & Therapy*, *12*(1), 32. DOI: 10.1186/s13287-020-02095-7 PMID: 33413661

Casarotti, G., & Tremolada, C. (2020). A new treatment of genito-urinary post-menopausal atrophy with autologous micro-fragmented fat tissue: A thirty-six months follow up case series. *European Review for Medical and Pharmacological Sciences*, *24*(13), 7420–7426. DOI: 10.26355/eurrev_202007_21910 PMID: 32706081

Casarotti, G. A., Chiodera, P., & Tremolada, C. (2018). Menopause: New frontiers in the treatment of urogenital atrophy. *European Review for Medical and Pharmacological Sciences*, *22*(2), 567–574. DOI: 10.26355/eurrev_201801_14211 PMID: 29424920

Crapo, P. M., Gilbert, T. W., & Badylak, S. F. (2011). An overview of tissue and whole organ decellularization processes. *Biomaterials*, *32*(12), 3233–3243. DOI: 10.1016/j.biomaterials.2011.01.057 PMID: 21296410 de Weerd, L., Weum, S., & Norderval, S. (2015). Novel treatment for recalcitrant rectovaginal fistulas: Fat injection. *International Urogynecology Journal*, *26*(1), 139–144. DOI: 10.1007/s00192-014-2497-4 PMID: 25199495

Diallo, A. B., Diallo, T. M. O., Bah, I., Bah, M. D., & Barry, M. (2015). Vesicovaginal Fistulas: Anatomical Clinical and Surgical Aspects in the Conakry University Hospital Center. *Open Journal of Urology*, *5*(12), 12. Advance online publication. DOI: 10.4236/oju.2015.512036

EAU Guidelines on Urethral Strictures—DEFINITION EPIDEMIOLOGY AETIOLOGY AND PREVENTION - Uroweb. (s. f.). Uroweb - European Association of Urology. Recuperado 11 de agosto de 2024, de https://uroweb.org/guidelines/urethral-strictures/chapter/definition-epidemiology-aetiology-and-prevention

el-Kassaby, A., AbouShwareb, T., & Atala, A. (2008). Randomized comparative study between buccal mucosal and acellular bladder matrix grafts in complex anterior urethral strictures. *The Journal of Urology*, *179*(4), 1432–1436. DOI: 10.1016/j.juro.2007.11.101 PMID: 18295282

Erceg Ivkošić, I., Fureš, R., Ćosić, V., Mikelin, N., Bulić, L., Dobranić, D., Brlek, P., & Primorac, D. (2023). Unlocking the Potential of Mesenchymal Stem Cells in Gynecology: Where Are We Now? *Journal of Personalized Medicine*, *13*(8), 1253. DOI: 10.3390/jpm13081253 PMID: 37623503

Eshtiaghi, P., & Sadownik, L. A. (2019). Fact or Fiction? Adipose-Derived Stem Cells and Platelet-Rich Plasma for the Treatment of Vulvar Lichen Sclerosus. *Journal of Lower Genital Tract Disease*, *23*(1), 65–70. DOI: 10.1097/LGT.0000000000000440 PMID: 30252710

Fiala, R., Vidlar, A., Vrtal, R., Belej, K., & Student, V. (2007). Porcine small intestinal submucosa graft for repair of anterior urethral strictures. *European Urology*, *51*(6), 1702–1708. DOI: 10.1016/j.eururo.2007.01.099 PMID: 17306922

Fu, W.-J., Zhang, X., Zhang, B.-H., Zhang, P., Hong, B.-F., Gao, J.-P., Meng, B., Kun, H., & Cui, F.-Z. (2009). Biodegradable urethral stents seeded with autologous urethral epithelial cells in the treatment of post-traumatic urethral stricture: A feasibility study in a rabbit model. *BJU International*, *104*(2), 263–268. DOI: 10.1111/j.1464-410X.2009.08366.x PMID: 19220257

García-Arranz, M., Herreros, M. D., González-Gómez, C., de la Quintana, P., Guadalajara, H., Georgiev-Hristov, T., Trébol, J., & Garcia-Olmo, D. (2016). Treatment of Crohn's-Related Rectovaginal Fistula With Allogeneic Expanded-Adipose Derived Stem Cells: A Phase I-IIa Clinical Trial. *Stem Cells Translational Medicine*, *5*(11), 1441–1446. DOI: 10.5966/sctm.2015-0356 PMID: 27412883

Gargett, C. E. (2004). Stem cells in gynaecology. *Australian and New Zealand Journal of Obstetrics and Gynaecology*, *44*(5), 380–386. DOI: 10.1111/j.1479-828X.2004.00290.x PMID: 15387855

Heinemann, C., & Reid, G. (2005). Vaginal microbial diversity among postmenopausal women with and without hormone replacement therapy. *Canadian Journal of Microbiology*, *51*(9), 777–781. DOI: 10.1139/w05-070 PMID: 16391657

Jack, G. S., Almeida, F. G., Zhang, R., Alfonso, Z. C., Zuk, P. A., & Rodríguez, L. V. (2005). Processed lipoaspirate cells for tissue engineering of the lower urinary tract: Implications for the treatment of stress urinary incontinence and bladder reconstruction. *The Journal of Urology*, *174*(5), 2041–2045. DOI: 10.1097/01.ju.0000176489.96993.84 PMID: 16217390

Jack, G. S., Zhang, R., Lee, M., Xu, Y., Wu, B. M., & Rodríguez, L. V. (2009). Urinary bladder smooth muscle engineered from adipose stem cells and a three dimensional synthetic composite. *Biomaterials*, *30*(19), 3259–3270. DOI: 10.1016/j.biomaterials.2009.02.035 PMID: 19345408

Jackson, S., James, M., & Abrams, P. (2002). The effect of oestradiol on vaginal collagen metabolism in postmenopausal women with genuine stress incontinence. *BJOG*, *109*(3), 339–344. DOI: 10.1111/j.1471-0528.2002.01052.x PMID: 11950190

Jin, M., Chen, Y., Zhou, Y., Mei, Y., Liu, W., Pan, C., & Hua, X. (2016). Transplantation of bone marrow-derived mesenchymal stem cells expressing elastin alleviates pelvic floor dysfunction. *Stem Cell Research & Therapy*, *7*(1), 51. DOI: 10.1186/s13287-016-0308-1 PMID: 27048404

Karon, M., & Chatterjee, S. (2019). Sacrocolpopexy: Patient Outcomes Support the Use of Non-Crosslinked Acellular Dermal Matrix as an Alternative to the Synthetic Polypropylene Mesh. *Journal of Gynecologic Surgery*, *35*(6), 337–344. DOI: 10.1089/gyn.2019.0010 PMID: 31871389

Krapf, J. M., Mitchell, L., Holton, M. A., & Goldstein, A. T. (2020). Vulvar Lichen Sclerosus: Current Perspectives. *International Journal of Women's Health*, *12*, 11–20. DOI: 10.2147/IJWH.S191200 PMID: 32021489

Kravvas, G., Muneer, A., Watchorn, R. E., Castiglione, F., Haider, A., Freeman, A., Hadway, P., Alnajjar, H., Lynch, M., & Bunker, C. B. (2022). Male genital lichen sclerosus, microincontinence and occlusion: Mapping the disease across the prepuce. *Clinical and Experimental Dermatology*, *47*(6), 1124–1130. DOI: 10.1111/ced.15127 PMID: 35150005

Kurniawati, E. M., Santoso, B., Rantam, F. A., Santoso, B. I., Widjiati, , Hadi, T. H. S., Hardianto, G., & Paraton, H. (2022). Does freeze dried amnion useful as scaffold for mesenchymal stem cell for repair of vesicovaginal fistula? An overview of proliferative and remodeling phase in the wound healing process. *European Journal of Obstetrics, Gynecology, and Reproductive Biology*, *274*, 113–116. DOI: 10.1016/j.ejogrb.2022.05.017 PMID: 35640439

Latini, J. M., McAninch, J. W., Brandes, S. B., Chung, J. Y., & Rosenstein, D. (2014). SIU/ICUD Consultation On Urethral Strictures: Epidemiology, etiology, anatomy, and nomenclature of urethral stenoses, strictures, and pelvic fracture urethral disruption injuries. *Urology*, *83*(3, Suppl), S1–S7. DOI: 10.1016/j.urology.2013.09.009 PMID: 24210733

Li, X., Wang, Y., Sharif-Afshar, A.-R., Uwamariya, C., Yi, A., Ishii, K., Hayward, S. W., Matusik, R. J., & Bhowmick, N. A. (2009). Urothelial transdifferentiation to prostate epithelia is mediated by paracrine TGF-beta signaling. *Differentiation. Differentiation*, *77*(1), 95–102. DOI: 10.1016/j.diff.2008.09.012 PMID: 19281768

Liang, R., Knight, K., Barone, W., Powers, R. W., Nolfi, A., Palcsey, S., Abramowitch, S., & Moalli, P. A. (2017). Extracellular matrix regenerative graft attenuates the negative impact of polypropylene prolapse mesh on vagina in rhesus macaque. *American Journal of Obstetrics and Gynecology*, *216*(2), 153.e1–153.e9. DOI: 10.1016/j.ajog.2016.09.073 PMID: 27615441

Liang, R., Knight, K., Easley, D., Palcsey, S., Abramowitch, S., & Moalli, P. A. (2017). Towards rebuilding vaginal support utilizing an extracellular matrix bioscaffold. *Acta Biomaterialia*, *57*, 324–333. DOI: 10.1016/j.actbio.2017.05.015 PMID: 28487243

Liang, R., Zong, W., Palcsey, S., Abramowitch, S., & Moalli, P. A. (2015). Impact of prolapse meshes on the metabolism of vaginal extracellular matrix in rhesus macaque. *American Journal of Obstetrics and Gynecology*, *212*(2), 174.e1–174.e7. DOI: 10.1016/j.ajog.2014.08.008 PMID: 25128444

Liu, X., Smith, L. A., Hu, J., & Ma, P. X. (2009). Biomimetic nanofibrous gelatin/apatite composite scaffolds for bone tissue engineering. *Biomaterials*, *30*(12), 2252–2258. DOI: 10.1016/j.biomaterials.2008.12.068 PMID: 19152974

Lu, H., Hoshiba, T., Kawazoe, N., & Chen, G. (2012). Comparison of decellularization techniques for preparation of extracellular matrix scaffolds derived from three-dimensional cell culture. *Journal of Biomedical Materials Research. Part A*, *100*(9), 2507–2516. DOI: 10.1002/jbm.a.34150 PMID: 22623317

Ma, H., Liu, M., Li, Y., Wang, W., Yang, K., Lu, L., He, M., Deng, T., Li, M., & Wu, D. (2020). Intrauterine transplantation of autologous menstrual blood stem cells increases endometrial thickness and pregnancy potential in patients with refractory intrauterine adhesion. *Journal of Obstetrics and Gynaecology Research*, *46*(11), 2347–2355. DOI: 10.1111/jog.14449 PMID: 32856391

Montorsi, F., Salonia, A., Centemero, A., Guazzoni, G., Nava, L., Da Pozzo, L. F., Cestari, A., Colombo, R., Barbagli, G., & Rigatti, P. (2002). Vestibular flap urethroplasty for strictures of the female urethra. Impact on symptoms and flow patterns. *Urologia Internationalis*, *69*(1), 12–16. DOI: 10.1159/000064353 PMID: 12119432

Mundy, A. R., & Andrich, D. E. (2011). Urethral trauma. Part I: Introduction, history, anatomy, pathology, assessment and emergency management. *BJU International*, *108*(3), 310–327. DOI: 10.1111/j.1464-410X.2011.10339.x PMID: 21771241

Murphy, C. M., O'Brien, F. J., Little, D. G., & Schindeler, A. (2013). Cell-scaffold interactions in the bone tissue engineering triad. *European Cells & Materials*, *26*, 120–132. DOI: 10.22203/eCM.v026a09 PMID: 24052425

Nappi, R. E., & Nijland, E. A. (2008). Women's perception of sexuality around the menopause: Outcomes of a European telephone survey. *European Journal of Obstetrics, Gynecology, and Reproductive Biology*, *137*(1), 10–16. DOI: 10.1016/j.ejogrb.2006.10.036 PMID: 17207567

Palminteri, E., Berdondini, E., Fusco, F., De Nunzio, C., & Salonia, A. (2012). Long-term results of small intestinal submucosa graft in bulbar urethral reconstruction. *Urology*, *79*(3), 695–701. DOI: 10.1016/j.urology.2011.09.055 PMID: 22245298

Ravindran, S., Song, Y., & George, A. (2010). Development of three-dimensional biomimetic scaffold to study epithelial-mesenchymal interactions. *Tissue Engineering. Part A*, *16*(1), 327–342. DOI: 10.1089/ten.tea.2009.0110 PMID: 19712044

Rider, D. A., Dombrowski, C., Sawyer, A. A., Ng, G. H. B., Leong, D., Hutmacher, D. W., Nurcombe, V., & Cool, S. M. (2008). Autocrine fibroblast growth factor 2 increases the multipotentiality of human adipose-derived mesenchymal stem cells. *Stem Cells (Dayton, Ohio)*, *26*(6), 1598–1608. DOI: 10.1634/stemcells.2007-0480 PMID: 18356575

Robinson, D., & Cardozo, L. (2011). Estrogens and the lower urinary tract. *Neurourology and Urodynamics*, *30*(5), 754–757. DOI: 10.1002/nau.21106 PMID: 21661025

Semmelink, H. J., de Wilde, P. C., van Houwelingen, J. C., & Vooijs, G. P. (1990). Histomorphometric study of the lower urogenital tract in pre- and post-menopausal women. *Cytometry*, *11*(6), 700–707. DOI: 10.1002/cyto.990110607 PMID: 2384029

Semmens, J. P., & Wagner, G. (1982). Estrogen deprivation and vaginal function in postmenopausal women. *Journal of the American Medical Association*, *248*(4), 445–448. DOI: 10.1001/jama.1982.03330040033027 PMID: 6283189

Shi, J.-G., Fu, W.-J., Wang, X.-X., Xu, Y.-D., Li, G., Hong, B.-F., Hu, K., Cui, F.-Z., Wang, Y., & Zhang, X. (2012). Transdifferentiation of human adipose-derived stem cells into urothelial cells: Potential for urinary tract tissue engineering. *Cell and Tissue Research*, *347*(3), 737–746. DOI: 10.1007/s00441-011-1317-0 PMID: 22290635

Singh, M., Kapoor, R., Kapoor, D., Kapoor, R., Srivastav, A., & Chipde, S. (2013). Dorsal onlay vaginal graft urethroplasty for female urethral stricture. *Indian Journal of Urology : IJU : Journal of the Urological Society of India*, *29*(2), 124–128. DOI: 10.4103/0970-1591.21041 PMID: 23956514

Tan, J., Li, P., Wang, Q., Li, Y., Li, X., Zhao, D., Xu, X., & Kong, L. (2016). Autologous menstrual blood-derived stromal cells transplantation for severe Asherman's syndrome. *Human Reproduction (Oxford, England)*, *31*(12), 2723–2729. DOI: 10.1093/humrep/dew235 PMID: 27664218

Tang, X., Zhang, X., Wu, Y., Yin, H., Du, Y., Zhang, X., Li, Q., Liu, S., & Xu, T. (2020). The Clinical Effects of Utilizing Allogeneic Acellular Dermal Matrix in the Surgical Therapy of Anterior Urethral Stricture. *Urologia Internationalis*, *104*(11-12), 933–938. DOI: 10.1159/000510317 PMID: 33022676

Tedesco, M., Bellei, B., Garelli, V., Caputo, S., Latini, A., Giuliani, M., Cota, C., Chichierchia, G., Romani, C., Foddai, M. L., Cristaudo, A., Morrone, A., & Migliano, E. (2020). Adipose tissue stromal vascular fraction and adipose tissue stromal vascular fraction plus platelet-rich plasma grafting: New regenerative perspectives in genital lichen sclerosus. *Dermatologic Therapy*, *33*(6), e14277. DOI: 10.1111/dth.14277 PMID: 32888384

Tian, H., Bharadwaj, S., Liu, Y., Ma, P. X., Atala, A., & Zhang, Y. (2010). Differentiation of human bone marrow mesenchymal stem cells into bladder cells: Potential for urological tissue engineering. *Tissue Engineering. Part A*, *16*(5), 1769–1779. DOI: 10.1089/ten.tea.2009.0625 PMID: 20020816

Tremolada, C.Image Regenerative Clinic. (2019). Mesenchymal Stromal Cells and Micro Fragmented Adipose Tissue: New Horizons of Effectiveness of Lipogems. *Journal of Stem Cells Research. HSOA Journal of Stem Cells Research, Development & Therapy*, *5*(1), 1–7. DOI: 10.24966/SRDT-2060/100017

Trounson, A., & DeWitt, N. D. (2016). Pluripotent stem cells progressing to the clinic. *Nature Reviews. Molecular Cell Biology*, *17*(3), 194–200. DOI: 10.1038/nrm.2016.10 PMID: 26908143

Tseng, L. H., Lin, Y.-H., Chang, S.-D., Lim, C.-K., Chiang, C.-C., & Huang, C.-H. (s. f.). *Three-dimensional printing vaginal pessary to treat pelvic organ prolapse: A pilot study*. Recuperado 6 de julio de 2025, de https://www.authorea.com/users/517755/articles/592178-three-dimensional-printing-vaginal-pessary-to-treat-pelvic-organ-prolapse-a-pilot-study

Urogenital tract fistulas in females—UpToDate. (s. f.). Recuperado 11 de agosto de 2024, de https://www.uptodate.com/contents/urogenital-tract-fistulas-in-females?search=urogenital-tractfistulas_in-women&source=search_result&selectedTitle=1~150&usage_type=default&display_rank=1

Wen, Y., Wani, P., Zhou, L., Baer, T., Phadnis, S. M., Reijo Pera, R. A., & Chen, B. (2013). Reprogramming of fibroblasts from older women with pelvic floor disorders alters cellular behavior associated with donor age. *Stem Cells Translational Medicine*, *2*(2), 118–128. DOI: 10.5966/sctm.2012-0092 PMID: 23341439

Xu, Y.-M., Fu, Q., Sa, Y.-L., Zhang, J., Song, L.-J., & Feng, C. (2013). Outcome of small intestinal submucosa graft for repair of anterior urethral strictures. *International Journal of Urology*, *20*(6), 622–629. DOI: 10.1111/j.1442-2042.2012.03230.x PMID: 23131085

Zambon, J. P., Batezini, N. S. S., Pinto, E. R. S., Skaff, M., Girotti, M. E., & Almeida, F. G. (2010). Do we need new surgical techniques to repair vesico-vaginal fistulas? *International Urogynecology Journal*, *21*(3), 337–342. DOI: 10.1007/s00192-009-1040-5 PMID: 19949771

Zaret, K. S. (2009). Using small molecules to great effect in stem cell differentiation. *Cell Stem Cell*, *4*(5), 373–374. DOI: 10.1016/j.stem.2009.04.012 PMID: 19427285

Zhang, Y., Ma, Y., Chen, J., Wang, M., Cao, Y., Li, L., Yang, H., Liu, X., Li, Y., & Zhu, L. (2021). Mesenchymal stem cell transplantation for vaginal repair in an ovariectomized rhesus macaque model. *Stem Cell Research & Therapy*, *12*(1), 406. DOI: 10.1186/s13287-021-02488-2 PMID: 34266489

Zhang, Y., Shi, L., Lin, X., Zhou, F., Xin, L., Xu, W., Yu, H., Li, J., Pan, M., Pan, Y., Dai, Y., Zhang, Y., Shen, J., Zhao, L., Lu, M., & Zhang, S. (2021). Unresponsive thin endometrium caused by Asherman syndrome treated with umbilical cord mesenchymal stem cells on collagen scaffolds: A pilot study. *Stem Cell Research & Therapy*, *12*(1), 420. DOI: 10.1186/s13287-021-02499-z PMID: 34294152

Zhao, Y., Luo, Q., Zhang, X., Qin, Y., Hao, J., Kong, D., Wang, H., Li, G., Gu, X., & Wang, H. (2020). Clinical Efficacy and Safety of Stem Cell-Based Therapy in Treating Asherman Syndrome: A System Review and Meta-Analysis. *Stem Cells International*, *2020*, 1–11. Advance online publication. DOI: 10.1155/2020/8820538 PMID: 33414830

Zhu, W.-D., Xu, Y.-M., Feng, C., Fu, Q., Song, L.-J., & Cui, L. (2010). Bladder reconstruction with adipose-derived stem cell-seeded bladder acellular matrix grafts improve morphology composition. *World Journal of Urology*, *28*(4), 493–498. DOI: 10.1007/s00345-010-0508-8 PMID: 20091038

Chapter 6
Stem Cell Therapy in Reproductive Medicine

ABSTRACT

Stem cell therapy holds immense promise for addressing infertility problems. This chapter delves into the promising application of regenerative technology to revolutionize reproductive medicine, in overcoming infertility challenges for both men and women. We explore the different clinical studies for potential of in vitro gametogenesis and spermatogenesis, as a groundbreaking approach to infertility treatment. Furthermore, the chapter examines the clinical implications of stem cell therapy in addressing testicular dysfunction and failure. While significant advancements have been made, we emphasize the importance of ongoing research to translate these findings into safe and effective clinical practices, but also acknowledges the complex ethical considerations surrounding stem cell research and its translation into clinical practice. It is imperative to balance scientific advancements with ethical principles to ensure responsible and beneficial utilization of this technology.

INTRODUCTION

The undifferentiated cells present in the different stages of life that give rise to differentiated and specialized cells to build tissue and organs are called the stem cells. (Volarevic et al., 2014) The main properties of the stem cells are its self-renewal ability, high proliferative potential and high differentiation potential. (Gargett, 2010; L. Li & Xie, 2005) Moreover, they serve as a reservoir of cells to regenerate tissue. (Simoni & Taylor, 2018) Based on developmental potency (table1), stem cells

DOI: 10.4018/979-8-3693-8829-7.ch006

were described as totipotent, pluripotent, multipotent, oligopotent and monopotent. (Zakrzewski et al., 2019)

Totipotent stem cells have the greatest potency with the ability to divide and differentiate into any type cells of the whole organism. (Zakrzewski et al., 2019) They are found in early stages of development, that is, a fertilized oocyte and the cells of the first two divisions are totipotent cells. They differentiate into both embryonic and extraembryonic tissues, forming the embryo and the placenta. (Rossant, 2001)

Table 1. Stem cell classification based on cell potency

Cell potency	Cell types capable of generating	Examples
Totipotent	Differentiate into any cell type (Whole organism)	Zygote (derived from early cells of fertilized egg), early morula
Pluripotent	Differentiate into cells from any of the three germ layers	embryonic stem cells (derived from inner cell mass of blastocyst), induced pluripotent stem cell
Multipotent	Differentiate into a limited range of cell types	Hematopoietic stem cells, neural stem cells, mesenchymal stem cells
Unipotent	Differentiate into a single cell type	Ovarian stem cell, spermatogonial stem cells

At the blastocyst stage of embryo development (approximately four days after fertilization), the inner cell mass (ICM)acquires pluripotency, that is, it can differentiate to every cell type of the whole organism, but not the placenta and umbilical cord. (Zakrzewski et al., 2019) Embryonic stem cells (ESCs) are pluripotent stem cells that can form cells of all germ layers but not extraembryonic structures. Induced pluripotent stem cells (iPSCs) are created artificially by inducing the specialized cells to express genetic markers that shows the characteristics and properties of ESC. ESCs and iPSCs may differentiate to form less potent -multi-, oligo- or monopotent-cells. (Zakrzewski et al., 2019)

Monopotent stem cells can produce only one cell type but has self-renewal property (e.g., spermatogonial stem cells). (Vassena et al., 2015) Diverse types of stem cells, their characteristics and possible application in reproductive medicine are widely studied. Stem cells are considered as potentially new therapeutic agents for the treatment of infertility because of unlimited source and high differentiation potential. (Kharazi & Badalzadeh, 2020; Volarevic et al., 2014)

Reproductive medicine specializes in diagnosing and treating infertility, fertility preservation, and other reproductive problems. (*Definition of reproductive medicine - NCI Dictionary of Cancer Terms - NCI*, s. f.) Infertility is "a disease of the reproductive system defined by the failure to achieve a clinical pregnancy after 12 months or more of regular unprotected sexual intercourse." (Zegers-Hochschild et al., 2009) Reproductive medicine research and subsequent clinical development have attained scientific growth and maturity in the field of human embryology, reproductive endocrinology, coagulation disorders, autoimmune disorders, reproductive tract microbiota, and reproductive genetics. (Aleixandre-Benavent et al., 2015)

The field has evolved rapidly as a new area of specialization in medicine where technology is of paramount importance. Comprehensive technological advances have been developed to treat many infertility cases includes recombinant gonadotropin therapy, in vitro fertilization and embryo culture, intracytoplasmic sperm injection, assisted hatching, and preimplantation genetic testing. (Zegers-Hochschild et al., 2009) Stem cell technology is widely studied and is emerging as promising therapy for infertility treatment. (Ahmadian et al., 2020)

Infertility effects the quality of life and well-being of affected individuals and couples. Mostly, couples wish to have their biological child. Human reproduction research and infertility treatment revolutionized in last 30 years with the development of treatments such as hormonal induction of follicular development and ovulation, in vitro fertilization, intracytoplasmic sperm injection, fresh and frozen embryo transfer, and cryopreservation of egg and embryo. (Wang & Sauer, 2006) Critical physiological barriers were bypassed positively with a combination of these advanced assisted conception procedures. (Moore et al., 2008) However, providing an effective treatment is often severely hampered when primary infertility is caused due to failure of germ cell production as a result of genetic problems, physiological problems, toxic exposure or any other unexplained factors. (Moore et al., 2008)

As a matter of fact, in cases in which there is absence of gametes (due to azoospermia in males and ovarian failure in females) or poor-quality gamete, the only option available clinically is gamete (egg or sperm) donation; techniques which are not universally accepted due to cultural, religious, ethical, legal, and personal concerns. (Cervelló et al., 2016) This again led to rise in scientific interest in the study of human germ line development and stem cell therapy. (Hudson et al., 2009) The possibility to derive artificial gametes from stem cells with the rapid development of technology may provide new therapeutic strategies to treat infertile couples. (Fang et al., 2018; Radhakrishnan, 2017)

In vitro gamete production, ovarian rejuvenation, endometrial regeneration, and testicular tissue reactivation are a few interesting areas of research in stem cell and reproductive medicine. One of the leading causes of infertility reported were due to the disruption or absence of germ cells. (Y. Hayashi et al., 2012) Pluripotent stem cells, gonadal stem cells and mesenchymal stem cells have been studied widely to obtain gametes that could be used in the treatment. (Bhartiya et al., 2014)

In females with infertility due to ovarian dysfunction, stem cell-based strategies for ovarian regeneration and oocyte production have been proposed as clinical therapies. (Volarevic et al., 2014) In males with spermatogenic dysfunction, use of stem cells such as spermatogonial stem cells (SSCs) has been studied in model organisms for the regeneration of the impaired or damaged spermatogenesis. (Abdelaal et al., 2021; Ibtisham & Honaramooz, 2020; Kharazi & Badalzadeh, 2020; Sharma et al., 2019) However, the challenges in the field include inefficient existing protocols associated with extensive in vitro manipulation (cell differentiation, epigenetic and genetic changes) in addition to the regulatory constraints. (Bhartiya et al., 2014)

Scientific research has been conducted to understand the basis of early germ cell development with the hope that treatments can be developed that stimulate de novo gamete development from the primary stem cells. Stem cells potential use in reproductive medicine were studied by stimulation them in vitro to develop specialized cells including male and female gametes. (Volarevic et al., 2014) In addition, this may also help to understand the precise molecular mechanisms of infertility. (Y. Hayashi et al., 2012) In vitro generation of functional gametes is a promising approach for treating infertility by understanding drug formulation and targeted application to treat infertility. (Zhou et al., 2016)

Theoretically, in vitro cultured stem cells could be utilized in vivo to restore faulty gametogenesis process that may develop into specialized gametocytes. (P. Y. Zhang et al., 2020) Moreover, studying the very earliest stages of germ cell commitment and differentiation provide precise knowledge to investigate some of the origins of human gonadal dysfunction (e.g., testicular failure, testicular cancer, early menopause), and impact of environmental factors on embryos developmental processes and early placentation. (Anway & Skinner, 2008)

In this chapter, we summarize recent advances of stem cell technology in the field of reproductive medicine. The strategies developed using in vivo and in vitro cultured stem cells to understand the pathophysiology of human infertility (diagram 1).

Figure 1. Causes of male and female infertility and proposed stem cell therapy in reproductive medicine

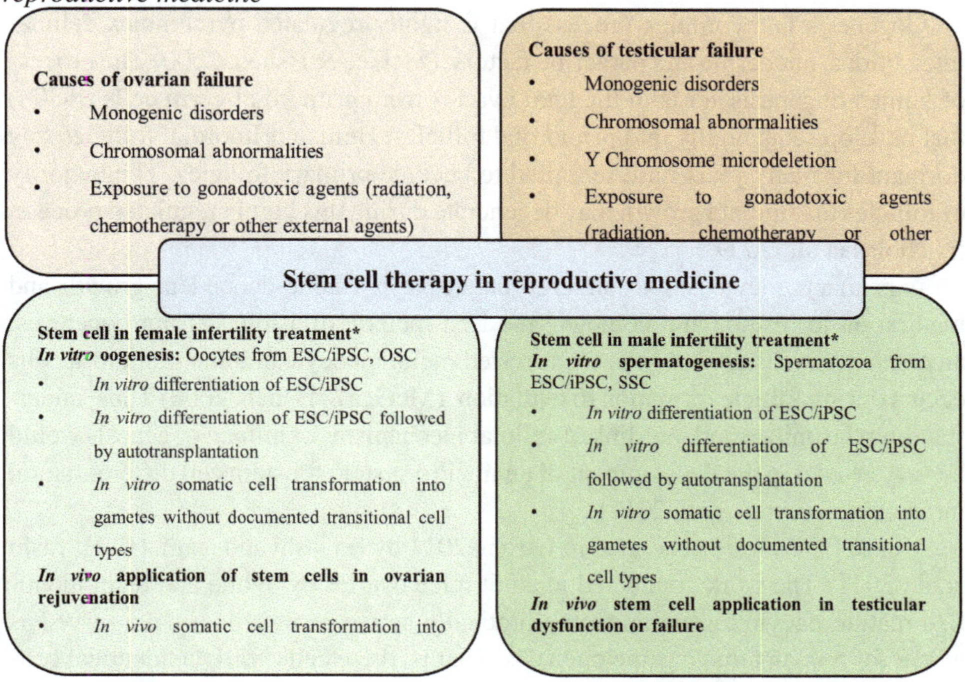

FEMALE INFERTILITY

Female reproductive diseases, such as primary ovarian insufficiency, polycystic ovary syndrome, fallopian tube obstruction, Asherman syndrome, uterine fibroids, autoimmune disorders and implantation failure may cause infertility. (Y. X. Zhao et al., 2019) Advanced clinical research and rapid development in translational medicine are focusing on the treatment of female reproductive disorders using novel stem cell therapies.

The existence of proliferative germ cells that sustain oocyte and follicle production in the postnatal mammalian ovary were reported in 2004, providing new directions in the study of mammalian female germline stem cells. (J. Johnson et al., 2004) In addition, related research studies had a significant clinical implication related to premature ovarian failure. (J. Johnson et al., 2004; Tilly et al., 2009; Zou et al., 2009) Stem cell-based therapies for ovarian dysfunction have been proposed for treating infertility in women. (White et al., 2012) Most studied in vitro and in vivo applications of stem-cell therapy in reproductive system are briefed in the sections below.

In Vitro Gametogenesis/Oogenesis

Oogenesis is a complex process that is tightly regulated by complex cellular interactions, endocrine and paracrine factors. (McGee & Hsueh, 2000) The process of human oogenesis starts in the fetal ovaries from primordial germ cells (PGCs) that develop sequentially into primordial follicles. Human primordial follicles stay dormant for many years until recruited to become primary follicles. The majority of follicles that initiate growth may degenerate during this highly regulated process. (Nicholas et al., 2009)

A primary oocyte has to undergo extensive hormone-dependent growth and maturation to resume and complete the first meiotic division. During oogenesis, an extensive crosstalk between the arrested meiotic oocyte and the granulosa cells occurs during follicle formation to ovulation. (McGee & Hsueh, 2000) Thus, understanding the unique and interlinked cellular mechanisms of human oogenesis would be very crucial in the development of an in vitro system for stem cell differentiation into oocytes. (Jung et al., 2017)

Mouse PGCs were generated in vitro in 2011 by Hayashi and team. (K. Hayashi et al., 2011) The work conducted in adult mice ovaries by White and team identified mature oocyte formation from mitotically-active germ cells in vivo served as model for human female gametogenesis. That is, these cells with the identical cell-surface expression in human ovaries may be purified and cultured in vitro to form oocytes. (White et al., 2012) Pluripotent cells (ESCsoriPSCs) differentiation into the primordial germ cell-like cells (PGC-LCs), its controlled initiation and completion of meiosis are the critical phases of ex vivo gametogenesis. These phases are fundamentally independent, and different in males and females. (K. Hayashi et al., 2011; Hikabe et al., 2016) In mice model experiments, PGC-LCs graft in mice ovary differentiated into mature oocytes under in vivo-like conditions. (K. Hayashi et al., 2012; K. Hayashi & Saitou, 2013)

Oocytes from ESC/iPSC

Human ESCs are pluripotent stem cells which can differentiate into the cells of all three germ layers. (Nicholas et al., 2009) The development of iPSCs from a patient's adult somatic cells by introducing the combinations of several transcription factors (e.g., OCT3/4, SOX2, KLF4, and MYC) enabled to generate strategies to develop in vitro disease models of infertility. (Y. Hayashi et al., 2012) Pluripotent cell-specific genes or markers of iPSCs are similar to ESCs which could imply that they exhibit similar morphology, gene expression, and epigenetic status.

The derivation of gametes from human ESCs or iPSCs may require cell regulators that functions in the exit pathway from pluripotency and the entry pathway into a germ cell-specific state.(Jung et al., 2017) iPSCs differentiate into all three germ layer cell types in vitro and form teratoma in vivo. (Y. Hayashi et al., 2012)

Many in vitro studies in mice models showed that pluripotent stem cells (PSCs) could be converted into germ cells such as sperm or oocytes. (K. Hayashi et al., 2011; Y. Lee & Kang, 2019) ESC or iPSC derived oocytes could help to restore fertility. (Nicholas et al., 2009) Controlled clinical trials showed scope of PSCs in regenerative medicine. (Vassena et al., 2015) Developmental process of germ cell and embryo in humans could be investigated with generation of various pluripotent stem cell lines which could be used for future therapies in the field of reproductive medicine. (Moore et al., 2008)

Research conducted using human pluripotent stem cells (hPSCs) differentiation into the female germ line has so lagged because ESCs were able to form follicle-like structures but were not able to progress into meiosis. (Hübner et al., 2003; Lacham-Kaplan et al., 2006) But this outcome was achieved in the mouse model by Hayashi and his research team in 2012 by making use of PGCs (obtained from female epiblast-like cells) which were subsequently aggregated with somatic cells from embryonic ovaries and transplanted under the ovarian bursa. (Y. Hayashi et al., 2012)

Another in vitro animal experiment demonstrated that PGC-derived immature oocytes that fertilized to form embryos were transplanted into foster mothers, to obtain healthy and fertile offspring. (Cyranoski, 2013) However, they reported abnormal pronucleus formation post fertilization of these artificial oocytes, which could be responsible for the lower rates of post-implantation development compared with controls. (Cyranoski, 2013) This finding may direct to defects in epigenetic reprogramming during the in vitro formation of PGCs. (Vassena et al., 2015) Anyhow, in vitro human oocytes development from pluripotent cells is still in research stage and in far prospect as far as clinical application is concerned. (Vassena et al., 2015)

Oocytes from OSC

There was a long-term belief on "fixed ovarian reserves" due to absence of renewable germ cells in mammalian females. However, mice studies challenged this theory of fixed ovarian reserve by suggesting that mouse ovaries contain mitotically active germ cells that might be able to generate oocytes throughout life. (J. Johnson et al., 2004, 2005) In addition, experiments demonstrated that ovaries of reproductive-aged women (similar to adult mice) may contain similar mitotically active germ cells that can be propagated in vitro that develop into some follicle-like structures containing oocytes. (White et al., 2012)

Ovarian stem cells (OSC) are defined as very small embryonic-like stem cells in the ovary which have the capacity to develop as germ cells under certain conditions. (Bhartiya et al., 2013; Lei & Spradling, 2013) OSC are derived from bi-potential stem cells in the ovarian cortex that can produce new stem cells or differentiate into ovarian reproductive cells and primitive granular cells. (Ye et al., 2017)

In vitro differentiated OSCs from adult human ovarian cortex expressed pluripotency markers. (Stimpfel et al., 2013) Zhang et al., 2012 clearly demonstrated that germline cells in post-natal female mice do not enter mitosis nor contribute to de novo folliculogenesis. (H. Zhang et al., 2012) While another research postulated that de novo oocyte production originates from circulating bone-marrow-derived germ cells (J. Johnson et al., 2005), this claim was refuted, with no proof of bone marrow cells, or any other circulating cells, contributing to the de novo production of oocytes. (Eggan et al., 2006) Although studies are very actively ongoing to understand the possibility of using OSC to treat some kinds of infertility; there is currently no consensus on their functionality and therapeutically application. (Vassena et al., 2015)

In Vivo Application of Stem Cells in Ovarian Rejuvenation

Evaluating ovarian reserve for precise therapeutic strategies is crucial for optimal result in infertility treatment. Conditions such as premature ovarian insufficiency (POI)characterized by amenorrhea, oligomenorrhea, hypoestrogenism, low-anti-Mullerian hormone or high gonadotrophin levels, is a common cause of infertility in women. (Gupta et al., 2018) For POI patient, autologous bone marrow-derived stem cell (BMSC) therapy for successful ovarian reactivation and regeneration has been reported. (Gupta et al., 2018)

SC therapy for POI-induced infertility is based on the possibility of renewal of damaged ovaries to regain its function. Mouse model of chemotherapy-induced POI study showed that the transplanted stem cells could reside in the ovarian tissue and rescue ovarian function, even though, this experiment necessitate further investigation. (H. J. Lee et al., 2007) In vitro produced oocytes from germline stem cells and fertilization-competent oocytes in vivo have been identified and isolated from adult mouse ovaries.(White et al., 2012)

APPLICATION OF MESENCHYMAL STEM CELLS TO TREAT OVARIAN FAILURE

Mesenchymal stem cells (MSCs) are multipotent adult stem cells derived from the mesoderm and isolated from fat tissue, bone marrow, amniotic fluid, umbilical cord or placenta. (Shah et al., 2018) MSCs are widely used in research and therapeutics because of their ability to differentiate to multiple lineages of cells, such as chondrocytes, osteocytes, adipocytes, hepatocytes, or neurocytes. (Samsonraj et al., 2017) In the treatment of ovarian and uterine diseases, four types of mesenchymal stem cells (bone marrow stromal cells, adipose-derived stem cells, menstrual blood mesenchymal stem cells, and umbilical cord mesenchymal stem cells) have been mainly studied. (Y. X. Zhao et al., 2019)

One major advantage of MSCs is that they do not raise much concern due to its low immunogenicity and low immunomodulatory function.(Na & Kim, 2020) In addition, abundant sources and convenient sampling make MSCs great choice in the clinic application studies related to infertility treatment. (Y. X. Zhao et al., 2019) However, the limitations may include the factors such as age and the invasive collection methods (bone marrow aspiration and liposuction) which affect its self-renewal potential. (Feng et al., 2019)

Application of Stem Cells to Treat Endometrial Disorders

During a woman's reproductive, endometrium undergoes more than 400 cycles of growth, differentiation, and shedding. In addition, the endometrium grows 4–10 mm each cycle under the influence estrogen during the proliferative phase of the menstrual cycle. Furthermore, high level of endometrial tissue regeneration also occurs after parturition, endometrial resection, and in postmenopausal women taking estrogen replacement therapy.

Adult stem cells maintain cellular production in this highly regenerating tissue. It was originally hypothesized that the adult stem or progenitor cells are responsible for the cyclic regeneration of the endometrial functional is layer each month and also these adult stem cells reside in the basalis layer, and are present in the atrophic endometrium of postmenopausal women.

However, as endometrial stem/progenitor cells are being identified with specific markers, it has become apparent that they may reside in both functional is and basalis layers. Endometrial stem/progenitor cells originating from the functional is and shed in menstrual debris may also contribute to eutopic endometrial regeneration from those fragments remaining within the uterine cavity. (Gargett & Ye, 2012) The progenitor stem cells in the endometrium can replace all of the endometrium during each menstrual cycle. In addition, when these progenitor stem cells are depleted, multipotent endometrial stem cells replenish them.

Recruitment of stem cells from outside of the uterus occurs in setting of increased demand such as ischemia or injury. For instance, bone marrow-derived multipotent stem cells are recruited to the uterus by estrogen or injury-induced expression of the chemokine. (Gil-Sanchis et al., 2015) In case of an injury, especially low estrogen levels are low, there may be insufficient stem cell recruitment to adequately repair the uterus resulting in conditions such as Asherman´s syndrome or other endometrial defects.

Asherman's Syndrome is an acquired gynecological disorder that seems difficult to treat. Better understanding of stem-cell mobilization, recruitment, and engraftment shaped the possibility of improved therapy for endometrial defects through enhanced manipulation of stem-cell trafficking. (Cervelló & Simón, 2009)

Adult stem cells have been identified in the highly regenerative human endometrium on the basis of their functional attributes. Furthermore, the normal endometrium is a rich source of multipotent stem cells which are studied for applications in regenerative medicine. (Simoni & Taylor, 2018) A study about stem cell-based therapy in treating Asherman's syndrome suggest that stem cell and hormone combination therapy was safe and more effective in improving menstruation duration, pregnancy outcome, and endometrial thickness. (Y. Zhao et al., 2020)

However, there is still limitations on application and stem cell-based therapy has been proposed as a novel strategy for management of Asherman's syndrome.(Y. Zhao et al., 2020) Characterization of endometrial MSC specific markers and epithelial progenitor cell markers enables their potential use in reconstructing endometrial tissue. (Gargett & Ye, 2012) Moreover, adult stem cells can reconstruct endometrial tissue in vivo suggesting their possible use in treating disorders associated with inadequate endometrium.

MALE INFERTILITY

Male infertility and gonadal dysfunction may lead to low sperm production, abnormal sperm function or blockages that prevent the delivery of sperm. Major factors such as genetics, environmental exposures, aging or gonadotoxic therapies (e.g. immunosuppression and anticancer treatment) may affect spermatogenesis (biological mechanism of sperm production). (Cassim & Mohamed, 2019) Male infertility accounts for approximately half of all cases of infertility and from that, many infertile males are diagnosed as idiopathic with unknown causes. (Ferlin et al., 2007)

Azoospermia is a condition of total absence of sperm in the ejaculate due to obstructive or non-obstructive process. In obstructive azoospermia, sperm may be found in testes but not in the ejaculate due to either congenital or acquired obstruction of the ejaculatory duct, whereas in non-obstructive azoospermia (NOA), no sperm is found in the ejaculate due to inability of the testes to produce sperm or failure of spermatogenesis. NOA may occur due to inadequate central stimulation of the testes or an abnormality within the testes. (Cocuzza et al., 2013) NOA have been reported as the most severe cause of male infertility. (Radhakrishnan, 2017)

Defect in spermatogenic processor NOA could be attributed to SSC or its niche, or both. Various treatments using stem cells are being explored with variable success rate to treat infertile male with azoospermia. (Cassim & Mohamed, 2019) Though therapeutic options for NOA are limited, the rapid advancement in the stem cell research and gene editing technologies could enable new treatment options. SSC, iPSC, testicular niche cells and mesenchymal stromal cells have been investigated for their potential use in male infertility. (Gauthier-Fisher et al., 2020)

In addition, human umbilical cord perivascular cells (HUCPVC) been reported as a scurce of cells that could be utilized in male infertility treatment methods. (Gauthier-Fisher et al., 2020) During past few years, a considerable progress in the derivation of male germ cells from pluripotent stem cells has been made. (Volarevic et al., 2014)

Significant progress has been made in ART for the treatment of infertility. Male factor infertility due to primary NOA remains a challenge for the couple, as well as for the attending infertility specialist. Current ART treatments methods are unable to help infertile patients with NOA (unless donor sperm is used) due to the fact that the mechanisms regulating spermatogenesis and sperm function in humans were poorly understood yet. (Y. Hayashi et al., 2012)

In Vitro Gametogenesis- Male/ Spermatogenesis

In the biological system, spermatogenesis is a tightly regulated process that takes place in the seminiferous tubules within the testis. (Parreira et al., 1998) In vitro spermatogenesis or the generation of functional sperm in vitro, has been a goal for almost a century. Researchers struggled to find and develop a micro-environmental condition that can support the process in vitro, because of the complexity of the cellular differentiation processes in a body or in vivo. Several cytological markers such as genes, transcription factors, cytokines, growth factors, proteins and micro

RNAs have been studied at distinct phases of spermatogenesis pathway from embryonic germ cells allocation to postnatal spermatogenesis. (Dissanayake, 2018)

Sato et al. reported successful in vitro production of functional sperm in neonatal mouse testes, thereby potentially opening a new era of reproductive biology. (Sato et al., 2011; Song & Wilkinson, 2012)

In vitro spermatogenesis procedures may have many applications in the clinic such as restoration of fertility in patients diagnosed with cancer (who require chemotherapy or radiotherapy). (Song & Wilkinson, 2012) Other promising application may include generating functional sperm from iPSC in patients with NOA.

Main approaches in achieving this goal include autologous testicular transplantation of stem cells, auto or xeno-grafting of testicular tissues, in vitro culture of stem cells with testicular tissues (organ culture systems) or without testicular tissues. However, major challenges faced in the in vitro spermatogenesis process are haploidization of stem cells or progression beyond pachytene stage, and inability to further differentiation of the few round spermatids in culture (especially up to formation of tail). (Dissanayake, 2018) Knowledge on cytological markers expressed or suppressed at different stages of spermatogenesis may play as key factors in the development strategies of in vitro spermatogenesis.

Spermatozoa from ESC/iPSC

In vitro studies reported differentiation of human embryonic stem cells (hESCs) and induced pluripotent stem cells (hiPSCs)into primordial germ cells (PGCs) but not into spermatogonia nor haploid spermatocytes or spermatids.(Easley et al., 2012) Two methods to produce sperm from the mouse ESCs in vitro were the monolayer differentiation and the embryoid body (cellular aggregate or mixture of cells from the endoderm, the ectoderm, and the mesoderm)formation. (P. Li et al., 2013; West et al., 2006)

Successful in vitro experiments have been reported demonstrating ESCs to form embryonic germ cells, based on the similarity of their gene and protein expression patterns to those of PGCs. The first reported differentiation of human ESCs into embryonic germ cells confirmed that the cultivation of human ESCs into embryoid bodies formed embryonic germ cells that express germ cell-specific genes. (Clark et al., 2004)

Male germ cells at different stages have been derived from hiPSCs with different protocols such as spontaneous differentiation, overexpression of germ cell regulators, addition of cytokines, co-culture with gonadal cells in vitro and xeno-transplantation. (Fang et al., 2018) In addition, differentiation medium was supplemented with the growth factors, regulator proteins, retinoid and hormones for optimal result. (K.

Hayashi et al., 2011; Vassena et al., 2015) Mouse iPS cells can differentiate into spermatogonial stem cells and late stages of male germ cells.

The overall developmental dynamics of human primordial germ cells are similar to that in mice, but accumulating evidence suggests that there are crucial differences between human and mouse PGC specification. Unlike mouse iPSCs (miPSCs) in naive state, hiPSCs exhibit a primed pluripotency which possess less potential for the germ cell fate. (Fang et al., 2018)

Hayashi and team made the remarkable finding that PGC-LCs could be obtained from mouse ESCs and mouse iPSCs.(Y. Hayashi et al., 2012) These PGC-LCs could be differentiated into spermatozoa in vivo resulting in the birth of healthy offspring via ICSI. (Radhakrishnan, 2017) Moreover, the overexpression of specific male germ cell genes has been used to guide differentiation toward spermatozoa. (Vassena et al., 2015)

In vitro produced male gametes from ESCs or iPSCs could provide better disease models to study male infertility. iPSCs resembles ESCs in cell morphology, expression of pluripotency marker and differentiation potential. hiPSCs may be superior to hESCs for reproductive medicine application because there are few ethical concerns, abundant source and it can be generated from patients' somatic cells that are immunocompatible. Therefore, the generation of patient-specific spermatozoa from hiPSCs may provide the hope for the future treatment of male infertility. However, hiPSCs may not faithfully recapitulate the characteristics of hESCs at both genetic and epigenetic levels.

Especially, hiPSCs are reported to keep some epigenetic marks of the donor cell type from which they were reprogrammed.(Panula et al., 2019; Radhakrishnan, 2017) So, it is necessary to reconstitute human germ line development in vitro, which could possibly be achieved by differentiation of hiPSCs . (Fang et al., 2018)

Comparison of patient-derived hiPSCs with normal hiPSCs for their germ line differentiation abilities may help identify abnormalities and decipher the molecular mechanisms of idiopathic male infertility involved in differentiation and maturation. (Durruthy et al., 2014) Furthermore, male germ cells differentiation from hiPSCs would be an invaluable tool to explore the human germ line specification, including transcriptional networks, signaling pathways and epigenetic reprogramming. (Durruthy et al., 2014).

Obtaining PGCs from pluripotent cells is the first step in the differentiation toward post-meiotic spermatozoa. To recreate the testicular niche in vitro for a better differentiation of pluripotent cells is also very crucial. The functional male gametes derived from autologous iPSCs could benefit azoospermia patients.

However, there are only few reliable studies and limited evidences on derivation of functional gametes from hiPSCs, which makes keeps it still in an immature research stage for clinical applications.(Nikolic et al., 2016) The germ cells go through dramatic epigenetic reprogramming during development). (Jung et al., 2017; Wen & Tang, 2016) Thus, correct epigenetic reprogramming is essential for the development of functional gametes. (Eguizabal et al., 2013; Kee et al., 2009; Medrano et al., 2012; Metcalfe, 2020; Panula et al., 2011; Park et al., 2009; Tilgner et al., 2008)

The evaluation of epigenetic status during germ cell differentiation from hiPSCs is informative but not fully evaluated in most published research studies. (Fang et al., 2018) hiPSCs may have tremendous potential for biological and therapeutic applications for male infertility. It is important to note that the molecular mechanisms underlying human male germ cell development are still poorly understood yet.

But, recent advances in generation of male germ cells from mouse iPSCs not only hold great promise for the establishment of in vitro human spermatogenesis models, but also provide better insights into the mechanism of PGC specification and spermatogenesis regulation. In addition, advanced genomic technology may help in precise understanding of molecular basis of reproductive biology and its possible applications in infertility research. (Fang et al., 2018)

Spermatozoa from SSC

SSCs are located at the basement membrane of seminiferous tubules and are the least differentiated germ cells. (Sharma et al., 2019) SSCs can undergo self-renewal to keep the stem cell pool intact, or differentiation resulting in the formation of sperm. SSCs differentiation ensures the continuous, lifelong production of approximately 100 million spermatozoa each day from puberty onwards in the normospermic man. (Oatley & Brinster, 2008) Long-term propagation of SSCs in vitro was developed for the first time in 2003 in a mouse model and thereafter in other animal model systems including rat, hamster, bovine, and humans. (Abdelaal et al., 2021; Kanatsu-Shinohara et al., 2003; Sadri-Ardekani et al., 2009) Even though SSC therapy in humans is not clinically applied yet, it is identified as a potential source for the development of stem cell-based therapies to treat male infertility.

In patients with cancer, SSCs therapy may include cryopreservation of a testis biopsy before gonadotoxic treatment, in vitro propagation of SSCs from the biopsy and auto-transplantation once the patient has recovered. (Sharma et al., 2019)

Alternatively, sperm could be generated from the immature germ cells in vitro or in a tissue graft. However, these methods are still experimental. Though, cryopreservation of testicular biopsies is now offered at various centers. Furthermore, it is not clear if sustained spermatogenesis will occur from SSCs using these methods. From the recent scientific literature, it seems that so far there is still the need of a natural cellular niche in order to obtain mature functional spermatozoa, and that in vitro production of spermatozoa is currently far prospect.

In Vivo Application in Testicular Dysfunction or Failure

The second critical phase in mammalian spermatogenesis (that is meiosis) is regulated by a highly specialized microenvironment in Sertoli cells. Thus, it is essential to mimic spermatogenic environment when spermatogenesis is recreated ex vivo. (Cai et al., 2013) An important proof of functionality of male germ cells from pluripotent cells could be studied by its transplantation into the testis of sterile individuals. (Y. Hayashi et al., 2012; Nakaki et al., 2013; West et al., 2006)

Recently, hiPSCs from azoospermic and normospermic males were transplanted into sterile mouse testis, partially colonizing the testicular niche and showing signs of early stage spermatogenesis.(Durruthy et al., 2014) Study shows that transplanting in vitro produced PGC-LC into the mouse testis could be used to obtain functional mouse sperm, capable of producing healthy and fertile offspring.(K. Hayashi et al., 2011)

Transplantation of SSCs was first described by Brinster and Zimmerman in 1994 when suspensions of testicular cells from fertile mice were transplanted into infertile mice resulting in fertility restoration and progeny. (Avarbock et al., 1996; Brinster & Zimmermann, 1994) It has been shown that cryopreservation of SSCs does not impair their ability to produce healthy spermatozoa capable of fertilization. (Wu et al., 2012) Currently, two strategies exist for transplantation of SSCs – testicular tissue harvest and grafting and SSC isolate injection. (Vij et al., 2018) The basic theory of SSC in vivo therapy is to induce spermatogenesis from the organism's own SSCs, after which these cells may be supported in vivo within the patient's testis or in xenograft or other ex vivo culture. The sperm obtained from this method could then be used for fertilization. (Forbes et al., 2017)

The concept of transplantation was investigated by taking cells from the testis of allogenic animals and transplanting them to an allogenic recipient that was treated with the chemotherapeutic compound busulfan or with local irradiation to empty the SSC niche in the seminiferous epithelium. The results demonstrated the effectiveness of the transplantation and chemotherapeutic treatment or irradiation does not seem to affect the ability of the testis to support spermatogenesis.(Hermann et al., 2012)

Other factor that may have role in efficiency of the SSC transplantation was reported as the number of stem cells injected. (Oatley & Brinster, 2008) These experiments showed that frozen SSCs are capable of restoring spermatogenesis and fertility upon transplantation in infertile mice. Thereafter, auto-transplantation has successfully been established in several other species including rat, bull, goat, sheep, dog and recently non-human primates. (Hermann et al., 2012)

Transplantation of SSCs into a member of the same species or back into the original host is a useful technique for preserving SSCs and sperm outside the donor. This has numerous theoretical applications, including restoring fertility after fertility-ablative treatments in pre-adolescents. (E. K. Johnson et al., 2017) Techniques of in vitro and combined in vitro and in vivo approaches may improve autologous transplantation of SSCs in larger mammals. (Forbes et al., 2017) Irradiated bovine testes restored morphology after autologous transplant. (Izadyar et al., 2003)

Autologous transplantation of germ cells into the rete testis of infertile monkeys showed growing testicular volume with sperm production. (Schlatt et al., 2002) SSC grafting offers clinically options in survivors from childhood cancer. SSCs transplantation shows promises for fertility preservation and establishment of reproductive potential in infertile patients. (Tournaye & Goossens, 2011)

Stem cell therapy for infertile recipient based on spermatogonia cells transplantation from a fertile donor has also been proposed. (Tournaye & Goossens, 2011) However, more studies are needed to clinically apply these methods in human infertile patients. (Nazm Bojnordi, 2017) The generation of healthy offspring with normal karyotypes and typical imprinting patterns, are important issues that must be considered by clinical and developmental biologist. (Tournaye & Goossens, 2011)

Application of Mesenchymal Stem Cells to Treat Testicular Failure

Mesenchymal stem cell (MSC) therapy offers a broad spectrum of treatment of different conditions, including male infertility. MSC therapies for azoospermia were studied in animal models and demonstrated good results. (Tamadon et al., 2019) Most of the studies were conducted using MSC derived from the adipose tissue, bone marrow, and umbilical cord. A clinical case study reported adipose derived MSC therapy for male patient with NOA, to restore spermatogenesis leading to the in vitro production of superior quality embryos. (Cassim & Mohamed, 2019)

These scientific approaches give us hope for its gradual introduction into clinical practice even though this type of treatment in humans is not clinically available yet. (Tamadon et al., 2019) MSCs derived from umbilical cord, adipose tissue, and the bone marrow appear more appropriate to recover the fertility due to better results. Although applying those methods into the human practice seems to be investigated before introducing this method into clinic. (Tamadon et al., 2019)

FUTURE PERSPECTIVES AND CONCLUSION

Stem cells-based therapies are not yet available for the public or in clinic applications though there are very promising ongoing clinical trials directed towards complementing or solving reproductive medicine issues. (Sharma et al., 2019) Research based on producing functional gametes in vitro or stimulating the resident stem cell population in the gonad to revive its function in vivo may contribute to infertility treatments and to improve reproductive function. Stem cell research based on iPSC lines and its introduction in male and female reproductive biological applications seems a breakthrough on the field. In addition, development of therapies from adult stem cells in the treatment of reproductive tract disorders (such as recovery the damaged endometrial lining) may complement the existing treatment methods.

Though the research studies reported in the literature or that are published until now are mostly based on animal models, there are few preliminary studies reports involving patients in experimental state. (Sharma et al., 2019) Also, gametes specification from different types of stem cells as well as propagation of stem cells from the reproductive system and from organized embryos have been successfully studied in some non-human mammals. (P. Y. Zhang et al., 2020)

The current state of the art in stem cell and reproductive medicine mandates that patients and physicians should be aware of unfounded claims of improvement of existing medical conditions because as of now, there is no solid evidences of its applications. (Vassena et al., 2015) In addition, limited studies are being performed on human gametes and embryos due to ethical concerns.

Production of artificial gametes signifies hope for basic research and clinical application in the field of reproductive biology and medicine. It will not only provide advantages in terms of therapeutic application but also will be a great platform for studying embryology. Germline stem cells produced from hESCs or from gonadal sources along with refined transplantation techniques and manipulation techniques may provide new opportunities to treat primary reproductive disorders. (Kubota & Brinster, 2006)

The stem cell field owes a great deal to the previous work conducted by embryologists and researchers devoted to reproductive medicine. The time is coming when this emerging field will pay off in the reproductive sciences by offering new avenues of understanding gametogenesis, early embryonic development and possible treatment approaches.

REFERENCES

Abdelaal, N. E., Tanga, B. M., Abdelgawad, M., Allam, S., Fathi, M., Saadeldin, I. M., Bang, S., & Cho, J. (2021). Cellular Therapy via Spermatogonial Stem Cells for Treating Impaired Spermatogenesis, Non-Obstructive Azoospermia. *Cells*, *10*(7), 1779. Advance online publication. DOI: 10.3390/cells10071779 PMID: 34359947

Ahmadian, S., Mahdipour, M., Pazhang, M., Sheshpari, S., Mobarak, H., Bedate, A. M., Rahbarghazi, R., & Nouri, M. (2020). Effectiveness of Stem Cell Therapy in the Treatment of Ovarian Disorders and Female Infertility: A Systematic Review. *Current Stem Cell Research & Therapy*, *15*(2), 173–186. DOI: 10.2174/1574888X14666191119122159 PMID: 31746298

Aleixandre-Benavent, R., Simon, C., & Fauser, B. C. J. M. (2015). Trends in clinical reproductive medicine research: 10 years of growth. *Fertility and Sterility*, *104*(1), 131–137.e5. DOI: 10.1016/j.fertnstert.2015.03.025 PMID: 25936232

Anway, M. D., & Skinner, M. K. (2008). Epigenetic programming of the germ line: Effects of endocrine disruptors on the development of transgenerational disease. *Reproductive Biomedicine Online*, *16*(1), 23–25. DOI: 10.1016/S1472-6483(10)60553-6 PMID: 18252044

Avarbock, M. R., Brinster, C. J., & Brinster, R. L. (1996). Reconstitution of spermatogenesis from frozen spermatogonial stem cells. *Nature Medicine*, *2*(6), 693–696. DOI: 10.1038/nm0696-693 PMID: 8640563

Bhartiya, D., Hinduja, I., Patel, H., & Bhilawadikar, R. (2014). Making gametes from pluripotent stem cells—A promising role for very small embryonic-like stem cells. *Reproductive biology and endocrinology. RB&E*, *12*(1). Advance online publication. DOI: 10.1186/1477-7827-12-114 PMID: 25421462

Bhartiya, D., Sriraman, K., Parte, S., & Patel, H. (2013). Ovarian stem cells: Absence of evidence is not evidence of absence. *Journal of Ovarian Research*, *6*(1), 65. Advance online publication. DOI: 10.1186/1757-2215-6-65 PMID: 24044496

Brinster, R. L., & Zimmermann, J. W. (1994). Spermatogenesis following male germ-cell transplantation. *Proceedings of the National Academy of Sciences of the United States of America*, *91*(24), 11298–11302. DOI: 10.1073/pnas.91.24.11298 PMID: 7972053

Cai, H., Xia, X., Wang, L., Liu, Y., He, Z., Guo, Q., & Xu, C. (2013). In vitro and in vivo differentiation of induced pluripotent stem cells into male germ cells. *Biochemical and Biophysical Research Communications*, *433*(3), 286–291. DOI: 10.1016/j.bbrc.2013.02.107 PMID: 23524261

Cassim, M. I., & Mohamed, T. (2019). *Case Report A Novel Therapy for the Treatment of Malefactor Infertility Due to Non-obstructive Azoospermia: A Case Report. 6*(1), 129-131.

Cervelló, I., Medrano, J. V., & Simón, C. (2016). Regenerative Medicine and Tissue Engineering in Reproductive Medicine: Future Clinical Applications in Human Infertility. *Translating Regenerative Medicine to the Clinic*, 139-151. DOI: 10.1016/B978-0-12-800548-4.00010-3

Cervelló, I., & Simón, C. (2009). Somatic stem cells in the endometrium. *Reproductive Sciences (Thousand Oaks, Calif.), 16*(2), 200–205. DOI: 10.1177/1933719108329955 PMID: 19208788

Clark, A. T., Bodnar, M. S., Fox, M., Rodriquez, R. T., Abeyta, M. J., Firpo, M. T., & Reijo Pera, R. A. (2004). Spontaneous differentiation of germ cells from human embryonic stem cells in vitro. *Human Molecular Genetics, 13*(7), 727–739. DOI: 10.1093/hmg/ddh088 PMID: 14962983

Cocuzza, M., Alvarenga, C., & Pagani, R. (2013). The epidemiology and etiology of azoospermia. *Clinics (Sao Paulo, Brazil), 68 Suppl 1*(Suppl 1), 15-26. DOI: 10.6061/clinics/2013(Sup01)03

Cyranoski, D. (2013). Stem cells: Egg engineers. *Nature, 500*(7463), 392–394. DOI: 10.1038/500392a PMID: 23969442

Definition of reproductive medicine—NCI Dictionary of Cancer Terms—NCI. (s. f.). Recuperado 19 de agosto de 2023, de https://www.cancer.gov/publications/dictionaries/cancer-terms/def/reproductive-medicine

Dissanayake, D. (2018). In Vitro Spermatogenesis; Past, Present, and Future. In *Spermatozoa—Facts and Perspectives*. InTech., DOI: 10.5772/intechopen.73505

Durruthy, J. D., Ramathal, C., Sukhwani, M., Fang, F., Cui, J., Orwig, K. E., & Reijo Pera, R. A. (2014). Fate of induced pluripotent stem cells following transplantation to murine seminiferous tubules. *Human Molecular Genetics, 23*(12), 3071–3084. DOI: 10.1093/hmg/ddu012 PMID: 24449759

Easley, C. A.IV, Phillips, B. T., McGuire, M. M., Barringer, J. M., Valli, H., Hermann, B. P., Simerly, C. R., Rajkovic, A., Miki, T., Orwig, K. E., & Schatten, G. P. (2012). Direct differentiation of human pluripotent stem cells into haploid spermatogenic cells. *Cell Reports, 2*(3), 440–446. DOI: 10.1016/j.celrep.2012.07.015 PMID: 22921399

Eggan, K., Jurga, S., Gosden, R., Min, I. M., & Wagers, A. J. (2006). Ovulated oocytes in adult mice derive from non-circulating germ cells. *Nature*, *441*(7097), 1109–1114. DOI: 10.1038/nature04929 PMID: 16799565

Eguizabal, C., Montserrat, N., Veiga, A., & Belmonte, J. I. (2013). Dedifferentiation, transdifferentiation, and reprogramming: Future directions in regenerative medicine. *Seminars in Reproductive Medicine*, *31*(1), 82–94. DOI: 10.1055/s-0032-1331802 PMID: 23329641

Fang, F., Li, Z., Zhao, Q., Li, H., & Xiong, C. (2018). Human induced pluripotent stem cells and male infertility: An overview of current progress and perspectives. *Human Reproduction (Oxford, England)*, *33*(2), 188–195. DOI: 10.1093/humrep/dex369 PMID: 29315416

Feng, P., Li, P., & Tan, J. (2019). Human Menstrual Blood-Derived Stromal Cells Promote Recovery of Premature Ovarian Insufficiency Via Regulating the ECM-Dependent FAK/AKT Signaling. *Stem Cell Reviews and Reports*, *15*(2), 241–255. DOI: 10.1007/s12015-018-9867-0 PMID: 30560467

Ferlin, A., Raicu, F., Gatta, V., Zuccarello, D., Palka, G., & Foresta, C. (2007). Male infertility: Role of genetic background. *Reproductive Biomedicine Online*, *14*(6), 734–745. DOI: 10.1016/S1472-6483(10)60677-3 PMID: 17579990

Forbes, C. M., Flannigan, R., & Schlegel, P. N. (2017). Spermatogonial stem cell transplantation and male infertility: Current status and future directions. *Arab Journal of Urology*, *16*(1), 171–180. DOI: 10.1016/j.aju.2017.11.015 PMID: 29713548

Gargett, C. E. (2010). Stem cells in human reproduction. *Reproduction (Cambridge, England)*, *140*(1), 1–2. DOI: 10.1530/REP-10-0225 PMID: 20584995

Gargett, C. E., & Ye, L. (2012). Endometrial reconstruction from stem cells. *Fertility and Sterility*, *98*(1), 11–20. DOI: 10.1016/j.fertnstert.2012.05.004 PMID: 22657248

Gauthier-Fisher, A., Kauffman, A., & Librach, C. L. (2020). Potential use of stem cells for fertility preservation. *Andrology*, *8*(4), 862–878. DOI: 10.1111/andr.12713 PMID: 31560823

Gil-Sanchis, C., Cervelló, I., Khurana, S., Faus, A., Verfaillie, C., & Simón, C. (2015). Contribution of different bone marrow-derived cell types in endometrial regeneration using an irradiated murine model. *Fertility and Sterility*, *103*(6), 1596–1605.e1. DOI: 10.1016/j.fertnstert.2015.02.030 PMID: 25813284

Gupta, S., Lodha, P., Karthick, M. S., & Tandulwadkar, S. (2018). Role of Autologous Bone Marrow-Derived Stem Cell Therapy for Follicular Recruitment in Premature Ovarian Insufficiency: Review of Literature and a Case Report of World's First Baby with Ovarian Autologous Stem Cell Therapy in a Perimenopausal Woman of Age 45 Year. *Journal of Human Reproductive Sciences*, *11*(2), 125–130. DOI: 10.4103/jhrs.JHRS_57_18 PMID: 30158807

Hayashi, K., Ogushi, S., Kurimoto, K., Shimamoto, S., Ohta, H., & Saitou, M. (2012). Offspring from oocytes derived from in vitro primordial germ cell-like cells in mice. *Science*, *338*(6109), 971–975. DOI: 10.1126/science.1226889 PMID: 23042295

Hayashi, K., Ohta, H., Kurimoto, K., Aramaki, S., & Saitou, M. (2011). Reconstitution of the mouse germ cell specification pathway in culture by pluripotent stem cells. *Cell*, *146*(4), 519–532. DOI: 10.1016/j.cell.2011.06.052 PMID: 21820164

Hayashi, K., & Saitou, M. (2013). Generation of eggs from mouse embryonic stem cells and induced pluripotent stem cells. *Nature Protocols*, *8*(8), 1513–1524. DOI: 10.1038/nprot.2013.090 PMID: 23845963

Hayashi, Y., Saitou, M., & Yamanaka, S. (2012). Germline development from human pluripotent stem cells toward disease modeling of infertility. *Fertility and Sterility*, *97*(6), 1250–1259. DOI: 10.1016/j.fertnstert.2012.04.037 PMID: 22656305

Hermann, B. P., Sukhwani, M., Winkler, F., Pascarella, J. N., Peters, K. A., Sheng, Y., Valli, H., Rodriguez, M., Ezzelarab, M., Dargo, G., Peterson, K., Masterson, K., Ramsey, C., Ward, T., Lienesch, M., Volk, A., Cooper, D. K., Thomson, A. W., Kiss, J. E., & Orwig, K. E. (2012). Spermatogonial stem cell transplantation into rhesus testes regenerates spermatogenesis producing functional sperm. *Cell Stem Cell*, *11*(5), 715–726. DOI: 10.1016/j.stem.2012.07.017 PMID: 23122294

Hikabe, O., Hamazaki, N., Nagamatsu, G., Obata, Y., Hirao, Y., Hamada, N., Shimamoto, S., Imamura, T., Nakashima, K., Saitou, M., & Hayashi, K. (2016). Reconstitution in vitro of the entire cycle of the mouse female germ line. *Nature*, *539*(7628), 299–303. DOI: 10.1038/nature20104 PMID: 27750280

Hübner, K., Fuhrmann, G., Christenson, L. K., Kehler, J., Reinbold, R., De La Fuente, R., Wood, J., Strauss, J. F.III, Boiani, M., & Schöler, H. R. (2003). Derivation of oocytes from mouse embryonic stem cells. *Science*, *300*(5623), 1251–1256. DOI: 10.1126/science.1083452 PMID: 12730498

Hudson, N., Culley, L., Rapport, F., Johnson, M., & Bharadwaj, A. (2009). «Public» perceptions of gamete donation: A research review. *Public Understanding of Science (Bristol, England)*, *18*(1), 61–77. DOI: 10.1177/0963662507078396 PMID: 19579535

Ibtisham, F., & Honaramooz, A. (2020). Spermatogonial Stem Cells for In Vitro Spermatogenesis and In Vivo Restoration of Fertility. *Cells*, *9*(3), 745. Advance online publication. DOI: 10.3390/cells9030745 PMID: 32197440

Izadyar, F., Den Ouden, K., Stout, T. A. E., Stout, J., Coret, J., Lankveld, D. P. K., Spoormakers, T. J. P., Colenbrander, B., Oldenbroek, J. K., Van der Ploeg, K. D., Woelders, H., Kal, H. B., & De Rooij, D. G. (2003). Autologous and homologous transplantation of bovine spermatogonial stem cells. *Reproduction (Cambridge, England)*, *126*(6), 765–774. DOI: 10.1530/rep.0.1260765 PMID: 14748695

Johnson, E. K., Finlayson, C., Rowell, E. E., Gosiengfiao, Y., Pavone, M. E., Lockart, B., Orwig, K. E., Brannigan, R. E., & Woodruff, T. K. (2017). Fertility Preservation for Pediatric Patients: Current State and Future Possibilities. *The Journal of Urology*, *198*(1), 186–194. DOI: 10.1016/j.juro.2016.09.159 PMID: 28189577

Johnson, J., Bagley, J., Skaznik-Wikiel, M., Lee, H. J., Adams, G. B., Niikura, Y., Tschudy, K. S., Tilly, J. C., Cortes, M. L., Forkert, R., Spitzer, T., Iacomini, J., Scadden, D. T., & Tilly, J. L. (2005). Oocyte generation in adult mammalian ovaries by putative germ cells in bone marrow and peripheral blood. *Cell*, *122*(2), 303–315. DOI: 10.1016/j.cell.2005.06.031 PMID: 16051153

Johnson, J., Canning, J., Kaneko, T., Pru, J. K., & Tilly, J. L. (2004). Germline stem cells and follicular renewal in the postnatal mammalian ovary. *Nature*, *428*(6979), 145–150. DOI: 10.1038/nature02316 PMID: 15014492

Jung, D., Xiong, J., Ye, M., Qin, X., Li, L., Cheng, S., Luo, M., Peng, J., Dong, J., Tang, F., Shen, W., Matzuk, M. M., & Kee, K. (2017). In vitro differentiation of human embryonic stem cells into ovarian follicle-like cells. *Nature Communications*, *8*(1), 15680. Advance online publication. DOI: 10.1038/ncomms15680 PMID: 28604658

Kanatsu-Shinohara, M., Ogonuki, N., Inoue, K., Miki, H., Ogura, A., Toyokuni, S., & Shinohara, T. (2003). Long-term proliferation in culture and germline transmission of mouse male germline stem cells. *Biology of Reproduction*, *69*(2), 612–616. DOI: 10.1095/biolreprod.103.017012 PMID: 12700182

Kee, K., Angeles, V. T., Flores, M., Nguyen, H. N., & Reijo Pera, R. A. (2009). Human DAZL, DAZ and BOULE genes modulate primordial germ-cell and haploid gamete formation. *Nature*, *462*(7270), 222–225. DOI: 10.1038/nature08562 PMID: 19865085

Kharazi, U., & Badalzadeh, R. (2020). A review on the stem cell therapy and an introduction to exosomes as a new tool in reproductive medicine. *Reproductive Biology*, *20*(4), 447–459. DOI: 10.1016/j.repbio.2020.07.002 PMID: 32900639

Kubota, H., & Brinster, R. L. (2006). Technology insight: In vitro culture of spermatogonial stem cells and their potential therapeutic uses. *Nature Clinical Practice. Endocrinology & Metabolism*, *2*(2), 99–108. DOI: 10.1038/ncpendmet0098 PMID: 16932264

Lacham-Kaplan, O., Chy, H., & Trounson, A. (2006). Testicular cell conditioned medium supports differentiation of embryonic stem cells into ovarian structures containing oocytes. *Stem Cells (Dayton, Ohio)*, *24*(2), 266–273. DOI: 10.1634/stemcells.2005-0204 PMID: 16109761

Lee, H. J., Selesniemi, K., Niikura, Y., Niikura, T., Klein, R., Dombkowski, D. M., & Tilly, J. L. (2007). Bone marrow transplantation generates immature oocytes and rescues long-term fertility in a preclinical mouse model of chemotherapy-induced premature ovarian failure. *Journal of Clinical Oncology*, *25*(22), 3198–3204. DOI: 10.1200/JCO.2006.10.3028 PMID: 17664466

Lee, Y., & Kang, E. (2019). Stem cells and reproduction. *BMB Reports*, *52*(8), 482–489. DOI: 10.5483/BMBRep.2019.52.8.141 PMID: 31234956

Lei, L., & Spradling, A. C. (2013). Female mice lack adult germ-line stem cells but sustain oogenesis using stable primordial follicles. *Proceedings of the National Academy of Sciences of the United States of America*, *110*(21), 8585–8590. DOI: 10.1073/pnas.1306189110 PMID: 23630252

Li, L., & Xie, T. (2005). Stem cell niche: Structure and function. *Annual Review of Cell and Developmental Biology*, *21*(1), 605–631. DOI: 10.1146/annurev.cellbio.21.012704.131525 PMID: 16212509

Li, P., Hu, H., Yang, S., Tian, R., Zhang, Z., Zhang, W., Ma, M., Zhu, Y., Guo, X., Huang, Y., He, Z., & Li, Z. (2013). Differentiation of induced pluripotent stem cells into male germ cells in vitro through embryoid body formation and retinoic acid or testosterone induction. *BioMed Research International*, *2013*, 1–9. Advance online publication. DOI: 10.1155/2013/608728 PMID: 23509752

McGee, E. A., & Hsueh, A. J. W. (2000). Initial and cyclic recruitment of ovarian follicles. *Endocrine Reviews*, *21*(2), 200–214. DOI: 10.1210/edrv.21.2.0394 PMID: 10782364

Medrano, J. V., Ramathal, C., Nguyen, H. A. N., Simon, C., & Reijo Pera, R. A. (2012). Divergent RNA-binding proteins, DAZL and VASA, induce meiotic progression in human germ cells derived in vitro. *Stem Cells (Dayton, Ohio)*, *30*(3), 441–451. DOI: 10.1002/stem.1012 PMID: 22162380

Metcalfe, S. M. (2020). Mesenchymal stem cells and management of COVID-19 pneumonia. *Medicine in Drug Discovery*, *5*, 100019. DOI: 10.1016/j.medidd.2020.100019 PMID: 32296777

Moore, H., Udayashankar, R., & Aflatoonian, B. (2008). Stem cells for reproductive medicine. *Molecular and Cellular Endocrinology*, *288*(1-2), 104–110. DOI: 10.1016/j.mce.2008.02.019 PMID: 18406514

Na, J., & Kim, G. J. (2020). Recent trends in stem cell therapy for premature ovarian insufficiency and its therapeutic potential: A review. *Journal of Ovarian Research*, *13*(1), 74. Advance online publication. DOI: 10.1186/s13048-020-00671-2 PMID: 32576209

Nakaki, F., Hayashi, K., Ohta, H., Kurimoto, K., Yabuta, Y., & Saitou, M. (2013). Induction of mouse germ-cell fate by transcription factors in vitro. *Nature*, *501*(7466), 222–226. DOI: 10.1038/nature12417 PMID: 23913270

Nazm Bojnordi, M. (2017). The applications and recovery outcome of spermatogonia stem cells in regenerative medicine. *Middle East Fertility Society Journal*, *22*(4), 246–250. DOI: 10.1016/j.mefs.2017.05.008

Nicholas, C. R., Haston, K. M., Grewall, A. K., Longacre, T. A., & Reijo Pera, R. A. (2009). Transplantation directs oocyte maturation from embryonic stem cells and provides a therapeutic strategy for female infertility. *Human Molecular Genetics*, *18*(22), 4376–4389. DOI: 10.1093/hmg/ddp393 PMID: 19696121

Nikolic, A., Volarevic, V., Armstrong, L., Lako, M., & Stojkovic, M. (2016). Primordial Germ Cells: Current Knowledge and Perspectives. *Stem Cells International*, *2016*(1), 1741072. Advance online publication. DOI: 10.1155/2016/1741072 PMID: 26635880

Oatley, J. M., & Brinster, R. L. (2008). Regulation of spermatogonial stem cell self-renewal in mammals. *Annual Review of Cell and Developmental Biology*, *24*(1), 263–286. DOI: 10.1146/annurev.cellbio.24.110707.175355 PMID: 18588486

Panula, S., Kurek, M., Kumar, P., Albalushi, H., Sánchez, S. P., Damdimopoulou, P., Olofsson, J. I., Hovatta, O., Lanner, F., & Stukenborg, J. B. (2019). Human induced pluripotent stem cells from two azoospermic patients with Klinefelter syndrome show similar X chromosome inactivation behavior to female pluripotent stem cells. *Human Reproduction (Oxford, England)*, *34*(11), 2297–2310. DOI: 10.1093/humrep/dez134 PMID: 31743397

Panula, S., Medrano, J. V., Kee, K., Bergström, R., Nguyen, H. N., Byers, B., Wilson, K. D., Wu, J. C., Simon, C., Hovatta, O., & Pera, R. A. R. (2011). Human germ cell differentiation from fetal- and adult-derived induced pluripotent stem cells. *Human Molecular Genetics*, *20*(4), 752–762. DOI: 10.1093/hmg/ddq520 PMID: 21131292

Park, T. S., Galic, Z., Conway, A. E., Lindgren, A., Van Handel, B. J., Magnusson, M., Richter, L., Teitell, M. A., Mikkola, H. K. A., Lowry, W. E., Plath, K., & Clark, A. T. (2009). Derivation of primordial germ cells from human embryonic and induced pluripotent stem cells is significantly improved by coculture with human fetal gonadal cells. *Stem Cells (Dayton, Ohio)*, *27*(4), 783–795. DOI: 10.1002/stem.13 PMID: 19350678

Parreira, G. G., Ogawa, T., Avarbock, M. R., França, L. R., Brinster, R. L., & Russell, L. D. (1998). Development of germ cell transplants in mice. *Biology of Reproduction*, *59*(6), 1360–1370. DOI: 10.1095/biolreprod59.6.1360 PMID: 9828179

Radhakrishnan, G. (2017). Stem cells—The new agents in infertility treatment: The light at the end of the tunnel? *Fertility Science and Research*, *4*(2), 70. DOI: 10.4103/fsr.fsr_16_18

Rossant, J. (2001). Stem cells from the Mammalian blastocyst. *Stem Cells (Dayton, Ohio)*, *19*(6), 477–482. DOI: 10.1634/stemcells.19-6-477 PMID: 11713338

Sadri-Ardekani, H., Mizrak, S. C., Van Daalen, S. K. M., Korver, C. M., Roepers-Gajadien, H. L., Koruji, M., Hovingh, S., De Reijke, T. M., De La Rosette, J. J. M. C. H., Van Der Veen, F., De Rooij, D. G., Repping, S., & Van Pelt, A. M. M. (2009). Propagation of human spermatogonial stem cells in vitro. *Journal of the American Medical Association*, *302*(19), 2127–2134. DOI: 10.1001/jama.2009.1689 PMID: 19920237

Samsonraj, R. M., Raghunath, M., Nurcombe, V., Hui, J. H., van Wijnen, A. J., & Cool, S. M. (2017). Concise Review: Multifaceted Characterization of Human Mesenchymal Stem Cells for Use in Regenerative Medicine. *Stem Cells Translational Medicine*, *6*(12), 2173–2185. DOI: 10.1002/sctm.17-0129 PMID: 29076267

Sato, T., Katagiri, K., Gohbara, A., Inoue, K., Ogonuki, N., Ogura, A., Kubota, Y., & Ogawa, T. (2011). In vitro production of functional sperm in cultured neonatal mouse testes. *Nature*, *471*(7339), 504–508. DOI: 10.1038/nature09850 PMID: 21430778

Schlatt, S., Foppiani, L., Rolf, C., Weinbauer, G. F., & Nieschlag, E. (2002). Germ cell transplantation into X-irradiated monkey testes. *Human Reproduction (Oxford, England)*, *17*(1), 55–62. DOI: 10.1093/humrep/17.1.55 PMID: 11756362

Shah, J. S., Sabouni, R., Cayton Vaught, K. C., Owen, C. M., Albertini, D. F., & Segars, J. H. (2018). Biomechanics and mechanical signaling in the ovary: A systematic review. *Journal of Assisted Reproduction and Genetics*, *35*(7), 1135–1148. DOI: 10.1007/s10815-018-1180-y PMID: 29691711

Sharma, S., Wistuba, J., Pock, T., Schlatt, S., & Neuhaus, N. (2019). Spermatogonial stem cells: Updates from specification to clinical relevance. *Human Reproduction Update*, *25*(3), 275–297. DOI: 10.1093/humupd/dmz006 PMID: 30810745

Simoni, M., & Taylor, H. S. (2018). Therapeutic strategies involving uterine stem cells in reproductive medicine. *Current Opinion in Obstetrics & Gynecology*, *30*(3), 209–216. DOI: 10.1097/GCO.0000000000000457 PMID: 29652725

Song, H.-W., & Wilkinson, M. F. (2012). In vitro spermatogenesis: A long journey to get tails. *Spermatogenesis*, *2*(4), 238–244. DOI: 10.4161/spmg.22069 PMID: 23248764

Stimpfel, M., Skutella, T., Cvjeticanin, B., Meznaric, M., Dovc, P., Novakovic, S., Cerkovnik, P., Vrtacnik-Bokal, E., & Virant-Klun, I. (2013). Isolation, characterization and differentiation of cells expressing pluripotent/multipotent markers from adult human ovaries. *Cell and Tissue Research*, *354*(2), 593–607. DOI: 10.1007/s00441-013-1677-8 PMID: 23820736

Tamadon, A., Zhan-byrbekuly, U., Kairgaliyev, I., Khoradmehr, A., Tamadon, A., Zhan-byrbekuly, U., Kairgaliyev, I., & Khoradmehr, A. (2019). *Mesenchymal Stem Cell Therapy of Male Infertility*. Male Reproductive Health., DOI: 10.5772/intechopen.88343

Tilgner, K., Atkinson, S. P., Golebiewska, A., Stojković, M., Lako, M., & Armstrong, L. (2008). Isolation of primordial germ cells from differentiating human embryonic stem cells. *Stem Cells (Dayton, Ohio)*, *26*(12), 3075–3085. DOI: 10.1634/stemcells.2008-0289 PMID: 18802037

Tilly, J. L., Niikura, Y., & Rueda, B. R. (2009). The current status of evidence for and against postnatal oogenesis in mammals: A case of ovarian optimism versus pessimism? *Biology of Reproduction*, *80*(1), 2–12. DOI: 10.1095/biolreprod.108.069088 PMID: 18753611

Tournaye, H., & Goossens, E. (2011). Spermatogonial stem cells: What does the future hold? *Facts, Views & Vision in ObGyn*, *3*(1), 36–40. PMID: 24753846

Vassena, R., Eguizabal, C., Heindryckx, B., Sermon, K., Simon, C., Van Pelt, A. M. M., Veiga, A., & Zambelli, F. (2015). Stem cells in reproductive medicine: Ready for the patient? *Human Reproduction (Oxford, England)*, *30*(9), 2014–2021. DOI: 10.1093/humrep/dev181 PMID: 26202914

Vij, S. C., Sabanegh, E.Jr, & Agarwal, A. (2018). Biological therapy for non-obstructive azoospermia. *Expert Opinion on Biological Therapy*, *18*(1), 19–23. DOI: 10.1080/14712598.2018.1380622 PMID: 28927307

Volarevic, V., Bojic, S., Nurkovic, J., Volarevic, A., Ljujic, B., Arsenijevic, N., Lako, M., & Stojkovic, M. (2014). Stem cells as new agents for the treatment of infertility: Current and future perspectives and challenges. *BioMed Research International*, *2014*, 1–8. Advance online publication. DOI: 10.1155/2014/507234 PMID: 24826378

Wang, J., & Sauer, M. V. (2006). In vitro fertilization (IVF): A review of 3 decades of clinical innovation and technological advancement. *Therapeutics and Clinical Risk Management*, *2*(4), 355–364. DOI: 10.2147/tcrm.2006.2.4.355 PMID: 18360648

Wen, L., & Tang, F. (2016). Single-cell sequencing in stem cell biology. *Genome Biology*, *17*(1), 71. Advance online publication. DOI: 10.1186/s13059-016-0941-0 PMID: 27083874

West, J. A., Park, I. H., Daley, G. Q., & Geijsen, N. (2006). In vitro generation of germ cells from murine embryonic stem cells. *Nature Protocols*, *1*(4), 2026–2036. DOI: 10.1038/nprot.2006.303 PMID: 17487192

White, Y. A. R., Woods, D. C., Takai, Y., Ishihara, O., Seki, H., & Tilly, J. L. (2012). Oocyte formation by mitotically active germ cells purified from ovaries of reproductive-age women. *Nature Medicine*, *18*(3), 413–421. DOI: 10.1038/nm.2669 PMID: 22366948

Wu, X., Goodyear, S. M., Abramowitz, L. K., Bartolomei, M. S., Tobias, J. W., Avarbock, M. R., & Brinster, R. L. (2012). Fertile offspring derived from mouse spermatogonial stem cells cryopreserved for more than 14 years. *Human Reproduction (Oxford, England)*, *27*(5), 1249–1259. DOI: 10.1093/humrep/des077 PMID: 22416011

Ye, H., Zheng, T., Li, W., Li, X., Fu, X., Huang, Y., Hu, C., Li, J., Huang, J., Liu, Z., Zheng, L., & Zheng, Y. (2017). Ovarian Stem Cell Nests in Reproduction and Ovarian Aging. *Cellular physiology and biochemistry : international journal of experimental cellular physiology, biochemistry, and pharmacology, 43*(5), 1917-1925. DOI: 10.1159/000484114

Zakrzewski, W., Dobrzyński, M., Szymonowicz, M., & Rybak, Z. (2019). Stem cells: Past, present, and future. *Stem Cell Research & Therapy*, *10*(1), 68. Advance online publication. DOI: 10.1186/s13287-019-1165-5 PMID: 30808416

Zegers-Hochschild, F., Adamson, G. D., De Mouzon, J., Ishihara, O., Mansour, R., Nygren, K., Sullivan, E., & Van Der Poel, S. (2009). The International Committee for Monitoring Assisted Reproductive Technology (ICMART) and the World Health Organization (WHO) Revised Glossary on ART Terminology, 2009. *Human Reproduction (Oxford, England)*, *24*(11), 2683–2687. DOI: 10.1093/humrep/dep343 PMID: 19801627

Zhang, H., Zheng, W., Shen, Y., Adhikari, D., Ueno, H., & Liu, K. (2012). Experimental evidence showing that no mitotically active female germline progenitors exist in postnatal mouse ovaries. *Proceedings of the National Academy of Sciences of the United States of America*, *109*(31), 12580–12585. DOI: 10.1073/pnas.1206600109 PMID: 22778414

Zhang, P. Y., Fan, Y., Tan, T., & Yu, Y. (2020). Generation of Artificial Gamete and Embryo From Stem Cells in Reproductive Medicine. *Frontiers in Bioengineering and Biotechnology*, *8*, 781. Advance online publication. DOI: 10.3389/fbioe.2020.00781 PMID: 32793569

Zhao, Y., Luo, Q., Zhang, X., Qin, Y., Hao, J., Kong, D., Wang, H., Li, G., Gu, X., & Wang, H. (2020). Clinical Efficacy and Safety of Stem Cell-Based Therapy in Treating Asherman Syndrome: A System Review and Meta-Analysis. *Stem Cells International*, *2020*, 1–11. Advance online publication. DOI: 10.1155/2020/8820538 PMID: 33414830

Zhao, Y. X., Chen, S. R., Su, P. P., Huang, F. H., Shi, Y. C., Shi, Q. Y., & Lin, S. (2019). Using Mesenchymal Stem Cells to Treat Female Infertility: An Update on Female Reproductive Diseases. *Stem Cells International*, *2019*, 1–10. Advance online publication. DOI: 10.1155/2019/9071720 PMID: 31885630

Zhou, Q., Wang, M., Yuan, Y., Wang, X., Fu, R., Wan, H., Xie, M., Liu, M., Guo, X., Zheng, Y., Feng, G., Shi, Q., Zhao, X. Y., Sha, J., & Zhou, Q. (2016). Complete Meiosis from Embryonic Stem Cell-Derived Germ Cells In Vitro. *Cell Stem Cell*, *18*(3), 330–340. DOI: 10.1016/j.stem.2016.01.017 PMID: 26923202

Zou, K., Yuan, Z., Yang, Z., Luo, H., Sun, K., Zhou, L., Xiang, J., Shi, L., Yu, Q., Zhang, Y., Hou, R., & Wu, J. (2009). Production of offspring from a germline stem cell line derived from neonatal ovaries. *Nature Cell Biology*, *11*(5), 631–636. DOI: 10.1038/ncb1869 PMID: 19363485

KEY TERMS AND DEFINITIONS

Asherman's syndrome- an acquired gynecological disorder of the uterus characterized by the bonding of scar tissue that lines the walls of the uterus: which decreases the volume of the uterine cavity.

Autoimmune disorders: - when the body's natural defense system can't tell the difference between your own cells and foreign cells, causing the body to mistakenly attack normal or ones' own cells.

Ex vivo- outside of the living body:

Gonadotoxic- temporary or permanent damage to ovaries or testes after exposure to certain substances or drugs:

Intracytoplasmic sperm injection (ICSI): an in vitro fertilization procedure in which a single sperm is injected directly into the cytoplasm of the egg

In vitro: - when research or work is performed outside of a living organism or in a in the test tube

In vitro fertilization (IVF): - complex series of procedures in assisted reproduction technology that involves the removal of eggs from ovary, then it is fertilizedwith sperm to form embryos in a petri-dish. The embryo is then transferred into the uterus in the hopes of achieving a pregnancy.

In vivo: - when research or work is done with or within an entire, living organism

Leydig cells: - testicular cells responsible for the biosynthesis and secretion of androgens.

Mitosis- type of cell division where a single cell divides into two identical daughter cells:

Meiosis: - type of cell division that reduces the number of chromosomes in the parent cell by half and produces four gamete cells

Oocyte- an immature egg (immature ovum):

Oogenesis: - process in which the primary egg cell (or ovum) becomes a mature ovum in the ovaries

Primordial follicle- small: dormant follicle with a single layer of flattened granulosa cells

Primary follicle: -large, mitotic follicle with cuboidal granulosa cells

Primary oocyte- oocyte that arises from the oogonium via the process of oogenesis: and gives rise to secondary oocyte and polar body after first meiotic division

Sertoli cells: - "nurse" cell of the testicles that is part of a seminiferous tubule and helps in the process of spermatogenesis

Spermatozoa (sperm): - male gamete

Spermatogenesis: - process by which haploid spermatozoa develop from spermatogonial stem cells in the seminiferous tubules of the testis

Chapter 7
Stem Cell Therapy for Renal Diseases

ABSTRACT

Acute and chronic kidney diseases pose significant global health challenges. While conventional treatments offer limited efficacy, SC therapy has emerged as a promising regenerative approach This chapter delves into the potential of stem cells in treating AKI and CKD. We explore various stem cell types, including embryonic stem cells, induced pluripotent stem cells, and mesenchymal stem cells, and their differentiation capacities toward renal lineages. The chapter discusses the mechanisms of action, focusing on both direct cell replacement and paracrine effects, as well as the immunomodulatory properties of stem cells. Preclinical studies demonstrating the efficacy of stem cell therapy in AKI and CKD models are reviewed, highlighting challenges and limitations. the chapter provides an outlook on the future of stem cell-based therapies and gene editing, including the need for standardized protocols, safety considerations, and the potential for combination therapies, to optimize experimental interventions

BACKGROUND

Kidney diseases encompass a diverse spectrum of disorders impairing kidney structure and function, presenting with varying clinical manifestations and progressions. Categorized as acute or chronic based on duration (less than or greater than three months, respectively).(Lazzeri, Romagnani, et al., 2015; National Kidney Foundation, 2002)

The introduction of standardized definitions and staging criteria for chronic kidney disease (CKD) in 2002 unveiled its widespread prevalence, affecting over 10% of the world's population and increasing dramatically with age, surpassing 20% in individuals over 60 and reaching 35% in those over 70. (Coresh et al., 2007;

DOI: 10.4018/979-8-3693-8829-7.ch007

James et al., 2010; Levey & Coresh, 2012) While therapeutic interventions targeting urinary protein excretion can effectively slow the progression of certain chronic nephropathies, kidney failure remains the ultimate and devastating consequence of CKD. Dialysis and transplantation currently represent the only treatment options, but these are hindered by organ shortages, rejection risks, and exorbitant costs. (Lazzeri, Romagnani, et al., 2015; Ruggenenti et al., 2008).

Kidney disease is a burgeoning global health crisis, driven by an aging population and the escalating prevalence of cardiovascular disease, hypertension, and diabetes. The emergence of SC as potential therapeutics has ignited intense research into mitigating the disease burden, and could offer a limitless supply of kidney progenitor cells and can be differentiated into three-dimensional kidney organoids, providing valuable models for studying kidney development and disease.

The traditional view of the human kidney as a static, non-regenerative organ has been challenged by recent findings indicating the kidney's capacity for self-repair through resident progenitor cell populations.

While the SC is promising new treatment for kidney disease documented in many animal studies, and a significant progress has been made demonstrating the potential of SC to restore renal function in preclinical models of acute kidney injury (AKI) and CKD, contradictory findings within the literature hinder the translation of these promising therapies into clinical practice.

These cells exert their beneficial effects by modulating fibrotic and inflammatory pathways and promoting endogenous regenerative processes. Despite these promising preclinical findings, the translation of SC therapies to the clinic has been hindered. A critical re-evaluation of the existing literature, with a meticulous focus on methodological details, is necessary to address this translational gap.

One of our groups already demonstrate the benefit of SC treatment in patients with AKI. During the COVID 19 pandemic, an autologous peripheral blood non-hematopoietic enriched SC cocktail was administered to severe COVID-19 cases, dividing patient into treated and control group, finding a significantly lower mortality rate (20% vs. 30%, respectively) and reduced incidence of sepsis (25% vs. 35%, p=0.0095) in the treated group, with an additionally lower AKI prevalence (25% vs. 35%), and a manifest reduction in inflammatory markers (C-Reactive Protein and IL-6) at 25 days post-treatment and were more likely to achieve normal neutrophil-to-lymphocyte ratio compared to controls (p=0.0108). (Torres-Zambrano et al., 2021)

One study from Wuhan, China, found that AKI was the most common extrapulmonary complication in critically ill patients admitted in the ICU, becoming more common than expected and could develop fatal complications associated with higher mortality (Yang et al., 2020). These findings suggest that the proposed SC therapy may promote healing and accelerate recovery in severe COVID-19 patients, potentially through an anti-inflammatory mechanism.

However, the staging of CKD in animals is less precisely defined than in humans, and therapy is rarely initiated at a late stage comparable to pre-dialysis in clinical practice. Furthermore, preventive therapy in animal models is often implemented immediately after or even prior to renal ablation or toxin administration, well before the onset of glomerulosclerosis or interstitial fibrosis. Consequently, the applicability of these findings to the clinical setting may be somewhat limited. Nonetheless, it is conceivable that in clinical trials, a highly precise categorization of CKD stage might not be necessary to observe significant effects of cell-based therapy (Papazova et al., 2015).

It is imperative that both human and animal studies adhere to the highest methodological standards. While randomization and blinding during inclusion may often be impractical in experimental settings, blinding for the analysis of functional and structural parameters is readily achievable.

Moreover, there is no justification for neglecting to report all pertinent animal characteristics, as well as the number of dropouts due to technical failures or premature (non-scheduled) death or sacrifice. The omission of such data is prevalent in animal studies, including numerous studies within our meta-analysis and as documented in other meta-analyses (van der Spoel et al., 2011; Wever et al., 2012). Authors must disclose these details, as they are likely to influence outcome parameters.

PROPOSED TYPE OF CELLS

A variety of progenitor cells have been isolated from mouse and rat kidneys using slow-cycling properties or SC markers. These cells have shown beneficial effects in preclinical models of acute and chronic kidney injury (Bruno et al., 2014). Furthermore, multiple studies have investigated resident MSC populations within the glomeruli and renal interstitium.

The efficacy of this therapy hinges on SCs' inherent ability to differentiate, organize, and integrate into damaged tissues, thereby restoring function, or alternatively, on their capacity to secrete renoprotective factors.

A variety of stem cell types, including mesenchymal SC MSCs) (Eirin & Lerman, 2014; Kuppe & Kramann, 2016), adipose-derived stem cells (ADSCs) (Wankhade et al., 2016), bone marrow mesenchymal SCs (BMSC), endothelial progenitor cells (EPCs) and renal progenitor cells (RPs), embryonic SCs (ESCs), induced pluripotent SCs (iPSCs), amniotic fluid stem cells (AFSCs) (Baulier et al., 2014; Sedrakyan et al., 2012), placenta (PDSCs), umbilical cord (UCSCs), and blood and renal progenitors (RPs) (Bussolati & Camussi, 2015; Pleniceanu et al., 2018), have demonstrated the capacity to stimulate renal repair in preclinical models of AKI and CKD.

The way of inducing AKI in the studies, could be by using various insults such as cisplatin, glycerol, sepsis, or ischemia-reperfusion. MSCs have shown protective and regenerative properties (Morigi et al., 2004). Experimental evidence supports the role of MSCs in reducing fibrosis and renal dysfunction in CKD. Notably, MSCs have attenuated CKD progression in 5/6 nephrectomy models and ameliorated glomerular pathology in diabetic nephropathy (Ezquer et al., 2008; R. H. Lee et al., 2006; Semedo et al., 2009)

Mesenchymal Stem Cells (MSCs)

MSCs exert potent immunomodulatory effects. They produce prostaglandin E2, which downregulates T-cell function by suppressing pro-inflammatory cytokines TNF-α and IFN-γ while promoting anti-inflammatory IL-10. This anti-inflammatory milieu is supported by elevated IL-10 levels in kidney tissues of MSC-treated animals following ischemia-reperfusion (I/R) injury (Semedo et al., 2007). MSC-derived prostaglandin E2 also enhances IL-10 production in monocytes/macrophages via EP2 and EP4 receptors, contributing to the overall immunosuppressive effects. Intravenous MSC administration successfully modulates the immune response in sepsis models, improving survival (Németh et al., 2009).

Furthermore, exposure to proinflammatory stimuli amplifies MSC immunosuppressive properties, leading to protection against natural killer cell-mediated cytotoxicity, inhibition of T cell proliferation, and induction of hepatocyte growth factor, TGF-β, and indoleamine 2,3-dioxygenase (Meisel et al., 2004; Shi et al., 2011; Spaggiari et al., 2006; Stagg, 2007).

MSC therapy demonstrated reductions in BUN and improvements in glomerulosclerosis and interstitial fibrosis. MSCs also ameliorated glomerular damage induced by Adriamycin, reducing proteinuria, inflammation, and microvascular rarefaction (S.-R. Lee et al., 2010; Morigi et al., 2004)

Renal Progenitors (RPS)

Diverse populations of both human and murine RPs, derived from embryonic and adult kidneys, have been explored for renal regeneration. Of particular interest, human CD133+ cells isolated from infant or adult kidneys have demonstrated protective effects in murine models of acute tubular and glomerular injury (Bussolati & Camussi, 2015). Administration of these cells has been shown to mitigate fibrosis following AKI and stimulate erythropoietin production (Aggarwal et al., 2016; Marcheque et al., 2019).

Compared to non-renal MSCs, resident renal MSCs exhibit a more nephrogenic gene expression profile, suggesting a potential for trans-differentiation or differentiation into renal cells under appropriate stimuli. Indeed, these cells can differentiate into renal epithelial cells, erythropoietin-producing fibroblasts, and renin-granule containing juxtaglomerular cells (Chen et al., 2008).

Nestin+ renal MSCs have demonstrated the ability to rescue AKI in mice by reducing serum creatinine, BUN levels, and apoptosis. Intravenous administration of these cells in AKI models resulted in improved renal function, as evidenced by decreased serum creatinine and BUN levels, and reduced cell apoptosis. Their paracrine effects, including VEGF signaling, are crucial for maintaining the glomerular basement membrane, protecting against ischemic renal failure (Jiang et al., 2015). However, due to depletion during CKD progression, autologous use of these kidney-derived cells is unlikely.

A progenitor cell derived from a specific tissue may be the optimal source for treating tissue-specific diseases. Such cells can be generated through the differentiation of ESCs or induced iPSCs. However, research on kidney regeneration strategies utilizing ESCs and iPSCs is still in its early stages. Ongoing efforts are essential to establish detailed differentiation protocols that guide pluripotent stem cells towards the formation of intermediate mesoderm, followed by renal progenitors, and ultimately mature, functional renal cells. Additionally, addressing the current challenges associated with these cell types, including heterogeneity in differentiation potential, interline variability, and low frequency and inefficient transduction of iPSCs, is crucial (Lazzeri, Romagnani, et al., 2015).

Embryonic SCs (ESCs)

While ESCs are theoretically ideal due to their ability to generate any kidney cell, their pluripotency also poses a significant safety risk: tumor formation. To mitigate this, researchers have developed protocols using retinoic acid, activin A, and bone morphogenic proteins to differentiate ESCs into renal lineages (Batchelder et al., 2009; Bruce et al., 2007; D. Kim & Dressler, 2005; Vigneau et al., 2007). Despite recent advancements increasing the efficiency of differentiating ESCs into intermediate mesoderm (Mae et al., 2013), the practical application of ESC-derived renal cells is hindered by teratoma formation, immune rejection, and ethical concerns.

Similar cell populations can be isolated from the urine of adult humans (Lazzeri, Ronconi, et al., 2015). Human fetal kidney-derived NCAM+ progenitors, when administered into the kidney parenchyma of a NOD/SCID mouse model of CKD induced by 5/6 nephrectomy, demonstrated engraftment, improved creatinine clearance, and reduced fibrosis (Harari-Steinberg et al., 2013). Intriguingly, RPs co-expressing stromal and epithelial markers have been identified in the urine of preterm neonates, offering a potentially accessible cell source (Arcolino et al., 2016).

Bone Marrow-Derived Mesenchymal Stem Cells (BMSC)

A 2015 meta-analysis of 71 preclinical studies evaluated the efficacy of cell-based therapy for CKD and identified factors influencing treatment outcomes, concluded that BMSCs administered intravenously or via renal artery injection, compared to iPSCs or organ-specific MSCs, were most effective in slowing CKD progression (Papazova et al., 2015). The analysis revealed that cell-based therapy consistently reduced CKD progression, particularly as measured by urinary protein and urea levels. Notably, the timing of treatment, cell origin, and dose did not impact efficacy. (Papazova et al., 2015)

The study highlighted the importance of considering disease models when interpreting results. While structural improvements were observed across different CKD models, functional outcomes varied based on disease pathogenesis. For instance, subtotal nephrectomy and toxic injury induced distinct CKD patterns. These findings emphasize the need f Papazova or tailored cell therapy approaches based on specific patient populations.

Adipose Derived Stem Cells (ADSC)

ADSCs have shown promising results in various AKI models (Sheashaa et al., 2016; Y. Song et al., 2017). Adipose-derived stromal vascular fraction (SVF) offers a potential advantage for autologous therapy as it can be directly administered without culturing, although isolation processes involving centrifugation and enzymatic digestion may classify it as "more than minimally manipulated" by FDA standards. SVF and cultured ADSCs exhibited comparable renoprotective effects in a rat I/R injury model (L. Zhou et al., 2017).

Umbilical Cord Stem Cell (UCSCs)

MSCs derived from UCSCs exhibit therapeutic effects similar to those of BMSCs in both AKI and CKD models (Morigi et al., 2010). While autologous MSCs from older renal failure patients offer a potential treatment option, their therapeutic efficacy may be compromised by age and disease-related factors. The complex pathological environment in these patients can impair MSC function, leading to reduced therapeutic potential compared to MSCs from healthy donors (Baban et al., 2016; Hickson et al., 2016)

Endothelial Progenitor Cells (EPCs)

EPCs derived from bone marrow or peripheral blood also hold therapeutic promise in renal disease. Primarily acting through paracrine signaling and in situ differentiation into mature endothelial cells, EPCs promote neovascularization. In porcine models, EPCs mitigated post-ischemic injury following renal artery stenosis (Zhu et al., 2013). However, direct comparisons have shown slightly inferior renoprotective effects compared to MSCs.

In rat models of glomerulosclerosis, unfractionated BMDCs, comprising MSCs and EPC precursors, demonstrated disease amelioration by engrafting into the glomerular microvasculature. This suggests that the endothelial differentiation potential of BMDCs is instrumental in their therapeutic efficacy (Goligorsky et al., 2009). Similar to MSCs, BMDCs derived from diseased animals can undergo phenotypic switching towards a smooth muscle/myofibroblast progenitor cell lineage, potentially limiting their effectiveness.

Optimizing EPC therapies necessitates balancing the generation of SPCs and EPCs while effectively mobilizing, recruiting, homing, and engrafting endothelial-committed cells (Goligorsky et al., 2009). Although regenerating the vasculature of an entire organ remains a formidable challenge, EPCs demonstrate the capacity to preserve existing microvasculature (Chade et al., 2009). The comparative advantages of lineage-restricted EPCs versus MSCs with broader differentiation potential are likely to be kidney disease-specific and require further elucidation.

Unfractionated BMDCs, comprising a mixture of MSCs and EPC precursors, have demonstrated efficacy in rat models of glomerulosclerosis by engrafting into the glomerular microvasculature (Goligorsky et al., 2009). This suggests that the endothelial differentiation potential of BMDCs contributes to their therapeutic benefit.

Similar to MSCs, BMDCs from diseased animals may exhibit a shift towards smooth muscle or myofibroblast lineages, diminishing their efficacy. Optimizing EPC-based therapies necessitates complex strategies to balance stem and endothelial progenitor cell populations, and to effectively mobilize, recruit, home, and integrate endothelial-committed cells.

While regenerating the vasculature of an entire organ remains challenging, EPCs excel at preserving existing microvasculature. The optimal cell type for kidney disease treatment, whether lineage-restricted EPCs or MSCs with broader differentiation potential, likely depends on the specific disease and requires further investigation.

Amniotic Fluid Stem Cells (AFSCs)

AFSCs exhibit characteristics intermediate between embryonic and adult SC. They appear to safeguard tubular and glomerular cells by stimulating endogenous repair mechanisms, thereby mitigating inflammation and fibrosis. Additional studies have demonstrated that AFSCs enhance recovery in 5/6 nephrectomy (Nx) models and cisplatin-induced toxicity. They possess the in vitro capacity to differentiate into renal lineages, expressing epithelial and podocyte markers when cultured under appropriate conditions (De Coppi et al., 2007; Perin et al., 2007; Siegel et al., 2009). In vivo studies have demonstrated that human AFSCs (hAFSCs) can survive and partially differentiate within injured kidneys following glycerol-induced acute kidney injury (AKI) (Perin et al., 2010).

Marcheque et al. demonstrated the renoprotective effects of AFSCs in both glycerol-induced AKI and CKD models (Baulier et al., 2014; Da Sacco et al., 2018). By stimulating endogenous repair mechanisms, AFSCs protect tubular and glomerular cells, reducing inflammation and fibrosis (Da Sacco et al., 2018). Other studies have shown AFSC efficacy in 5/6 nephrectomy and cisplatin-induced toxicity models (Al-Husseiny et al., 2016).

Early administration of hAFSCs during AKI correlates with improved renal function, increased tubular cell proliferation, and decreased apoptosis (Perin et al., 2010). These beneficial effects are likely attributed to both direct cell-based repair and immunomodulatory properties, creating a favorable microenvironment for tissue regeneration. Similar findings have been reported in other AKI models, with hAFSCs demonstrating a stronger anti-apoptotic effect compared to mesenchymal stem cells (MSCs) (Hauser et al., 2010).

While hAFSCs can engraft in the kidney, their primary therapeutic mechanism appears to be paracrine signaling. Studies have shown that hAFSCs improve renal function and limit tubular damage without significant differentiation into renal tissue (Rota et al., 2012). Moreover, AFSCs have demonstrated potential in chronic kidney disease (CKD) models, prolonging survival, ameliorating kidney function decline, and delaying the progression of fibrosis (Sedrakyan et al., 2012; Sun et al., 2013).

Induced Pluripotent Stem Cells (iPSCs)

iPSCs offer a potentially patient-specific cell source for treating kidney disease. While undifferentiated murine iPSCs can mitigate ischemia-reperfusion induced AKI when administered intra-arterially through a paracrine mechanism (P.-Y. Lee et al., 2012), their pluripotency, as with ESCs, raises safety concerns. Protocols for differentiating iPSCs into nephrogenic cells, renal proximal tubular cells, and podocytes have been established (Araoka et al., 2014; Hendry et al., 2013; Lam et al., 2014; Lin et al., 2010; B. Song et al., 2012; Taguchi et al., 2014; Takasato et al., 2014). However, human pre-differentiated iPSCs have shown no benefit in AKI animal models. Morigi's group reported that human iPSCs, differentiated into nephrogenic intermediate mesoderm and then metanephric mesenchyme, engrafted and promoted recovery from cisplatin-induced AKI (Imberti et al., 2015).

Similar benefits were observed in a mouse renal I/R model. Human iPSC-derived RPs, transplanted under the kidney capsule, reduced renal dysfunction and fibrosis (Toyohara et al., 2015). Although less studied in renal failure, encapsulated mouse ESCs transplanted onto the omentum slowed disease progression in a rat 5/6 nephrectomy model(Al-Husseiny et al., 2016).

A recent breakthrough by Imberti et al. challenges this paradigm. Using an efficient two-step protocol, they differentiated iPSCs into RPCs that robustly engrafted into damaged tubules, restoring renal function and structure in cisplatin-induced AKI. Notably, undifferentiated iPSCs exerted no protective effects. Importantly, no evidence of aberrant differentiation or tumorigenesis from iPSC-derived RPCs was found (Imberti et al., 2015).

While this review provides an overview of stem cell-based approaches to preclinical kidney disease, it is important to note that this is not an ecd133stive analysis. The presented studies highlight the diverse range of stem cell sources, characterizations, applications, and mechanisms of action explored, as well as the variety of species and disease models utilized. Table 1 summarizes the types of stem cells employed, delivery methods, disease models, and species across these experimental renal disease therapies.

ADVANCING CLINICAL TRIALS OF STEM CELLS FOR RENAL DISEASE

The majority of SC clinical trials for renal disease have employed MSCs, which do not represent a truly regenerative approach. The inherent variability of MSCs among individuals, laboratories, and tissue sources, coupled with the heterogeneity of kidney diseases, precludes a robust meta-analysis of MSC trials for renal disease (Peired et al., 2016). Nevertheless, these studies provide valuable insights.

Early trials using BMSC established their safety profile. Administration of MSCs was well-tolerated in cardiac surgery patients at risk of postoperative AKI. Autologous adipose-derived MSCs (AdMSCs) delivered via the renal artery to patients with renovascular disease were safe and improved renal blood flow and oxygenation at three months compared to medical therapy (n=14) (Saad et al., 2017).

The first placebo-controlled trial involving allogeneic bone marrow-derived mesenchymal precursor cells in diabetic nephropathy patients (n=10/group) demonstrated safety and a potential trend towards efficacy at 12 weeks, although the study was underpowered (Packham et al., 2016). A study of six patients with autosomal dominant polycystic kidney disease who received autologous BMSCs intravenously confirmed the safety of the treatment, but renal function remained unchanged at one year (Makhlough et al., 2017)

A study involving 30 patients with diverse CKDs, including 10 renal transplant recipients, reported improved renal function six months' post-treatment with autologous BMSCs (El-Ansary et al., 2012). However, complications associated with these seemingly "benign" cells have emerged. For instance, UCSCs administration to two renal transplant patients resulted in peripheral vein thrombosis likely due to inadequate prevention of cell clumping (Wu et al., 2017). A case report described worsened renal function in a CKD patient following AdMSC treatment, with a subsequent renal biopsy revealing a significant inflammatory response, including cells expressing markers of the administered stem cell product (J.-S. Kim et al., 2017). Notably, a large, well-designed international trial investigating MSCs for AKI in cardiac surgery patients was prematurely halted due to worse outcomes in the treatment group compared to controls (Swaminathan et al., 2018)

Table 1. Clinical trials of stem cells for renal disease

Model	Species	Cell	Route
Acute kidney injury			
I/R (Ranghino et al., 2017)	Mouse	Glomerular progenitors and tubular progenitors	IV
I/R (Chen et al., 2008)	Mouse	Mouse clonal renal MSC	IV
I/R (Jiang et al., 2015)	Mouse	Nestin+ kidney MSC	IV
I/R (Behr et al., 2009)	Sheep	Autologous BM-MSC	Renal artery
Kidney autotx (I/R) (Baulier et al., 2014)	Pig	Autologous AF-MSC	Renal artery
I/R (Sheashaa et al., 2016)	Rat	AdMSC	IV
I/R (L. Zhou et al., 2017)	Rat	hAdSVF, hAdMSC	Renal parenchyma
I/R (Toyohara et al., 2015)	Mouse	hiPSC-RPs (OSR1+ SIX2+)	Renal capsule

continued on following page

Table 1. Continued

Model	Species	Cell	Route
I/R (Gomes et al., 2018)	Rat	c-kit+ renal cells	IP, intraaortic
Ureter obstruction (Y. Song et al., 2017)	Rat	rADSC	IV
Sepsis (Sung et al., 2016)	Rat	Autologous AdMSC	IV
Glycerol (Aggarwal et al., 2016)	Mouse	hCD133+ RPs	IV
Glycerol (Lazzeri, Ronconi, et al., 2015)	Mouse	hCD133+CD24+ progenitors	IV
Glycerol (Perin et al., 2010)	Mouse	hAFSC	Renal parenchyma
Glycerol (Bruno et al., 2009)	Mouse	hMSC-microvesicles	IV
Glycerol (Herrera et al., 2007)	Mouse	hBM-MSC	IV
Cisplatin (Morigi et al., 2004)	Mouse	mBM-MSC	IV
Cisplatin (Morigi et al., 2010)	Mouse	hUC-MSC	IV
Cisplatin (Imberti et al., 2015)	Mouse	hiPSC-RPs	IV
Cisplatin (Al-Husseiny et al., 2016)	Rat	hAF-MSC	IV
Cisplatin (Y. Zhou et al., 2013)	Rat	hUC-MSC exosomes	Renal capsule
Cisplatin (Ashour et al., 2016)	Rat	Allo-rBM-MSC	IV
		hADSC	
		hAFSC	
Cisplatin (Santeramo et al., 2017)	Rat	hCD133+ kidney cells	IV
Cisplatin (Moghadasali et al., 2015)	Macaque	Autologous BM-MSC	Intraarterial
Adriamycin (Ratliff & Goligorsky, 2013)	Mouse EPC (in hydrogel)	SC in ear	Renal capsule
Adriamycin (Zoja et al., 2012)	Rat	rBM-MSC	IV
Gentamycin (S.-J. Lee et al., 2017)	Dog	cUC-MSC	Renal parenchyma
Chronic kidney disease			
Renal artery stenosis (Zhu et al., 2013)	Pig	Pig MSC, EPC	Renal artery
Renal artery stenosis (Chade et al., 2009)	Pig	Autologous EPC	Renal artery
RAS with metabolic syndrome (Eirin et al., 2018)	Pig	Autologous MSC-EV	Renal artery
5/6 Nx (Harari-Steinberg et al., 2013)	Mouse	Human NCAM+ RPs	Renal parenchyma

continued on following page

Table 1. Continued

Model	Species	Cell	Route
5/6 Nx (S.-R. Lee et al., 2010; Semedo et al., 2009)	Rat	rBM-MSC	IV
5/6 Nx (Geng et al., 2016)	Rat	mESC	Gel-on renal remnant Gel-on omentum
5/6 Nx (Tarng et al., 2016)	Rat	rBM-MSC, hiPSC	Renal parenchyma
Alport syndrome (Wankhade et al., 2016)	Mouse	AFSC	Intracardiac
Alport syndrome (Sedrakyan et al., 2017)	Mouse	AFSC-EV	Intracardiac
Lupus nephritis (Thiel et al., 2015)	Mouse	hESC-MSC	IV
Spontaneous CKD (Quimby et al., 2016)	Cat	Allo-AdMSC	IV
Diabetic nephropathy (Feng et al., 2018)	Mouse	mAFSC	Renal parenchyma
Diabetic nephropathy (R. H. Lee et al., 2006)	Mouse	hBM-MSC	Intracardiac
Diabetic nephropathy (Ezquer et al., 2008)	Mouse	mBM-MSC	IV
Diabetic nephropathy (Wang et al., 2013)	Rat	rBM-MSC	Renal artery

I/R: ischemia reperfusion; autotx: autotransplantation; IP: intraperitoneal; RAS: renal artery stenosis; Nx: nephrectomy; MSC: mesenchymal SC; BM: bone marrow; AF: amniotic fluid; Ad: adipose; SVF, stromal vascular fraction; iPSC: induced pluripotent SC; ADSC: adipose-derived SC; UC: umbilical cord; EPC: endothelial progenitor cell; SC: stem cells; EV: extracellular vesicles; RPs: renal progenito; AdMSC: adipose-derived MSC; AFSC: amniotic fluid SC Before a stem cell type; h: human; m: mouse; c: canine; r: rat; allo: allogeneic

In summary, MSC clinical trials for kidney diseases exhibit significant heterogeneity in patient populations and study designs. The limited sample sizes of these trials have thus far precluded definitive conclusions regarding the clinical benefits of MSC therapy for AKI or CKD.

Long-term follow-up data from stem cell therapies in renal disease are limited. One study of autologous BMSCs in 30 CKD patients demonstrated sustained renal function benefits up to 18 months (Makhlough et al., 2018). This review underscores the critical need to minimize confounding variables in cell sources, preclinical models, and human disease characterization to enhance data interpretability and comparability across studies. Proprietary concerns often hinder rigorous cross-study analysis due to insufficient methodological detail.

To address these challenges, the establishment of standardized cell lines and derivation protocols under a national funding agency's oversight is proposed. While implementation may be complex and face resistance from biotechnology companies, the long-term benefits for research and patient care are substantial. In the interim, peer-reviewed journals should mandate comprehensive reporting of cell authentication, isolation, culture, and donor characteristics.

This includes detailed protocols, reagent information, population doubling time, cell density, clonal status, and culture media composition. Such transparency is essential for both academic research and successful translation to clinical applications (Marcheque et al., 2019).

The recent call for a standardized the use of MSC, to serve as a common calibration tool underscores the need for a baseline characterization of the type of cells across various parameters, including phenotype, karyotype, gene expression, cytokine secretion, and functional readouts (Tanavde et al., 2015). Establishing this standard will require comprehensive data collection from diverse MSC sources, such as large cell material sets, single-donor cell bank lines, immortalized cell lines, and iPS-derived lines (Viswanathan et al., 2014).

While the creation and maintenance of such a repository demands significant resources and oversight, it is essential for advancing the field. A central cell bank would facilitate data comparability across research laboratories and foster collaborative efforts. However, the limited utilization of the NIH's 2012 clinical-grade marrow-derived stromal cell product bank highlights the challenges in establishing and maintaining such resources (Sabatino et al., 2012).

Identifying optimal human pluripotent stem cell (hPSC) lines for preclinical kidney disease research is crucial. Given the variability in differentiation potential among ESC and induced iPSC lines (Osafune et al., 2008), selecting lines with a strong propensity to generate RPs for multiple kidney disease models is essential.

Recognizing that stem cell behavior can vary across different disease contexts, application-specific efficacy and potency testing is imperative. While molecular characterization provides valuable insights, it alone is insufficient to predict clinical outcomes. Co-culture experiments with specific renal cell types and mixed lymphocyte reactions can offer additional characterization beyond expression analysis.

In vitro assays assessing tubular epithelial cell proliferation, cell survival, and lymphocyte proliferation can further elucidate stem cell mechanisms of action. For instance, an in vitro model quantifying IL-10 release from blood cells has been proposed as a potency assay for MSC immunomodulation (de Wolf et al., 2017).

The immunomodulatory effects of MSCs can be assessed by analyzing their influence on cytokine profiles in blood mononuclear cells (Chinnadurai et al., 2018; Jiao et al., 2011). Co-culture systems designed to mimic specific biological processes, such as tube formation assays for vasculogenesis, can provide additional mechanistic insights.

3D MODELS

The use of SC to create 3D renal models present several promising perspectives in biomedical research and clinical applications. Creating 3D renal models that fully replicate the complexity and functionality of the human kidney remains a challenge. Developing strategies to promote vascularization within 3D renal models is essential for their long-term survival and function.

The use of SC raises ethical concerns, particularly regarding their source and potential for unintended consequences. Generating an autologous organ derived from patients, can offer the possibility of offer personalized medicine, allowing for the creation of models that accurately mimic their specific disease conditions. That will allow the researches to use this 3D renal models to test the efficacy and toxicity of potential drugs before clinical trials, and may be possible to predict individual susceptibility to kidney damage, reducing the risk of adverse side effects, and avoiding animal harm.

SC can be differentiated into kidney cell types and potentially used to regenerate damaged kidney tissue, and in the future, could offer an alternative for transplantation, addressing the shortage of donor organs.

Also, these models can provide insights into the complex processes involved in kidney development and maturation, by studying tem, can help researchers better understand the underlying causes of kidney diseases and identify potential therapeutic targets, assess the potential nephrotoxic effects of new chemicals, drugs, or environmental pollutants.

While ESCs and iPSCs differentiated into renal phenotypes have shown promise in generating three-dimensional kidney structures in vitro (Lin et al., 2010; Taguchi et al., 2014; Takasato et al., 2014; Xia et al., 2013), their capacity to form functional kidneys in vivo remains unproven. AFSCs offer an intermediate profile between ESCs and adult stem cells (De Coppi et al., 2007), with potential for renal lineage differentiation in vitro (Perin et al., 2007; Siegel et al., 2009).

In vivo studies have demonstrated that human AFSCs (hAFSCs) can survive and partially differentiate within injured kidneys. When administered early in the course of glycerol-induced AKI, hAFSCs improved renal function, stimulated tubular cell proliferation, and reduced apoptosis (Perin et al., 2010). These effects were attributed to both direct cell-based repair and immunomodulatory properties, creating a favorable environment for tissue regeneration. Similar benefits were observed in other AKI models (Hauser et al., 2010), with hAFSCs showing a stronger anti-apoptotic effect compared to MSCs.

However, hAFSCs primarily exert their therapeutic effects through paracrine mechanisms, as demonstrated by their limited engraftment and differentiation into renal tissue (Rota et al., 2012). Nonetheless, their ability to ameliorate kidney function and delay disease progression has been reported in CKD models (Sedrakyan et al., 2012; Sun et al., 2013).

Despite these challenges, the potential benefits of using SC for 3D renal models are substantial. Continued research and development in this area may lead to significant advancements in kidney disease research, treatment, and prevention.

GENOME EDITING TECHNOLOGY

Genome editing is a powerful technology that allows scientists to precisely modify the genetic material of cells. This technique holds significant promise for treating kidney failure by harnessing the potential of SC (NIH, 2024).

SC, often derived from bone marrow or embryonic tissue, are isolated and cultured in the laboratory, and with the help of genome editing tools, such as CRISPR-Cas9, it is possible to introduce specific genetic changes into the SC. These changes can correct mutations that cause kidney disease or enhance the cells' ability to differentiate into kidney-specific cell types (NHI, 2023).

These genetically modified SC could be then guided to differentiate into kidney cells, including podocytes, tubular epithelial cells, and endothelial cells, and be transplanted into a damaged kidney or injected directly into the affected area to promote regeneration.

Genome editing technology has emerged as a powerful tool for biomedical research, offering exceptional versatility and feasibility. These technologies can be tailored to address the specific genetic causes of kidney disease in individual patients. By correcting disease-causing mutations and promoting cell differentiation, genome editing may help restore the function of damaged kidney tissue. Autologous SC, derived from the patient's own body, are less likely to be rejected by the immune system, compared to transplanted organs.

In 2012, Jinek et al. demonstrated that a specific bacterial defense system known as clustered regularly interspaced short palindromic repeats (CRISPR)-associated protein-9 nucleases (Cas9) could be adapted into an RNA-guided genome editing tool (Jinek et al., 2012). It is a revolutionary gene editing technology that has been making waves in the scientific world. This technique was subsequently shown to be capable of modifying the human genome (Cong et al., 2013; Mali et al., 2013). Since then, the rapid advancement of CRISPR-Cas9 engineering has significantly propelled progress in related fields, including SC research for disease modeling, drug screening, and potential regenerative medicine (Avior et al., 2017; Gupta et al., 2018; Lau et al., 2018; Luo et al., 2016; Mandegar et al., 2016).

A specific DNA sequence is chosen as the target for editing. A short Ribonucleic acid (RNA) molecule, called a guide RNA, is designed to match the target DNA sequence. The guide RNA binds to the target DNA sequence, guiding the Cas9 enzyme to the exact location, who will cut the DNA at the target location, and then insert, delete, or replace specific DNA sequences at the cut site (Le Page, s. f.).

The genome editing in humans involves the accurate manipulation of DNA using engineered nucleases. This technique is more specific than the indiscriminate DNA damage caused by various agents, such as harmful cellular metabolic byproducts, environmental radiation, and replication errors, or other methods discussed previously (Hafezparast et al., 1993; Jeggo et al., 2016; Thacker & Wilkinson, 1991).

In contrast to the creation of multiple repair sites at undesired genomic locations using restriction endonucleases,14–16 site-specific double-strand breaks (DSBs) can be introduced in a targeted manner through the fusion of a DNA-recognizing component with an engineered nuclease (Anders et al., 2014; Jinek et al., 2012; Rouet et al., 1994).

Double-strand breaks can then be repaired via either non-homologous end joining (NHEJ) or homology-directed repair (HDR), resulting in targeted gene editing. In mammals, nicks or double-strand damages in DNA induced by ionizing radiation or oxidative stress are frequently repaired through the NHEJ pathway in all cell cycles (Cary et al., 1997; Hoeijmakers, 2001; Pardo et al., 2009; Rudin et al., 1989; Takata et al., 1998).

Physiologically, NHEJ is utilized to trigger V(D)J recombination for the maturation of T and B lymphocytes in the peripheral blood system (Lieber et al., 2004).

Takahashi et al, published the groundbreaking study, demonstrating that the reprogramming of adult cells iPSCs, has presented an unparalleled opportunity to investigate disease mechanisms, conduct drug and toxicology screening, and explore potential cell replacement therapy in regenerative medicine (Takahashi et al., 2007).

One of the studied cells are the fibroblasts derived from polycystic kidney disease (PKD) patients with diverse mutations, they can be isolated and reprogrammed into iPSCs, yielding disease-specific, patient-matched cell lines capable of indefinite division (Freedman et al., 2013; Thatava et al., 2011). Freedman et al. confirmed the retention of the PKD1 mutation in patient-derived iPSC lines following the establishment of pluripotency, subsequent differentiation into embryonic germ layers, and teratoma formation (Freedman et al., 2013).

Three ADPKD and two autosomal-recessive PKD iPSC lines were generated and differentiated into somatic epithelial cells and hepatoblasts, cell types previously implicated in the formation of liver cysts in PKD patients. Immunoblotting revealed a reduction of polycystin-2 in the cilia of non-differentiated ADPKD iPSCs, somatic epithelial cells, and hepatoblasts. These findings suggest that iPSC-based in vitro disease modeling corroborates the potential role of polycystin-1 in the regulation of polycystin-2 within the pathogenesis of PKD (Freedman et al., 2013).

The theory guide to mapped PKD mutation, that could potentially be corrected using CRISPR-Cas9 following cell reprogramming. This would allow for the retention of the corrected mutation in the desired kidney cell type for disease modeling or, ultimately, cell transplantation. The progeny of genetically corrected cells would be autologous to the patient.

Moreover, CRISPR-Cas9 has been employed to develop a potential therapeutic intervention for Fabry disease (FD), a rare X-linked lysosomal storage disorder leading to premature death in FD patients.77 In this approach, a disease model of FD was generated by correcting the galactosidase alpha gene in human immortalized podocytes (Pereira et al., 2016).

It is important to remember that genome editing tools can sometimes make unintended changes to the genome, which could have negative consequences, raising ethical questions about the potential for unintended consequences and the long-term implications for future generations.

Translating genome editing technology from the laboratory to clinical practice requires careful research and development to ensure safety and efficacy. While there are challenges to overcome, the potential benefits of genome editing and stem cells for treating kidney failure are significant. Ongoing research and development in this field may lead to new and innovative therapies for patients with kidney disease.

Pluripotent SC (PSC) and Kidney Organoids

The development of kidney organoids from PSC represents a significant advance in regenerative medicine and has the potential to revolutionize the treatment of kidney diseases.

Kidney organoids are 3D structures that mimic the development and function of the human kidney. They are created by culturing pluripotent SC under specific conditions that promote their differentiation into kidney cell types. In the future, it may be possible to create kidney organoids that can be transplanted into patients with kidney failure.

PSC are carefully cultured in a controlled environment, exposed to a precise sequence of growth factors and signaling molecules that guide their differentiation into specific kidney cell types, including podocytes, tubular epithelial cells, and endothelial cells. These differentiated cells subsequently self-organize into intricate 3D structures that closely resemble the complex architecture of the human kidney. Through a process of maturation, these organoids acquire functional characteristics, such as the ability to filter blood and produce urine, mirroring the physiological processes of a real kidney.

iPSCs have been utilized to generate kidney progenitor cells through directed differentiation, guided by a variety of specific cell markers previously identified. Several studies have also reported the development of kidney organoids from nephron progenitor cells (Kang & Han, 2014; Lam et al., 2014; Mae et al., 2013; Morizane et al., 2015; Taguchi et al., 2014, 2014; Takasato et al., 2015; Toyohara et al., 2015).

In addition to nephron progenitor cells, studies have generated kidney organoids from differentiated mesenchyme and intermediate mesoderm. This research has significantly advanced our understanding of kidney progenitors and nephrogenesis, while also providing valuable tools for disease modeling. iPSC-derived kidney organoids may also be therapeutically beneficial for identifying potential drug candidates, for targeting personalized therapies, in this casae for renal injuries, but also, can offer a tailored tumor treatments (Freedman et al., 2015; Morizane et al., 2015; Takasato et al., 2015).

Freedman et al. successfully generated PKD pathogenic kidney organoids by selectively knocking out PKD1 or PKD2 with CRISPR-Cas9, resulting in cyst formation in kidney tubules (Freedman et al., 2015).

A nephrin-GFP PSC line was created using TALEN, which demonstrated the ability to form vascularized glomeruli when transplanted into immunodeficient NOD/SCID mice (Taguchi et al., 2014). CRISPR-Cas9 was also successfully used to knockout podocalyxin in human PSC-derived kidney organoids, recapitulating the capillary loop stage of podocyte development. The results revealed defects in podocalyxin -/- kidney organoids, as assessed by the development of microvilli and cell junctions in the gene-edited podocyte-like cells (Y. K. Kim et al., 2017).

Furthermore, Forbes et alt, retrieve fibroblast tissue from a child diagnosed with nephronophthisis, for reprogramming into iPSCs, and then, these cells were then differentiated into kidney organoids using a modified version of a previous method. The specific form of nephronophthisis in this case was caused by a point mutation c.634G>A in exon 6, resulting in a defect in the IFT140 locus (encoding a key protein for maintaining ciliary integrity in the human kidney) (OMIM: 614620). The iPSC-derived kidney organoids displayed a relatively low messenger RNA (mRNA) and protein expression of IFT140 compared to the gene-corrected group (Forbes et al., 2018).

Through CRISPR-Cas9 gene correction, a repair template was introduced to rectify the c.634G>A point mutation in the IFT140 locus. This gene correction successfully rescued the expression of IFT140. The same cell model was subsequently used to analyze cilia phenotype, RNA sequencing, and epithelial cell polarity following the sorting of EpCam+ cells. This study demonstrates the incremental value of gene editing in the application of SC disease modeling. In vitro disease modeling serves as a foundation for functional genomics, and there is a growing number of comparable studies targeting other kidney diseases such as PKD (Forbes et al., 2018).

CONCLUSION

SC hold significant promise as a therapeutic approach for kidney injury, AKI and CKD. Their ability to differentiate into various cell types, including renal epithelial cells, endothelial cells, and fibroblasts, offers potential for regenerating damaged kidney tissue.

3D models have emerged as valuable tools for studying kidney development and disease, providing a more physiologically relevant environment for evaluating SC-based therapies. These models can be used to investigate the interactions between SC and kidney cells, as well as to assess the efficacy of different differentiation protocols and therapeutic strategies.

While animal models offer valuable insights, they do not fully recapitulate the complexity of human kidney disease. Diverse animal models exhibit distinct pathophysiological features, requiring tailored endpoint assessments. Technical limitations, including difficulties in frequent blood sampling, surgical procedures, and continuous physiological monitoring in small models, further complicate data collection. The inherent heterogeneity of human disease often surpasses the capabilities of murine models to predict clinical outcomes. Consequently, relying solely on murine data to justify clinical translation of novel therapies is insufficient. Moreover, interspecies variations in renal structure and function necessitate cautious interpretation and extrapolation of findings across animal models and to humans.

Given these limitations, large animal models remain indispensable for translational research. Comprehensive characterization by renal physiology and veterinary disease experts is crucial. Until advanced virtual models become available, sadly, large animals will continue to be essential for determining optimal dosage, scalability, safety, toxicity, and delivery routes, but it is in the hands of the researches to evolve and improve this cruel experiments.

While large animal models are complex and costly, collaborative efforts between researchers and veterinarians can optimize resource utilization and enhance study outcomes. A comprehensive characterization of animal models for acute and chronic kidney disease is essential to advance the field. Detailed comparative studies across different mouse strains, including assessments of disease kinetics, blood markers, proteinuria, and histological progression, are crucial but underfunded.

To address these challenges, a multidisciplinary committee of experts should be convened to establish best practices for preclinical study design, encompassing stem cell sources, disease models, isolation, culture, administration, and analysis methods.

Professional societies and funding agencies can play a pivotal role in fostering such initiatives and promoting standardized approaches.

REFERENCES

Aggarwal, S., Grange, C., Iampietro, C., Camussi, G., & Bussolati, B. (2016). Human CD133+ Renal Progenitor Cells Induce Erythropoietin Production and Limit Fibrosis After Acute Tubular Injury. *Scientific Reports*, *6*(1), 37270. DOI: 10.1038/srep37270 PMID: 27853265

Al-Husseiny, F., Sobh, M. A., Ashour, R. H., Foud, S., Medhat, T., El-Gilany, A.-H., Elghannam, D., Abdel-Ghaffar, H., Saad, M.-A., & Sobh, M. (2016). Amniotic Fluid-Derived Mesenchymal Stem Cells Cut Short the Acuteness of Cisplatin-Induced Nephrotoxicity in Sprague-Dawley Rats. *International Journal of Stem Cells*, *9*(1), 70–78. DOI: 10.15283/ijsc.2016.9.1.70 PMID: 27426088

Anders, C., Niewoehner, O., Duerst, A., & Jinek, M. (2014). Structural basis of PAM-dependent target DNA recognition by the Cas9 endonuclease. *Nature*, *513*(7519), 569–573. DOI: 10.1038/nature13579 PMID: 25079318

Araoka, T., Mae, S., Kurose, Y., Uesugi, M., Ohta, A., Yamanaka, S., & Osafune, K. (2014). Efficient and rapid induction of human iPSCs/ESCs into nephrogenic intermediate mesoderm using small molecule-based differentiation methods. *PLoS One*, *9*(1), e84881. DOI: 10.1371/journal.pone.0084881 PMID: 24454758

Arcolino, F. O., Zia, S., Held, K., Papadimitriou, E., Theunis, K., Bussolati, B., Raaijmakers, A., Allegaert, K., Voet, T., Deprest, J., Vriens, J., Toelen, J., van den Heuvel, L., & Levtchenko, E. (2016). Urine of Preterm Neonates as a Novel Source of Kidney Progenitor Cells. *Journal of the American Society of Nephrology*, *27*(9), 2762–2770. DOI: 10.1681/ASN.2015060664 PMID: 26940093

Ashour, R. H., Saad, M.-A., Sobh, M.-A., Al-Husseiny, F., Abouelkheir, M., Awad, A., Elghannam, D., Abdel-Ghaffar, H., & Sobh, M. (2016). Comparative study of allogenic and xenogeneic mesenchymal stem cells on cisplatin-induced acute kidney injury in Sprague-Dawley rats. *Stem Cell Research & Therapy*, *7*(1), 126. DOI: 10.1186/s13287-016-0386-0 PMID: 27585525

Avior, Y., Lezmi, E., Yanuka, D., & Benvenisty, N. (2017). Modeling Developmental and Tumorigenic Aspects of Trilateral Retinoblastoma via Human Embryonic Stem Cells. *Stem Cell Reports*, *8*(5), 1354–1365. DOI: 10.1016/j.stemcr.2017.03.005 PMID: 28392220

Baban, B., Liu, J. Y., Payne, S., Abebe, W., Yu, J. C., & Mozaffari, M. S. (2016). Status of stem cells in diabetic nephropathy: Predictive and preventive potentials. *The EPMA Journal*, *7*(1), 21. DOI: 10.1186/s13167-016-0070-6 PMID: 27729946

Batchelder, C. A., Lee, C. C. I., Matsell, D. G., Yoder, M. C., & Tarantal, A. F. (2009). Renal ontogeny in the rhesus monkey (Macaca mulatta) and directed differentiation of human embryonic stem cells towards kidney precursors. *Differentiation. Differentiation*, *78*(1), 45–56. DOI: 10.1016/j.diff.2009.05.001 PMID: 19500897

Baulier, E., Favreau, F., Le Corf, A., Jayle, C., Schneider, F., Goujon, J.-M., Feraud, O., Bennaceur-Griscelli, A., Hauet, T., & Turhan, A. G. (2014). Amniotic fluid-derived mesenchymal stem cells prevent fibrosis and preserve renal function in a preclinical porcine model of kidney transplantation. *Stem Cells Translational Medicine*, *3*(7), 809–820. DOI: 10.5966/sctm.2013-0186 PMID: 24797827

Behr, L., Hekmati, M., Lucchini, A., Houcinet, K., Faussat, A.-M., Borenstein, N., Noel, L.-H., Lelievre-Pegorier, M., & Laborde, K. (2009). Evaluation of the effect of autologous mesenchymal stem cell injection in a large-animal model of bilateral kidney ischaemia reperfusion injury. *Cell Proliferation*, *42*(3), 284–297. DOI: 10.1111/j.1365-2184.2009.00591.x PMID: 19438896

Bruce, S. J., Rea, R. W., Steptoe, A. L., Busslinger, M., Bertram, J. F., & Perkins, A. C. (2007). In vitro differentiation of murine embryonic stem cells toward a renal lineage. *Differentiation. Differentiation*, *75*(5), 337–349. DOI: 10.1111/j.1432-0436.2006.00149.x PMID: 17286599

Bruno, S., Chiabotto, G., & Camussi, G. (2014). Concise review: Different mesenchymal stromal/stem cell populations reside in the adult kidney. *Stem Cells Translational Medicine*, *3*(12), 1451–1455. DOI: 10.5966/sctm.2014-0142 PMID: 25355731

Bruno, S., Grange, C., Deregibus, M. C., Calogero, R. A., Saviozzi, S., Collino, F., Morando, L., Busca, A., Falda, M., Bussolati, B., Tetta, C., & Camussi, G. (2009). Mesenchymal stem cell-derived microvesicles protect against acute tubular injury. *Journal of the American Society of Nephrology*, *20*(5), 1053–1067. DOI: 10.1681/ASN.2008070798 PMID: 19389847

Bussolati, B., & Camussi, G. (2015). Therapeutic use of human renal progenitor cells for kidney regeneration. *Nature Reviews. Nephrology*, *11*(12), 695–706. DOI: 10.1038/nrneph.2015.126 PMID: 26241019

Cary, R. B., Peterson, S. R., Wang, J., Bear, D. G., Bradbury, E. M., & Chen, D. J. (1997). DNA looping by Ku and the DNA-dependent protein kinase. *Proceedings of the National Academy of Sciences of the United States of America*, *94*(9), 4267–4272. DOI: 10.1073/pnas.94.9.4267 PMID: 9113978

Chade, A. R., Zhu, X., Lavi, R., Krier, J. D., Pislaru, S., Simari, R. D., Napoli, C., Lerman, A., & Lerman, L. O. (2009). Endothelial progenitor cells restore renal function in chronic experimental renovascular disease. *Circulation*, *119*(4), 547–557. DOI: 10.1161/CIRCULATIONAHA.108.788653 PMID: 19153272

Chen, J., Park, H.-C., Addabbo, F., Ni, J., Pelger, E., Li, H., Plotkin, M., & Goligorsky, M. S. (2008). Kidney-derived mesenchymal stem cells contribute to vasculogenesis, angiogenesis and endothelial repair. *Kidney International*, *74*(7), 879–889. DOI: 10.1038/ki.2008.304 PMID: 18596729

Chinnadurai, R., Rajan, D., Qayed, M., Arafat, D., Garcia, M., Liu, Y., Kugathasan, S., Anderson, L. J., Gibson, G., & Galipeau, J. (2018). Potency Analysis of Mesenchymal Stromal Cells Using a Combinatorial Assay Matrix Approach. *Cell Reports*, *22*(9), 2504–2517. DOI: 10.1016/j.celrep.2018.02.013 PMID: 29490284

Cong, L., Ran, F. A., Cox, D., Lin, S., Barretto, R., Habib, N., Hsu, P. D., Wu, X., Jiang, W., Marraffini, L. A., & Zhang, F. (2013). Multiplex genome engineering using CRISPR/Cas systems. *Science*, *339*(6121), 819–823. DOI: 10.1126/science.1231143 PMID: 23287718

Coresh, J., Selvin, E., Stevens, L. A., Manzi, J., Kusek, J. W., Eggers, P., Van Lente, F., & Levey, A. S. (2007). Prevalence of chronic kidney disease in the United States. *Journal of the American Medical Association*, *298*(17), 2038–2047. DOI: 10.1001/jama.298.17.2038 PMID: 17986697

Da Sacco, S., Perin, L., & Sedrakyan, S. (2018). Amniotic fluid cells: Current progress and emerging challenges in renal regeneration. *Pediatric Nephrology (Berlin, Germany)*, *33*(6), 935–945. DOI: 10.1007/s00467-017-3711-7 PMID: 28620747

De Coppi, P., Bartsch, G.Jr, Siddiqui, M. M., Xu, T., Santos, C. C., Perin, L., Mostoslavsky, G., Serre, A. C., Snyder, E. Y., Yoo, J. J., Furth, M. E., Soker, S., & Atala, A. (2007). Isolation of amniotic stem cell lines with potential for therapy. *Nature Biotechnology*, *25*(1), 100–106. DOI: 10.1038/nbt1274 PMID: 17206138

de Wolf, C., van de Bovenkamp, M., & Hoefnagel, M. (2017). Regulatory perspective on in vitro potency assays for human mesenchymal stromal cells used in immunotherapy. *Cytotherapy*, *19*(7), 784–797. DOI: 10.1016/j.jcyt.2017.03.076 PMID: 28457740

Eirin, A., & Lerman, L. O. (2014). Mesenchymal stem cell treatment for chronic renal failure. *Stem Cell Research & Therapy*, *5*(4), 83. DOI: 10.1186/scrt472 PMID: 25158205

Eirin, A., Zhu, X.-Y., Jonnada, S., Lerman, A., van Wijnen, A. J., & Lerman, L. O. (2018). Mesenchymal Stem Cell-Derived Extracellular Vesicles Improve the Renal Microvasculature in Metabolic Renovascular Disease in Swine. *Cell Transplantation, 27*(7), 1080–1095. DOI: 10.1177/0963689718780942 PMID: 29954220

El-Ansary, M., Saadi, G., & Abd El-Hamid, S. M. (2012). Mesenchymal stem cells are a rescue approach for recovery of deteriorating kidney function. *Nephrology (Carlton, Vic.), 17*(7), 650–657. DOI: 10.1111/j.1440-1797.2012.01622.x PMID: 22640266

Ezquer, F. E., Ezquer, M. E., Parrau, D. B., Carpio, D., Yañez, A. J., & Conget, P. A. (2008). Systemic administration of multipotent mesenchymal stromal cells reverts hyperglycemia and prevents nephropathy in type 1 diabetic mice. *Biology of Blood and Marrow Transplantation, 14*(6), 631–640. DOI: 10.1016/j.bbmt.2008.01.006 PMID: 18489988

Feng, J., Lu, C., Dai, Q., Sheng, J., & Xu, M. (2018). SIRT3 Facilitates Amniotic Fluid Stem Cells to Repair Diabetic Nephropathy Through Protecting Mitochondrial Homeostasis by Modulation of Mitophagy. *Cellular Physiology and Biochemistry: International Journal of Experimental Cellular Physiology. Biochemical Pharmacology, 46*(4), 1508–1524. DOI: 10.1159/000489194 PMID: 29689547

Forbes, T. A., Howden, S. E., Lawlor, K., Phipson, B., Maksimovic, J., Hale, L., Wilson, S., Quinlan, C., Ho, G., Holman, K., Bennetts, B., Crawford, J., Trnka, P., Oshlack, A., Patel, C., Mallett, A., Simons, C., & Little, M. H. (2018). Patient-iPSC-Derived Kidney Organoids Show Functional Validation of a Ciliopathic Renal Phenotype and Reveal Underlying Pathogenetic Mechanisms. *American Journal of Human Genetics, 102*(5), 816–831. DOI: 10.1016/j.ajhg.2018.03.014 PMID: 29706353

Freedman, B. S., Brooks, C. R., Lam, A. Q., Fu, H., Morizane, R., Agrawal, V., Saad, A. F., Li, M. K., Hughes, M. R., Werff, R. V., Peters, D. T., Lu, J., Baccei, A., Siedlecki, A. M., Valerius, M. T., Musunuru, K., McNagny, K. M., Steinman, T. I., Zhou, J., & Bonventre, J. V. (2015). Modelling kidney disease with CRISPR-mutant kidney organoids derived from human pluripotent epiblast spheroids. *Nature Communications, 6*(1), 8715. DOI: 10.1038/ncomms9715 PMID: 26493500

Freedman, B. S., Lam, A. Q., Sundsbak, J. L., Iatrino, R., Su, X., Koon, S. J., Wu, M., Daheron, L., Harris, P. C., Zhou, J., & Bonventre, J. V. (2013). Reduced ciliary polycystin-2 in induced pluripotent stem cells from polycystic kidney disease patients with PKD1 mutations. *Journal of the American Society of Nephrology, 24*(10), 1571–1586. DOI: 10.1681/ASN.2012111089 PMID: 24009235

Geng, X.-D., Zheng, W., Wu, C.-M., Wang, S.-Q., Hong, Q., Cai, G.-Y., Chen, X.-M., & Wu, D. (2016). Embryonic Stem Cells-loaded Gelatin Microcryogels Slow Progression of Chronic Kidney Disease. *Chinese Medical Journal, 129*(4), 392–398. DOI: 10.4103/0366-6999.176088 PMID: 26879011

Goligorsky, M. S., Kuo, M.-C., Patschan, D., & Verhaar, M. C. (2009). Review article: Endothelial progenitor cells in renal disease. *Nephrology (Carlton, Vic.), 14*(3), 291–297. DOI: 10.1111/j.1440-1797.2009.01112.x PMID: 19444962

Gomes, S. A., Hare, J. M., & Rangel, E. B. (2018). Kidney-Derived c-Kit+ Cells Possess Regenerative Potential. *Stem Cells Translational Medicine, 7*(4), 317–324. DOI: 10.1002/sctm.17-0232 PMID: 29575816

Gupta, S., Lodha, P., Karthick, M. S., & Tandulwadkar, S. (2018). Role of Autologous Bone Marrow-Derived Stem Cell Therapy for Follicular Recruitment in Premature Ovarian Insufficiency: Review of Literature and a Case Report of World's First Baby with Ovarian Autologous Stem Cell Therapy in a Perimenopausal Woman of Age 45 Year. *Journal of Human Reproductive Sciences, 11*(2), 125–130. DOI: 10.4103/jhrs.JHRS_57_18 PMID: 30158807

Hafezparast, M., Kaur, G. P., Zdzienicka, M., Athwal, R. S., Lehmann, A. R., & Jeggo, P. A. (1993). Subchromosomal localization of a gene (XRCC5) involved in double strand break repair to the region 2q34-36. *Somatic Cell and Molecular Genetics, 19*(5), 413–421. DOI: 10.1007/BF01233246 PMID: 8291019

Harari-Steinberg, O., Metsuyanim, S., Omer, D., Gnatek, Y., Gershon, R., Pri-Chen, S., Ozdemir, D. D., Lerenthal, Y., Noiman, T., Ben-Hur, H., Vaknin, Z., Schneider, D. F., Aronow, B. J., Goldstein, R. S., Hohenstein, P., & Dekel, B. (2013). Identification of human nephron progenitors capable of generation of kidney structures and functional repair of chronic renal disease. *EMBO Molecular Medicine, 5*(10), 1556–1568. DOI: 10.1002/emmm.201201584 PMID: 23996934

Hauser, P. V., De Fazio, R., Bruno, S., Sdei, S., Grange, C., Bussolati, B., Benedetto, C., & Camussi, G. (2010). Stem cells derived from human amniotic fluid contribute to acute kidney injury recovery. *American Journal of Pathology, 177*(4), 2011–2021. DOI: 10.2353/ajpath.2010.091245 PMID: 20724594

Hendry, C. E., Vanslambrouck, J. M., Ineson, J., Suhaimi, N., Takasato, M., Rae, F., & Little, M. H. (2013). Direct transcriptional reprogramming of adult cells to embryonic nephron progenitors. *Journal of the American Society of Nephrology, 24*(9), 1424–1434. DOI: 10.1681/ASN.2012121143 PMID: 23766537

Herrera, M. B., Bussolati, B., Bruno, S., Morando, L., Mauriello-Romanazzi, G., Sanavio, F., Stamenkovic, I., Biancone, L., & Camussi, G. (2007). Exogenous mesenchymal stem cells localize to the kidney by means of CD44 following acute tubular injury. *Kidney International*, *72*(4), 430–441. DOI: 10.1038/sj.ki.5002334 PMID: 17507906

Hickson, L. J., Eirin, A., & Lerman, L. O. (2016). Challenges and opportunities for stem cell therapy in patients with chronic kidney disease. *Kidney International*, *89*(4), 767–778. DOI: 10.1016/j.kint.2015.11.023 PMID: 26924058

Hoeijmakers, J. H. (2001). Genome maintenance mechanisms for preventing cancer. *Nature*, *411*(6835), 366–374. DOI: 10.1038/35077232 PMID: 11357144

Imberti, B., Tomasoni, S., Ciampi, O., Pezzotta, A., Derosas, M., Xinaris, C., Rizzo, P., Papadimou, E., Novelli, R., Benigni, A., Remuzzi, G., & Morigi, M. (2015). Renal progenitors derived from human iPSCs engraft and restore function in a mouse model of acute kidney injury. *Scientific Reports*, *5*(1), 8826. DOI: 10.1038/srep08826 PMID: 25744951

James, M. T., Hemmelgarn, B. R., & Tonelli, M. (2010). Early recognition and prevention of chronic kidney disease. *Lancet*, *375*(9722), 1296–1309. DOI: 10.1016/S0140-6736(09)62004-3 PMID: 20382326

Jeggo, P. A., Pearl, L. H., & Carr, A. M. (2016). DNA repair, genome stability and cancer: A historical perspective. *Nature Reviews. Cancer*, *16*(1), 35–42. DOI: 10.1038/nrc.2015.4 PMID: 26667849

Jiang, M. H., Li, G., Liu, J., Liu, L., Wu, B., Huang, W., He, W., Deng, C., Wang, D., Li, C., Lahn, B. T., Shi, C., & Xiang, A. P. (2015). Nestin(+) kidney resident mesenchymal stem cells for the treatment of acute kidney ischemia injury. *Biomaterials*, *50*, 56–66. DOI: 10.1016/j.biomaterials.2015.01.029 PMID: 25736496

Jiao, J., Milwid, J. M., Yarmush, M. L., & Parekkadan, B. (2011). A mesenchymal stem cell potency assay. *Methods in Molecular Biology (Clifton, N.J.)*, *677*, 221–231. DOI: 10.1007/978-1-60761-869-0_16 PMID: 20941614

Jinek, M., Chylinski, K., Fonfara, I., Hauer, M., Doudna, J. A., & Charpentier, E. (2012). A programmable dual-RNA-guided DNA endonuclease in adaptive bacterial immunity. *Science*, *337*(6096), 816–821. DOI: 10.1126/science.1225829 PMID: 22745249

Kang, M., & Han, Y.-M. (2014). Differentiation of human pluripotent stem cells into nephron progenitor cells in a serum and feeder free system. *PLoS One*, *9*(4), e94888. DOI: 10.1371/journal.pone.0094888 PMID: 24728509

Kim, D., & Dressler, G. R. (2005). Nephrogenic factors promote differentiation of mouse embryonic stem cells into renal epithelia. *Journal of the American Society of Nephrology*, *16*(12), 3527–3534. DOI: 10.1681/ASN.2005050544 PMID: 16267156

Kim, J.-S., Lee, J.-H., Kwon, O., Cho, J.-H., Choi, J.-Y., Park, S.-H., Kim, C.-D., Kim, Y.-J., & Kim, Y.-L. (2017). Rapid deterioration of preexisting renal insufficiency after autologous mesenchymal stem cell therapy. *Kidney Research and Clinical Practice*, *36*(2), 200–204. DOI: 10.23876/j.krcp.2017.36.2.200 PMID: 28680828

Kim, Y. K., Refaeli, I., Brooks, C. R., Jing, P., Gulieva, R. E., Hughes, M. R., Cruz, N. M., Liu, Y., Churchill, A. J., Wang, Y., Fu, H., Pippin, J. W., Lin, L. Y., Shankland, S. J., Vogl, A. W., McNagny, K. M., & Freedman, B. S. (2017). Gene-Edited Human Kidney Organoids Reveal Mechanisms of Disease in Podocyte Development. *Stem Cells (Dayton, Ohio)*, *35*(12), 2366-2378. DOI: 10.1002/stem.2707

Kuppe, C., & Kramann, R. (2016). Role of mesenchymal stem cells in kidney injury and fibrosis. *Current Opinion in Nephrology and Hypertension*, *25*(4), 372–377. DOI: 10.1097/MNH.0000000000000230 PMID: 27191350

Lam, A. Q., Freedman, B. S., Morizane, R., Lerou, P. H., Valerius, M. T., & Bonventre, J. V. (2014). Rapid and efficient differentiation of human pluripotent stem cells into intermediate mesoderm that forms tubules expressing kidney proximal tubular markers. *Journal of the American Society of Nephrology*, *25*(6), 1211–1225. DOI: 10.1681/ASN.2013080831 PMID: 24357672

Lau, R. W., Wang, B., & Ricardo, S. D. (2018). Gene editing of stem cells for kidney disease modelling and therapeutic intervention. *Nephrology (Carlton, Vic.)*, *23*(11), 981–990. DOI: 10.1111/nep.13410 PMID: 29851168

Lazzeri, E., Romagnani, P., & Lasagni, L. (2015). Stem cell therapy for kidney disease. *Expert Opinion on Biological Therapy*, *15*(10), 1455–1468. DOI: 10.1517/14712598.2015.1067300 PMID: 26160420

Lazzeri, E., Ronconi, E., Angelotti, M. L., Peired, A., Mazzinghi, B., Becherucci, F., Conti, S., Sansavini, G., Sisti, A., Ravaglia, F., Lombardi, D., Provenzano, A., Manonelles, A., Cruzado, J. M., Giglio, S., Roperto, R. M., Materassi, M., Lasagni, L., & Romagnani, P. (2015). Human Urine-Derived Renal Progenitors for Personalized Modeling of Genetic Kidney Disorders. *Journal of the American Society of Nephrology*, *26*(8), 1961–1974. DOI: 10.1681/ASN.2014010057 PMID: 25568173

Le Page, M. (s. f.). What is CRISPR? *NewScientist*. https://www.newscientist.com/definition/what-is-crispr/#:~:text=A%20technology%20that%20can%20be,of%20DNA%20inside%20a%20cell

Lee, P.-Y., Chien, Y., Chiou, G.-Y., Lin, C.-H., Chiou, C.-H., & Tarng, D.-C. (2012). Induced pluripotent stem cells without c-Myc attenuate acute kidney injury via down-regulating the signaling of oxidative stress and inflammation in ischemia-reperfusion rats. *Cell Transplantation*, *21*(12), 2569–2585. DOI: 10.3727/096368912X636902 PMID: 22507855

Lee, R. H., Seo, M. J., Reger, R. L., Spees, J. L., Pulin, A. A., Olson, S. D., & Prockop, D. J. (2006). Multipotent stromal cells from human marrow home to and promote repair of pancreatic islets and renal glomeruli in diabetic NOD/scid mice. *Proceedings of the National Academy of Sciences of the United States of America*, *103*(46), 17438–17443. DOI: 10.1073/pnas.0608249103 PMID: 17088535

Lee, S.-J., Ryu, M.-O., Seo, M.-S., Park, S.-B., Ahn, J.-O., Han, S.-M., Kang, K.-S., Bhang, D.-H., & Youn, H.-Y. (2017). Mesenchymal Stem Cells Contribute to Improvement of Renal Function in a Canine Kidney Injury Model. *In Vivo (Athens, Greece)*, *31*(6), 1115–1124. DOI: 10.21873/invivo.11177 PMID: 29102933

Lee, S.-R., Lee, S.-H., Moon, J.-Y., Park, J.-Y., Lee, D., Lim, S. J., Jeong, K.-H., Park, J.-K., Lee, T.-W., & Ihm, C.-G. (2010). Repeated administration of bone marrow-derived mesenchymal stem cells improved the protective effects on a remnant kidney model. *Renal Failure*, *32*(7), 840–848. DOI: 10.3109/0886022X.2010.494803 PMID: 20662698

Levey, A. S., & Coresh, J. (2012). Chronic kidney disease. *Lancet*, *379*(9811), 165–180. DOI: 10.1016/S0140-6736(11)60178-5 PMID: 21840587

Lieber, M. R., Ma, Y., Pannicke, U., & Schwarz, K. (2004). The mechanism of vertebrate nonhomologous DNA end joining and its role in V(D)J recombination. *DNA Repair*, *3*(8-9), 817–826. DOI: 10.1016/j.dnarep.2004.03.015 PMID: 15279766

Lin, S. A., Kolle, G., Grimmond, S. M., Zhou, Q., Doust, E., Little, M. H., Aronow, B., Ricardo, S. D., Pera, M. F., Bertram, J. F., & Laslett, A. L. (2010). Subfractionation of differentiating human embryonic stem cell populations allows the isolation of a mesodermal population enriched for intermediate mesoderm and putative renal progenitors. *Stem Cells and Development*, *19*(10), 1637–1648. DOI: 10.1089/scd.2010.0017 PMID: 20143954

Luo, Y., Xu, X., An, X., Sun, X., Wang, S., & Zhu, D. (2016). Targeted Inhibition of the miR-199a/214 Cluster by CRISPR Interference Augments the Tumor Tropism of Human Induced Pluripotent Stem Cell-Derived Neural Stem Cells under Hypoxic Condition. *Stem Cells International*, *3598542*(1), 3598542. Advance online publication. DOI: 10.1155/2016/3598542 PMID: 27965712

Mae, S.-I., Shono, A., Shiota, F., Yasuno, T., Kajiwara, M., Gotoda-Nishimura, N., Arai, S., Sato-Otubo, A., Toyoda, T., Takahashi, K., Nakayama, N., Cowan, C. A., Aoi, T., Ogawa, S., McMahon, A. P., Yamanaka, S., & Osafune, K. (2013). Monitoring and robust induction of nephrogenic intermediate mesoderm from human pluripotent stem cells. *Nature Communications*, *4*(1), 1367. DOI: 10.1038/ncomms2378 PMID: 23340407

Makhlough, A., Shekarchian, S., Moghadasali, R., Einollahi, B., Dastgheib, M., Janbabaee, G., Hosseini, S. E., Falah, N., Abbasi, F., Baharvand, H., & Aghdami, N. (2018). Bone marrow-mesenchymal stromal cell infusion in patients with chronic kidney disease: A safety study with 18 months of follow-up. *Cytotherapy*, *20*(5), 660–669. DOI: 10.1016/j.jcyt.2018.02.368 PMID: 29580865

Makhlough, A., Shekarchian, S., Moghadasali, R., Einollahi, B., Hosseini, S. E., Jaroughi, N., Bolurieh, T., Baharvand, H., & Aghdami, N. (2017). Safety and tolerability of autologous bone marrow mesenchymal stromal cells in ADPKD patients. *Stem Cell Research & Therapy*, *8*(1), 116. DOI: 10.1186/s13287-017-0557-7 PMID: 28535817

Mali, P., Yang, L., Esvelt, K. M., Aach, J., Guell, M., DiCarlo, J. E., Norville, J. E., & Church, G. M. (2013). RNA-guided human genome engineering via Cas9. *Science*, *339*(6121), 823–826. DOI: 10.1126/science.1232033 PMID: 23287722

Mandegar, M. A., Huebsch, N., Frolov, E. B., Shin, E., Truong, A., Olvera, M. P., Chan, A. H., Miyaoka, Y., Holmes, K., Spencer, C. I., Judge, L. M., Gordon, D. E., Eskildsen, T. V., Villalta, J. E., Horlbeck, M. A., Gilbert, L. A., Krogan, N. J., Sheikh, S. P., Weissman, J. S., & Conklin, B. R. (2016). CRISPR Interference Efficiently Induces Specific and Reversible Gene Silencing in Human iPSCs. *Cell Stem Cell*, *18*(4), 541–553. DOI: 10.1016/j.stem.2016.01.022 PMID: 26971820

Marcheque, J., Bussolati, B., Csete, M., & Perin, L. (2019). Concise Reviews: Stem Cells and Kidney Regeneration: An Update. *Stem Cells Translational Medicine*, *8*(1), 82–92. DOI: 10.1002/sctm.18-0115 PMID: 30302937

Meisel, R., Zibert, A., Laryea, M., Göbel, U., Däubener, W., & Dilloo, D. (2004). Human bone marrow stromal cells inhibit allogeneic T-cell responses by indoleamine 2,3-dioxygenase-mediated tryptophan degradation. *Blood*, *103*(12), 4619–4621. DOI: 10.1182/blood-2003-11-3909 PMID: 15001472

Moghadasali, R., Hajinasrollah, M., Argani, H., Nassiri, S. M., Najarasl, M., Sodeifi, N., Baharvand, H., & Aghdami, N. (2015). Autologous transplantation of mesenchymal stromal cells tends to prevent progress of interstitial fibrosis in a rhesus Macaca mulatta monkey model of chronic kidney disease. *Cytotherapy*, *17*(11), 1495–1505. DOI: 10.1016/j.jcyt.2015.06.006 PMID: 26341479

Morigi, M., Imberti, B., Zoja, C., Corna, D., Tomasoni, S., Abbate, M., Rottoli, D., Angioletti, S., Benigni, A., Perico, N., Alison, M., & Remuzzi, G. (2004). Mesenchymal stem cells are renotropic, helping to repair the kidney and improve function in acute renal failure. *Journal of the American Society of Nephrology*, *15*(7), 1794–1804. DOI: 10.1097/01.ASN.0000128974.07460.34 PMID: 15213267

Morigi, M., Rota, C., Montemurro, T., Montelatici, E., Lo Cicero, V., Imberti, B., Abbate, M., Zoja, C., Cassis, P., Longaretti, L., Rebulla, P., Introna, M., Capelli, C., Benigni, A., Remuzzi, G., & Lazzari, L. (2010). Life-sparing effect of human cord blood-mesenchymal stem cells in experimental acute kidney injury. *Stem Cells (Dayton, Ohio)*, *28*(3), 513–522. DOI: 10.1002/stem.293 PMID: 20049901

Morizane, R., Lam, A. Q., Freedman, B. S., Kishi, S., Valerius, M. T., & Bonventre, J. V. (2015). Nephron organoids derived from human pluripotent stem cells model kidney development and injury. *Nature Biotechnology*, *33*(11), 1193–1200. DOI: 10.1038/nbt.3392 PMID: 26458176

National Kidney Foundation. (2002). K/DOQI clinical practice guidelines for chronic kidney disease: Evaluation, classification, and stratification. *American Journal of Kidney Diseases*, *39*(2, Suppl 1), S1–S266. PMID: 11904577

Németh, K., Leelahavanichkul, A., Yuen, P. S. T., Mayer, B., Parmelee, A., Doi, K., Robey, P. G., Leelahavanichkul, K., Koller, B. H., Brown, J. M., Hu, X., Jelinek, I., Star, R. A., & Mezey, E. (2009). Bone marrow stromal cells attenuate sepsis via prostaglandin E(2)-dependent reprogramming of host macrophages to increase their interleukin-10 production. *Nature Medicine*, *15*(1), 42–49. DOI: 10.1038/nm.1905 PMID: 19098906

NHI. (2023, noviembre 16). *The CRISPR Revolution*. https://www.nih.gov/about-nih/what-we-do/nih-turning-discovery-into-health/transformative-technologies/crispr-revolution#:~:text=NIH%20scientists%20are%20getting%20close,even%20removing%20the%20DNA%20as

NIH. (2024, agosto 16). CRISPR. National Human Genome Research Institute. https://www.genome.gov/genetics-glossary/CRISPR#:~:text=CRISPR%20(short%20for%20%E2%80%9Cclustered%20regularly,selectively%20modify%20the%20DNA%20of

Osafune, K., Caron, L., Borowiak, M., Martinez, R. J., Fitz-Gerald, C. S., Sato, Y., Cowan, C. A., Chien, K. R., & Melton, D. A. (2008). Marked differences in differentiation propensity among human embryonic stem cell lines. *Nature Biotechnology*, *26*(3), 313–315. DOI: 10.1038/nbt1383

Packham, D. K., Fraser, I. R., Kerr, P. G., & Segal, K. R. (2016). Allogeneic Mesenchymal Precursor Cells (MPC) in Diabetic Nephropathy: A Randomized, Placebo-controlled, Dose Escalation Study. *EBioMedicine*, *12*, 263–269. DOI: 10.1016/j.ebiom.2016.09.011 PMID: 27743903

Papazova, D. A., Oosterhuis, N. R., Gremmels, H., van Koppen, A., Joles, J. A., & Verhaar, M. C. (2015). Cell-based therapies for experimental chronic kidney disease: A systematic review and meta-analysis. *Disease Models & Mechanisms*, *8*(3), 281–293. DOI: 10.1242/dmm.017699 PMID: 25633980

Pardo, B., Gómez-González, B., & Aguilera, A. (2009). DNA repair in mammalian cells: DNA double-strand break repair: how to fix a broken relationship. *Cellular and Molecular Life Sciences : CMLS*, *66*(6), 1039–1056. DOI: 10.1007/s00018-009-8740-3 PMID: 19153654

Peired, A. J., Sisti, A., & Romagnani, P. (2016). Mesenchymal Stem Cell-Based Therapy for Kidney Disease: A Review of Clinical Evidence. *Stem Cells International*, *4798639*(1), 4798639. Advance online publication. DOI: 10.1155/2016/4798639 PMID: 27721835

Pereira, E. M., Labilloy, A., Eshbach, M. L., Roy, A., Subramanya, A. R., Monte, S., Labilloy, G., & Weisz, O. A. (2016). Characterization and phosphoproteomic analysis of a human immortalized podocyte model of Fabry disease generated using CRISPR/Cas9 technology. *American Journal of Physiology. Renal Physiology*, *311*(5), F1015–F1024. DOI: 10.1152/ajprenal.00283.2016 PMID: 27681560

Perin, L., Giuliani, S., Jin, D., Sedrakyan, S., Carraro, G., Habibian, R., Warburton, D., Atala, A., & De Filippo, R. E. (2007). Renal differentiation of amniotic fluid stem cells. *Cell Proliferation*, *40*(6), 936–948. DOI: 10.1111/j.1365-2184.2007.00478.x PMID: 18021180

Perin, L., Sedrakyan, S., Giuliani, S., Da Sacco, S., Carraro, G., Shiri, L., Lemley, K. V., Rosol, M., Wu, S., Atala, A., Warburton, D., & De Filippo, R. E. (2010). Protective effect of human amniotic fluid stem cells in an immunodeficient mouse model of acute tubular necrosis. *PLoS One*, *5*(2), e9357. DOI: 10.1371/journal.pone.0009357 PMID: 20195358

Pleniceanu, O., Omer, D., Harari-Steinberg, O., & Dekel, B. (2018). Renal lineage cells as a source for renal regeneration. *Pediatric Research, 83*(1-2), 267–274. DOI: 10.1038/pr.2017.255 PMID: 28985199

Quimby, J. M., Webb, T. L., Randall, E., Marolf, A., Valdes-Martinez, A., & Dow, S. W. (2016). Assessment of intravenous adipose-derived allogeneic mesenchymal stem cells for the treatment of feline chronic kidney disease: A randomized, placebo-controlled clinical trial in eight cats. *Journal of Feline Medicine and Surgery, 18*(2), 165–171. DOI: 10.1177/1098612X15576980 PMID: 25784460

Ranghino, A., Bruno, S., Bussolati, B., Moggio, A., Dimuccio, V., Tapparo, M., Biancone, L., Gontero, P., Frea, B., & Camussi, G. (2017). The effects of glomerular and tubular renal progenitors and derived extracellular vesicles on recovery from acute kidney injury. *Stem Cell Research & Therapy, 8*(1), 24. DOI: 10.1186/s13287-017-0478-5 PMID: 28173878

Ratliff, B. B., & Goligorsky, M. S. (2013). Delivery of EPC embedded in HA-hydrogels for treatment of acute kidney injury. *Biomatter, 3*(1), e23284. DOI: 10.4161/biom.23284 PMID: 23507925

Rota, C., Imberti, B., Pozzobon, M., Piccoli, M., De Coppi, P., Atala, A., Gagliardini, E., Xinaris, C., Benedetti, V., Fabricio, A. S. C., Squarcina, E., Abbate, M., Benigni, A., Remuzzi, G., & Morigi, M. (2012). Human amniotic fluid stem cell preconditioning improves their regenerative potential. *Stem Cells and Development, 21*(11), 1911–1923. DOI: 10.1089/scd.2011.0333 PMID: 22066606

Rouet, P., Smih, F., & Jasin, M. (1994). Introduction of double-strand breaks into the genome of mouse cells by expression of a rare-cutting endonuclease. *Molecular and Cellular Biology, 14*(12), 8096–8106. DOI: 10.1128/mcb.14.12.8096-8106.1994 PMID: 7969147

Rudin, N., Sugarman, E., & Haber, J. E. (1989). Genetic and physical analysis of double-strand break repair and recombination in Saccharomyces cerevisiae. *Genetics, 122*(3), 519–534. DOI: 10.1093/genetics/122.3.519 PMID: 2668114

Ruggenenti, P., Perticucci, E., Cravedi, P., Gambara, V., Costantini, M., Sharma, S. K., Perna, A., & Remuzzi, G. (2008). Role of remission clinics in the longitudinal treatment of CKD. *Journal of the American Society of Nephrology, 19*(6), 1213–1224. DOI: 10.1681/ASN.2007090970 PMID: 18354029

Saad, A., Dietz, A. B., Herrmann, S. M. S., Hickson, L. J., Glockner, J. F., McKusick, M. A., Misra, S., Bjarnason, H., Armstrong, A. S., Gastineau, D. A., Lerman, L. O., & Textor, S. C. (2017). Autologous Mesenchymal Stem Cells Increase Cortical Perfusion in Renovascular Disease. *Journal of the American Society of Nephrology, 28*(9), 2777–2785. DOI: 10.1681/ASN.2017020151 PMID: 28461553

Sabatino, M., Ren, J., David-Ocampo, V., England, L., McGann, M., Tran, M., Kuznetsov, S. A., Khuu, H., Balakumaran, A., Klein, H. G., Robey, P. G., & Stroncek, D. F. (2012). The establishment of a bank of stored clinical bone marrow stromal cell products. *Journal of Translational Medicine, 10*(1), 23. DOI: 10.1186/1479-5876-10-23 PMID: 22309358

Santeramo, I., Herrera Perez, Z., Illera, A., Taylor, A., Kenny, S., Murray, P., Wilm, B., & Gretz, N. (2017). Human Kidney-Derived Cells Ameliorate Acute Kidney Injury Without Engrafting into Renal Tissue. *Stem Cells Translational Medicine, 6*(5), 1373–1384. DOI: 10.1002/sctm.16-0352 PMID: 28375556

Sedrakyan, S., Da Sacco, S., Milanesi, A., Shiri, L., Petrosyan, A., Varimezova, R., Warburton, D., Lemley, K. V., De Filippo, R. E., & Perin, L. (2012). Injection of amniotic fluid stem cells delays progression of renal fibrosis. *Journal of the American Society of Nephrology, 23*(4), 661–673. DOI: 10.1681/ASN.2011030243 PMID: 22302195

Sedrakyan, S., Villani, V., Da Sacco, S., Tripuraneni, N., Porta, S., Achena, A., Lavarreda-Pearce, M., Petrosyan, A., Soloyan, H., Filippo, R. E. D., Bussolati, B., & Perin, L. (2017). Amniotic fluid stem cell-derived vesicles protect from VEGF-induced endothelial damage. *Scientific Reports, 7*(1), 16875. DOI: 10.1038/s41598-017-17061-2 PMID: 29203902

Semedo, P., Correa-Costa, M., Antonio Cenedeze, M., Maria Avancini Costa Malheiros, D., Antonia dos Reis, M., Shimizu, M. H., Seguro, A. C., Pacheco-Silva, A., & Saraiva Camara, N. O. (2009). Mesenchymal stem cells attenuate renal fibrosis through immune modulation and remodeling properties in a rat remnant kidney model. *Stem Cells (Dayton, Ohio), 27*(12), 3063–3073. DOI: 10.1002/stem.214 PMID: 19750536

Semedo, P., Wang, P. M., Andreucci, T. H., Cenedeze, M. A., Teixeira, V. P. A., Reis, M. A., Pacheco-Silva, A., & Câmara, N. O. S. (2007). Mesenchymal stem cells ameliorate tissue damages triggered by renal ischemia and reperfusion injury. *Transplantation Proceedings, 39*(2), 421–423. DOI: 10.1016/j.transproceed.2007.01.036 PMID: 17362746

Sheashaa, H., Lotfy, A., Elhusseini, F., Aziz, A. A., Baiomy, A., Awad, S., Alsayed, A., El-Gilany, A.-H., Saad, M.-A. A. A., Mahmoud, K., Zahran, F., Salem, D. A., Sarhan, A., Ghaffar, H. A., & Sobh, M. (2016). Protective effect of adipose-derived mesenchymal stem cells against acute kidney injury induced by ischemia-reperfusion in Sprague-Dawley rats. *Experimental and Therapeutic Medicine*, *11*(5), 1573–1580. DOI: 10.3892/etm.2016.3109 PMID: 27168775

Shi, M., Liu, Z.-W., & Wang, F.-S. (2011). Immunomodulatory properties and therapeutic application of mesenchymal stem cells. *Clinical and Experimental Immunology*, *164*(1), 1–8. DOI: 10.1111/j.1365-2249.2011.04327.x PMID: 21352202

Siegel, N., Valli, A., Fuchs, C., Rosner, M., & Hengstschläger, M. (2009). Induction of mesenchymal/epithelial marker expression in human amniotic fluid stem cells. *Reproductive Biomedicine Online*, *19*(6), 838–846. DOI: 10.1016/j.rbmo.2009.09.015 PMID: 20031026

Song, B., Smink, A. M., Jones, C. V., Callaghan, J. M., Firth, S. D., Bernard, C. A., Laslett, A. L., Kerr, P. G., & Ricardo, S. D. (2012). The directed differentiation of human iPS cells into kidney podocytes. *PLoS One*, *7*(9), e46453. DOI: 10.1371/journal.pone.0046453 PMID: 23029522

Song, Y., Peng, C., Lv, S., Cheng, J., Liu, S., Wen, Q., Guan, G., & Liu, G. (2017). Adipose-derived stem cells ameliorate renal interstitial fibrosis through inhibition of EMT and inflammatory response via TGF-β1 signaling pathway. *International Immunopharmacology*, *44*, 115–122. DOI: 10.1016/j.intimp.2017.01.008 PMID: 28092863

Spaggiari, G. M., Capobianco, A., Becchetti, S., Mingari, M. C., & Moretta, L. (2006). Mesenchymal stem cell-natural killer cell interactions: Evidence that activated NK cells are capable of killing MSCs, whereas MSCs can inhibit IL-2-induced NK-cell proliferation. *Blood*, *107*(4), 1484–1490. DOI: 10.1182/blood-2005-07-2775 PMID: 16239427

Stagg, J. (2007). Immune regulation by mesenchymal stem cells: Two sides to the coin. *Tissue Antigens*, *69*(1), 1–9. DOI: 10.1111/j.1399-0039.2006.00739.x PMID: 17212702

Sun, D., Bu, L., Liu, C., Yin, Z., Zhou, X., Li, X., & Xiao, A. (2013). Therapeutic effects of human amniotic fluid-derived stem cells on renal interstitial fibrosis in a murine model of unilateral ureteral obstruction. *PLoS One*, *8*(5), e65042. DOI: 10.1371/journal.pone.0065042 PMID: 23724119

Sung, P.-H., Chiang, H.-J., Chen, C.-H., Chen, Y.-L., Huang, T.-H., Zhen, Y.-Y., Chang, M.-W., Liu, C.-F., Chung, S.-Y., Chen, Y.-L., Chai, H.-T., Sun, C.-K., & Yip, H.-K. (2016). Combined Therapy With Adipose-Derived Mesenchymal Stem Cells and Ciprofloxacin Against Acute Urogenital Organ Damage in Rat Sepsis Syndrome Induced by Intrapelvic Injection of Cecal Bacteria. *Stem Cells Translational Medicine*, *5*(6), 782–792. DOI: 10.5966/sctm.2015-0116 PMID: 27075767

Swaminathan, M., Stafford-Smith, M., Chertow, G. M., Warnock, D. G., Paragamian, V., Brenner, R. M., Lellouche, F., Fox-Robichaud, A., Atta, M. G., Melby, S., Mehta, R. L., Wald, R., Verma, S., & Mazer, C. D.ACT-AKI investigators. (2018). Allogeneic Mesenchymal Stem Cells for Treatment of AKI after Cardiac Surgery. *Journal of the American Society of Nephrology*, *29*(1), 260–267. DOI: 10.1681/ASN.2016101150 PMID: 29038286

Taguchi, A., Kaku, Y., Ohmori, T., Sharmin, S., Ogawa, M., Sasaki, H., & Nishinakamura, R. (2014). Redefining the in vivo origin of metanephric nephron progenitors enables generation of complex kidney structures from pluripotent stem cells. *Cell Stem Cell*, *14*(1), 53–67. DOI: 10.1016/j.stem.2013.11.010 PMID: 24332837

Takahashi, K., Tanabe, K., Ohnuki, M., Narita, M., Ichisaka, T., Tomoda, K., & Yamanaka, S. (2007). Induction of pluripotent stem cells from adult human fibroblasts by defined factors. *Cell*, *131*(5), 861–872. DOI: 10.1016/j.cell.2007.11.019 PMID: 18035408

Takasato, M., Er, P. X., Becroft, M., Vanslambrouck, J. M., Stanley, E. G., Elefanty, A. G., & Little, M. H. (2014). Directing human embryonic stem cell differentiation towards a renal lineage generates a self-organizing kidney. *Nature Cell Biology*, *16*(1), 118–126. DOI: 10.1038/ncb2894 PMID: 24335651

Takasato, M., Er, P. X., Chiu, H. S., Maier, B., Baillie, G. J., Ferguson, C., Parton, R. G., Wolvetang, E. J., Roost, M. S., Chuva de Sousa Lopes, S. M., & Little, M. H. (2015). Kidney organoids from human iPS cells contain multiple lineages and model human nephrogenesis. *Nature*, *526*(7574), 564–568. DOI: 10.1038/nature15695 PMID: 26444236

Takata, M., Sasaki, M. S., Sonoda, E., Morrison, C., Hashimoto, M., Utsumi, H., Yamaguchi-Iwai, Y., Shinohara, A., & Takeda, S. (1998). Homologous recombination and non-homologous end-joining pathways of DNA double-strand break repair have overlapping roles in the maintenance of chromosomal integrity in vertebrate cells. *The EMBO Journal*, *17*(18), 5497–5508. DOI: 10.1093/emboj/17.18.5497 PMID: 9736627

Tanavde, V., Vaz, C., Rao, M. S., Vemuri, M. C., & Pochampally, R. R. (2015). Research using Mesenchymal Stem/Stromal Cells: Quality metric towards developing a reference material. *Cytotherapy*, *17*(9), 1169–1177. DOI: 10.1016/j.jcyt.2015.07.008 PMID: 26276001

Tarng, D.-C., Tseng, W.-C., Lee, P.-Y., Chiou, S.-H., & Hsieh, S.-L. (2016). Induced Pluripotent Stem Cell-Derived Conditioned Medium Attenuates Acute Kidney Injury by Downregulating the Oxidative Stress-Related Pathway in Ischemia-Reperfusion Rats. *Cell Transplantation*, *25*(3), 517–530. DOI: 10.3727/096368915X688542 PMID: 26132529

Thacker, J., & Wilkinson, R. E. (1991). The genetic basis of resistance to ionising radiation damage in cultured mammalian cells. *Mutation Research*, *254*(2), 135–142. DOI: 10.1016/0921-8777(91)90004-9 PMID: 2002810

Thatava, T., Armstrong, A. S., De Lamo, J. G., Edukulla, R., Khan, Y. K., Sakuma, T., Ohmine, S., Sundsbak, J. L., Harris, P. C., Kudva, Y. C., & Ikeda, Y. (2011). Successful disease-specific induced pluripotent stem cell generation from patients with kidney transplantation. *Stem Cell Research & Therapy*, *2*(6), 48. DOI: 10.1186/scrt89 PMID: 22142803

Thiel, A., Yavanian, G., Nastke, M.-D., Morales, P., Kouris, N. A., Kimbrel, E. A., & Lanza, R. (2015). Human embryonic stem cell-derived mesenchymal cells preserve kidney function and extend lifespan in NZB/W F1 mouse model of lupus nephritis. *Scientific Reports*, *5*(1), 17685. DOI: 10.1038/srep17685 PMID: 26628350

Torres-zambrano, G. M., Villegas-valverde, C. A., Bencomo-hernandez, A. A., Hadi, L. A., Adukkadukkam, D., Rivero-jimenez, R. A., Ahmed, Y. M., Castillo-aleman, Y. M., Ventura-carmenate, Y., Torres-zambrano, G. M., Villegas-valverde, C. A., Bencomo-hernandez, A. A., Hadi, L. A., Rivero-jimenez, R. A., Ahmed, Y. M., Castillo-aleman, Y. M., & Renal, Y. V. (2021). *Renal Involvement and Outcomes in Severe COVID-19 Patients After Stem Cell Jet-Nebulization To cite this article : 10*(6), 231-237. DOI: 10.11648/j.cmr.20211006.19

Toyohara, T., Mae, S.-I., Sueta, S.-I., Inoue, T., Yamagishi, Y., Kawamoto, T., Kasahara, T., Hoshina, A., Toyoda, T., Tanaka, H., Araoka, T., Sato-Otsubo, A., Takahashi, K., Sato, Y., Yamaji, N., Ogawa, S., Yamanaka, S., & Osafune, K. (2015). Cell Therapy Using Human Induced Pluripotent Stem Cell-Derived Renal Progenitors Ameliorates Acute Kidney Injury in Mice. *Stem Cells Translational Medicine*, *4*(9), 980–992. DOI: 10.5966/sctm.2014-0219 PMID: 26198166 van der Spoel, T. I. G., Jansen of Lorkeers, S. J., Agostoni, P., van Belle, E., Gyongyosi, M., Sluijter, J. P. G., Cramer, M. J., Doevendans, P. A., & Chamuleau, S. A. J. (2011). Human relevance of pre-clinical studies in stem cell therapy: Systematic review and meta-analysis of large animal models of ischaemic heart disease. *Cardiovascular Research*, *91*(4), 649–658. DOI: 10.1093/cvr/cvr113 PMID: 21498423

Vigneau, C., Polgar, K., Striker, G., Elliott, J., Hyink, D., Weber, O., Fehling, H.-J., Keller, G., Burrow, C., & Wilson, P. (2007). Mouse embryonic stem cell-derived embryoid bodies generate progenitors that integrate long term into renal proximal tubules in vivo. *Journal of the American Society of Nephrology*, *18*(6), 1709–1720. DOI: 10.1681/ASN.2006101078 PMID: 17475814

Viswanathan, S., Keating, A., Deans, R., Hematti, P., Prockop, D., Stroncek, D. F., Stacey, G., Weiss, D. J., Mason, C., & Rao, M. S. (2014). Soliciting strategies for developing cell-based reference materials to advance mesenchymal stromal cell research and clinical translation. *Stem Cells and Development*, *23*(11), 1157–1167. DOI: 10.1089/scd.2013.0591 PMID: 24422625

Wang, S., Li, Y., Zhao, J., Zhang, J., & Huang, Y. (2013). Mesenchymal stem cells ameliorate podocyte injury and proteinuria in a type 1 diabetic nephropathy rat model. *Biology of Blood and Marrow Transplantation*, *19*(4), 538–546. DOI: 10.1016/j.bbmt.2013.01.001 PMID: 23295166

Wankhade, U. D., Shen, M., Kolhe, R., & Fulzele, S. (2016). Advances in Adipose-Derived Stem Cells Isolation, Characterization, and Application in Regenerative Tissue Engineering. *Stem Cells International*, *3206807*(1), 3206807. Advance online publication. DOI: 10.1155/2016/3206807 PMID: 26981130

Wever, K. E., Menting, T. P., Rovers, M., van der Vliet, J. A., Rongen, G. A., Masereeuw, R., Ritskes-Hoitinga, M., Hooijmans, C. R., & Warlé, M. (2012). Ischemic preconditioning in the animal kidney, a systematic review and meta-analysis. *PLoS One*, *7*(2), e32296. DOI: 10.1371/journal.pone.0032296 PMID: 22389693

Wu, Z., Zhang, S., Zhou, L., Cai, J., Tan, J., Gao, X., Zeng, Z., & Li, D. (2017). Thromboembolism Induced by Umbilical Cord Mesenchymal Stem Cell Infusion: A Report of Two Cases and Literature Review. *Transplantation Proceedings*, *49*(7), 1656–1658. DOI: 10.1016/j.transproceed.2017.03.078 PMID: 28838459

Xia, Y., Nivet, E., Sancho-Martinez, I., Gallegos, T., Suzuki, K., Okamura, D., Wu, M.-Z., Dubova, I., Esteban, C. R., Montserrat, N., Campistol, J. M., & Izpisua Belmonte, J. C. (2013). Directed differentiation of human pluripotent cells to ureteric bud kidney progenitor-like cells. *Nature Cell Biology*, *15*(12), 1507–1515. DOI: 10.1038/ncb2872 PMID: 24240476

Yang, X., Yu, Y., Xu, J., Shu, H., Xia, J., Liu, H., Wu, Y., Zhang, L., Yu, Z., Fang, M., Yu, T., Wang, Y., Pan, S., Zou, X., Yuan, S., & Shang, Y. (2020). Clinical course and outcomes of critically ill patients with SARS-CoV-2 pneumonia in Wuhan, China: A single-centered, retrospective, observational study. *The Lancet. Respiratory Medicine*, *8*(5), 475–481. DOI: 10.1016/S2213-2600(20)30079-5 PMID: 32105632

Zhou, L., Song, Q., Shen, J., Xu, L., Xu, Z., Wu, R., Ge, Y., Zhu, J., Wu, J., Dou, Q., & Jia, R. (2017). Comparison of human adipose stromal vascular fraction and adipose-derived mesenchymal stem cells for the attenuation of acute renal ischemia/reperfusion injury. *Scientific Reports*, *7*(1), 44058. DOI: 10.1038/srep44058 PMID: 28276451

Zhou, Y., Xu, H., Xu, W., Wang, B., Wu, H., Tao, Y., Zhang, B., Wang, M., Mao, F., Yan, Y., Gao, S., Gu, H., Zhu, W., & Qian, H. (2013). Exosomes released by human umbilical cord mesenchymal stem cells protect against cisplatin-induced renal oxidative stress and apoptosis in vivo and in vitro. *Stem Cell Research & Therapy*, *4*(2), 34. DOI: 10.1186/scrt194 PMID: 23618405

Zhu, X.-Y., Urbieta-Caceres, V., Krier, J. D., Textor, S. C., Lerman, A., & Lerman, L. O. (2013). Mesenchymal stem cells and endothelial progenitor cells decrease renal injury in experimental swine renal artery stenosis through different mechanisms. *Stem Cells (Dayton, Ohio)*, *31*(1), 117–125. DOI: 10.1002/stem.1263 PMID: 23097349

Zoja, C., Garcia, P. B., Rota, C., Conti, S., Gagliardini, E., Corna, D., Zanchi, C., Bigini, P., Benigni, A., Remuzzi, G., & Morigi, M. (2012). Mesenchymal stem cell therapy promotes renal repair by limiting glomerular podocyte and progenitor cell dysfunction in adriamycin-induced nephropathy. *American Journal of Physiology. Renal Physiology*, *303*(9), F1370–F1381. DOI: 10.1152/ajprenal.00057.2012 PMID: 22952284

Chapter 8
Stem Cell Applications in Pediatric Urology

ABSTRACT

Pediatric urology is a rapidly evolving field with significant advancements in surgical techniques and medical management. Stem cell, with their unique properties of self-renewal and differentiation, have emerged as promising therapeutic agents for various pediatric urological conditions, giving hope for the treatment of challenge diseases. This review aims to explore the current applications and potential future directions of stem cell therapy in pediatric urology. Current and emerging applications of stem cells in pediatric urology, including treatment of vesicoureteral reflux, neurogenic bladder, hypospadias, and other congenital anomalies, thanks to the role of stem cell in tissue regeneration, modulation of inflammation, and immunomodulation. It's important to note that while stem cell therapy shows promise, more research is needed to fully understand its benefits and risks. Clinical trials are ongoing to evaluate the safety and efficacy of stem cell therapy for various pediatric urological conditions.

INTRODUCTION

Pediatric urology is a specialized field that addresses congenital and acquired disorders of the genitourinary system in children. Numerous congenital and acquired anomalies affect the genitourinary tract, often requiring surgical intervention. These conditions can range from relatively common issues, such as vesicoureteral reflux, to more complex and rare disorders, like bladder exstrophy, neurogenic bladder, hypospadias, epispadias, posterior urethral valves, myelomeningocele, bladder carcinoma, urethral stricture disease, stress urinary incontinence, pelvic organ

DOI: 10.4018/979-8-3693-8829-7.ch008

prolapse, vesicoureteral reflux, and traumatic urinary tract injuries are among the most common (S. Zhang, Li, Li, XuminXie, et al., 2022; S. Zhang et al., 2024).

Despite advances in surgical techniques and medical management, many of these conditions present significant challenges, often requiring multiple surgeries, long-term medication, and ongoing monitoring. Furthermore, the impact of these conditions on a child's quality of life and long-term health can be profound.

Surgical repair typically involves autologous or heterologous grafts, such as skin, oral mucosa, or bowel, to replace or augment missing tissue. While these materials can restore basic anatomy, they often fail to fully restore normal function. Moreover, postoperative complications, particularly in bladder augmentation or neobladder creation using autologous bowel, are frequent.

Unlocking the molecular mechanisms behind bladder and renal tissue injury, repair, or prevention holds transformative potential for these conditions. To address these limitations, tissue engineering has emerged as a promising approach. By utilizing autologous materials like mature bladder cells, bone marrow-derived stem cells, and adipose tissue, engineered tissues can potentially offer superior functional outcomes and minimize immunogenicity (Kollhoff et al., 2011). Unlike muscular dystrophy or sickle cell disease—where single-gene mutations guide gene-based therapies—common pediatric urologic injuries are responses to external stimuli. Their molecular underpinnings are polygenic or environmentally driven rather than mutational. (Aitken et al., 2024)

The advent of SC therapy has opened up new possibilities for treating pediatric urological conditions that were previously difficult or impossible to manage effectively. SC, with their ability to differentiate into various cell types, promote tissue regeneration, and modulate the immune system, offer the potential for regenerative medicine to make a significant impact in pediatric urology.

Congenital anomalies of the kidney and urinary tract (CAKUT), the leading global cause of childhood renal failure, exhibit significant phenotypic diversity (Mulder et al., 2020). With the advent of induced pluripotent stem cell (iPSC)-derived kidney organoids and their demonstrated utility in CAKUT research using CRISPR/Cas9-mutated control iPSCs, there is a growing need for patient-specific CAKUT iPSCs (Morizane et al., 2015; Nicolaou et al., 2015; A. Taguchi et al., 2014; Takasato et al., 2015).

Given the high prevalence of CAKUT in infants and young children, a reliable and acceptable somatic cell source is essential to ensure successful cell isolation and patient or caregiver consent.

Traditional iPSC generation protocols rely on invasive collection of somatic cells, such as skin fibroblasts and peripheral blood cells, which can deter research participation. Urine-derived stem cells (UDSCs), obtained non-invasively, have been shown to be reprogrammable. While previous studies demonstrated the feasibility of generating UDSCs-iPSCs from pediatric patients, their applicability to infants and young children remains uncertain.

In this age group, urine collection can be challenging, often requiring alternative methods like urine bags that are prone to contamination. Additionally, obtaining sufficient urine volumes for UDSCs culture and implementing measures to enhance urine yield, such as increasing fluid intake, may be difficult. Compared to peripheral blood collection, which requires skilled phlebotomists, urine collection can be performed by patients, caregivers, or non-medical research personnel.

To address these challenges, Mulder et al. evaluated the feasibility of using urine as a reliable source of reprogrammable cells for iPSC generation and subsequent kidney organoid development in infants and children. Their results demonstrate that UDSCs could be successfully isolated from 71% of remnant urine samples collected routinely in a pediatric nephrology clinic. Even small-volume, bag-collected infant urine specimens yielded three successful UDSCs isolations, that additionally, can be frozen and safely transported, and subsequent used kidney organoid generation or other regenerative therapies.

Under research conditions with immediate urine sample processing, UDSCs isolation was achieved in 100% of target patients using at most two urine collections. These findings highlight the potential of urine as a valuable and minimally invasive source of reprogrammable cells for iPSC-based research in pediatric populations (Mulder et al., 2020).

APPLICATIONS OF SC IN PEDIATRIC UROLOGY

Conditions affecting bladder function in children and adults, such as myelomeningocele, posterior urethral valves, bladder exstrophy, or spinal cord injury, often necessitate urinary diversion or augmentation cystoplasty. If left untreated, these conditions can lead to severe bladder dysfunction and kidney failure (Schäfer & Stehr, 2018; Smolar et al., 2016).

Currently, the gold standard treatment for end-stage bladder disease refractory to conservative management is enterocystoplasty, a surgical enlargement of the bladder using intestinal tissue. While enterocystoplasty offers functional improvement, it is associated with significant long-term complications, including recurrent urinary tract infections, metabolic abnormalities, stone formation, and malignancies (Kropp, 1998; Rigaud & Le Normand, 2004; Y. Wang & Liao, 2014).

Consequently, there is a pressing clinical need for alternative therapies for these reconstructive procedures. SC-based tissue engineering emerges as one of the most promising future strategies to address this unmet need.

SC therapy offers potential benefits for children with urological conditions due to their ability to differentiate into various cell types and modulate immune responses. The application of SC in pediatric urology is an emerging field, with research focused on a range of congenital and acquired conditions. The following sections discuss some of the key areas where stem cells are being explored for their therapeutic potential. Most of the pathologies were discussed in the previous chapters.

Tissue Regeneration

- Vesicoureteral Reflux (VUR): Stem cells can potentially repair damaged ureters and bladder tissue, reducing the need for surgery in some cases.
- Neurogenic Bladder: Stem cells may help regenerate damaged nerves and muscle tissue, improving bladder function in children with neurological conditions.
- Hypospadias: Stem cells could be used to repair the urethra and create a new opening in the correct location.
- Urethral Strictures: Stem cells may help regenerate damaged urethral tissue and improve urine flow.

Immunomodulation

- Interstitial Cystitis (IC): Stem cells can potentially reduce inflammation and improve symptoms in children with IC.
- Urinary Tract Infections (UTIs): Stem cells may help modulate the immune response and reduce the frequency of UTIs.

Other Potential Applications

- Bladder Cancer: Stem cells could be used to create new bladder tissue, reducing the need for radical cystectomy in some cases.
- Kidney Disease: Stem cells may help regenerate damaged kidney tissue and improve kidney function.

HYPOSPADIAS AND POSTERIOR URETHRAL VALVES (PUV)

Hypospadias is the most common genital malformation in male children, occurring in approximately 1 out of 125 to 300 newborn boys, with a rising incidence. This condition not only affects urination, sexual function, and reproductive ability but also has significant psychological consequences for children and their families (Bouty et al., 2015; Kiss et al., 2011).

The only available treatment for hypospadias is surgery. While numerous surgical techniques have been described, postoperative complications like urethral stricture and urinary fistula remain common, necessitating additional procedures and highlighting the challenges associated with treating this condition (Mouravas et al., 2014).

A key factor contributing to the high incidence of complications in hypospadias reconstruction and repair is the inadequate angiogenic capacity following injury, which hinders normal tissue healing (D. Wang et al., 2015). MSCs and their secreted factors have been shown to promote tissue repair and angiogenesis. However, MSCs derived from different tissues vary in terms of accessibility, quantity, proliferation, immunomodulatory properties, cytokine secretion, and therapeutic potential for specific diseases (Preynat-Seauve & Krause, 2011; Qi et al., 2016; Zou et al., 2023)

Zhang et al. demonstrated that urethral mesenchymal stem-like cells (PU-MSCs) may be an ideal repair material due to their easy accessibility, describing that both PU-MSCs and their culture supernatant can significantly enhance urethral injury repair (S. Zhang, Li, Li, XuminXie, et al., 2022). The specific properties of PU-MSCs and the underlying mechanisms involved in urethral reconstruction and repair using PU-MSCs remain unclear. Recent research has revealed that MSCs can secrete small extracellular vesicles (sEVs) containing biologically active substances, such as proteins, lipids, mRNAs, and microRNAs, which facilitate communication between cells and play diverse biological roles. MSC cell-derived sEVs have demonstrated efficacy in tissue damage repair, immune regulation, and angiogenesis promotion (Liu et al., 2021; Lu et al., 2022).

The UCSCs exosomes can significantly enhance the proliferation and migration of pediatric urethral smooth muscle cells (PUSMCs) by delivering CD73 molecules, leading to adenosine production and activation of the PI3K/AKT signaling pathway in PUSMCs. Based on these observations, we hypothesized that PU-MSC-sEVs might exert their beneficial effects on urethral repair through a pro-angiogenic mechanism. We postulated that CD73 molecules and ADO, a CD73 metabolite, could play a crucial role in promoting angiogenesis (S. Zhang, Li, Li, Xie, et al., 2022).

This study aimed to investigate the pro-angiogenic effects of PU-MSC-sEVs on vascular endothelial cells (VECs) and elucidate the underlying molecular mechanisms, providing valuable theoretical support for urethral reconstruction and repair.

Posterior urethral valves (PUV) are a congenital condition characterized by the presence of abnormal folds of tissue in the urethra, which obstruct urine flow. This obstruction can lead to severe bladder dysfunction, hydronephrosis, and renal damage if not treated promptly. The standard treatment for PUV is endoscopic ablation of the valves, but long-term management can be challenging due to persistent bladder dysfunction and the risk of chronic kidney disease.

Stem cell therapy is being investigated as a potential treatment for the bladder dysfunction associated with PUV. MSCs have shown promise in promoting bladder muscle regeneration and improving bladder compliance, which could mitigate the long-term complications of PUV. Preclinical studies have demonstrated that MSCs can differentiate into smooth muscle cells and contribute to the repair of the bladder wall, potentially restoring normal bladder function.

In one study, MSCs were injected into the bladder wall of a rat model of bladder outlet obstruction. The results showed that the MSC-treated group had improved bladder compliance and reduced fibrosis compared to the control group, suggesting that MSCs could play a role in preventing the bladder dysfunction that often follows PUV treatment. While clinical trials are still needed to evaluate the safety and efficacy of this approach in human patients, these findings offer hope for a new therapeutic option for children with PUV (Smolar et al., 2016).

Under these experimental conditions, PU-MSC-sEVs significantly enhanced the proliferation, migration, and tube-forming abilities of HUVECs. The studies suggest that PU-MSC-sEVs may promote angiogenesis by activating the CD73/ADO/A2AR signaling axis.

URETHRAL STRICTURE

Urethral strictures, or the narrowing of the urethra, can occur in children due to trauma, surgery, or infection. The condition can lead to difficulty urinating, recurrent urinary tract infections, and bladder dysfunction. Traditional treatments, such as urethral dilation or surgical reconstruction, are often associated with high recurrence rates and significant morbidity, leading to ongoing challenges in management.

Stem cell-based therapies are being explored as a potential treatment for urethral strictures. MSCs have been shown to promote tissue regeneration and reduce scarring in preclinical models of urethral injury. By delivering stem cells directly to the site of the stricture, researchers aim to regenerate healthy urethral tissue and reduce the likelihood of recurrence.

Angiogenesis is a critical process in urethral injury repair, and stem cells and their secreted factors can significantly influence this process.

In our previous work, we isolated pediatric urethral mesenchymal stem-like cells (PU-MSCs). This study focused on investigating the pro-angiogenic potential of PU-MSC-derived small extracellular vesicles (PU-MSC-sEVs) and the underlying mechanisms.

We found that PU-MSC-sEVs markedly stimulated the proliferation, migration, and tube formation of human umbilical vein endothelial cells (HUVECs). Mechanistically, we discovered that PU-MSC-sEVs delivered CD73 molecules to HUVECs, promoting angiogenesis. The extracellular adenosine (ADO) produced by CD73 further enhanced the angiogenic capacity of HUVECs. Moreover, PU-MSC-sEVs activated the adenosine receptor A2A (A2AR), contributing to the observed pro-angiogenic effects.

These findings suggest that PU-MSC-sEVs may promote angiogenesis through the activation of the CD73/ADO/A2AR signaling axis. This mechanism could potentially be harnessed for therapeutic interventions in urethral injury repair.

MSCs from pediatric urethra (PU-MSCs) offer significant promise for cell therapy and tissue engineering due to their multipotent differentiation capacity, robust proliferation, low immunogenicity, and immunomodulatory properties. In this study, we successfully isolated and characterized a population of PU-MSCs. These cells exhibited a spindle-shaped, fibroblast-like morphology consistent with MSCs derived from other tissues. Flow cytometry analysis confirmed their mesenchymal phenotype by revealing high expression of surface markers CD29, CD73, CD90, and CD105, while negative for hematopoietic markers CD45 and HLA-DR. Additionally, PU-MSCs demonstrated multi-lineage differentiation potential into adipocytes, osteocytes, and chondrocytes under appropriate in vitro conditions (Jiang et al., 2015).

Importantly, these cells maintained a normal karyotype and exhibited no tumorigenic potential during prolonged culture. Studies suggest that PU-MSCs represent a valuable source of stem cells for regenerative medicine applications, particularly in the context of urethral tissue reconstruction (Mulder et al., 2020).

Preclinical validation of scaffold-based technologies is essential for assessing wound healing efficacy in animal models that mimic target patient populations. In one study, MSCs were applied to a biodegradable scaffold and implanted at the site of a urethral injury in a rabbit model, exploring the potential of bilayer silk fibroin (BLSF) scaffolds for urethral repair. The results demonstrated that the MSC-treated group had significantly less scarring and improved urethral patency compared to the control group. Two weeks after electrocoagulation-induced urethral stricture, rabbits underwent urethroplasty with BLSF scaffolds. Gross tissue assessments at 1 and 3 months post-surgery revealed complete host tissue ingrowth. Retrograde urethrography (RUG) analysis at 3 months demonstrated a 39% reduction in urethral stenosis compared to pre-operative measurements. The MSCs appeared to promote the regeneration of normal urethral tissue, reducing the formation of fibrotic tissue that typically leads to stricture recurrence (Algarrahi et al., 2018; Kapetanos et al., 2022).

Histological and immunohistochemical analyses confirmed the formation of well-vascularized and innervated neotissues with smooth muscle bundles and epithelial covering at the graft sites.

These results suggest that BLSF matrices are promising candidates for supporting the repair of damaged urethral tissues (Sack et al., 2016)

The potential for SC therapy to reduce the need for repeated surgical interventions and improve long-term outcomes for children with urethral strictures is an exciting area of research. Clinical trials will be necessary to determine the safety and efficacy of this approach in pediatric patients, but the preclinical data suggest that stem cell therapy could offer a new treatment option for this challenging condition.

NEUROGENIC BLADDER

Neurogenic bladder, a condition resulting from nerve damage that affects bladder control, is commonly associated with congenital conditions such as spina bifida and acquired injuries such as spinal cord trauma. Children with neurogenic bladder often face lifelong challenges, including urinary incontinence, recurrent urinary tract infections, and kidney damage. Traditional management strategies, such as clean intermittent catheterization and anticholinergic medications, aim to manage symptoms but do not address the underlying nerve damage (de Sèze et al., 2007; Salehi-Pourmehr et al., 2020).

SC, particularly MSCs and NSCs, are being explored for their potential to repair damaged nerves and restore bladder function in neurogenic bladder. NSCs have the ability to differentiate into neurons and glial cells, offering the possibility of regenerating the neural pathways that control bladder function. Additionally, MSCs can support nerve regeneration through their paracrine effects, which include the secretion of neurotrophic factors that promote nerve growth and survival (Mahajan et al., 2016).

Preclinical studies have shown promising results in animal models of neurogenic bladder. In one study, MSCs were injected into the bladder wall of rats with spinal cord injury-induced neurogenic bladder. The results demonstrated improved bladder function, including increased bladder capacity and reduced voiding pressure, suggesting that MSCs could help restore normal bladder function. Additionally, the MSC-treated rats showed signs of nerve regeneration, as evidenced by increased expression of nerve growth factor (NGF) and other markers of neural repair.

In the past two decades, various natural and synthetic materials have been explored for bladder tissue engineering applications. These include natural matrices like amniotic membrane (Jerman et al., 2014), bladder acellular matrix (el-Kassaby et al., 2008), collagen and silk (Mauney et al., 2011; Sack et al., 2016), as well as synthetic polymers such as polylactic-co-glycolic acid, polyurethane (Yao et al., 2013), and poly-carbonate-urethane-urea (Sivaraman et al., 2019). Recent advancements have focused on (multi-)layered hybrid scaffolds, which often provide a more tailored microenvironment for each cellular layer, leading to enhanced regeneration and improved biomechanical properties of the engineered construct (Franck et al., 2013; Horst et al., 2013; X. Wu et al., 2016).

While these findings are encouraging, translating them into clinical practice will require further research, including the development of safe and effective delivery methods for stem cells, as well as long-term studies to assess the durability of the treatment.

BLADDER RECONSTRUCTION AND REGENERATION

The urinary bladder is a complex, hollow organ composed of multiple cell and ECM layers. These layers interact to enable proper bladder function, including acting as an effective barrier against urine, as well as its accumulation, storage, and controlled release. This process is coordinated by afferent and efferent neuronal pathways that transmit sensory signals to the central nervous system and relay responses back to the bladder tissue, resulting in coordinated detrusor muscle contraction and voiding (Aldamanhori & Chapple, 2017; Chai et al., 2016; Ochodnicky et al., 2012).

The bladder wall is organized into distinct histological compartments. The mucosa consists of several layers of transitional uroepithelium and is separated from the submucosa by the basal lamina. Above the highly vascularized and innervated lamina propria lies the primary functional bladder component, composed of three smooth muscle layers that facilitate coordinated micturition. The adventitia, a connective tissue layer, covers the urinary bladder and contributes to protection, vascularization, and innervation of the bladder wall (Jost et al., 1989, p. 1).

The unique anatomical structure of the bladder imposes specific biomechanical requirements on any adequate substitute, representing a significant challenge in developing a functional bioengineered bladder wall equivalent (Hanczar et al., 2021).

Congenital anomalies like bladder exstrophy and neurogenic bladder due to spina bifida present significant challenges in pediatric urology. Traditional bladder reconstruction techniques, such as enterocystoplasty (using a segment of the bowel to enlarge the bladder), carry risks such as mucus production, stone formation, and an increased risk of malignancy. Additionally, these procedures do not fully restore bladder function, leading to ongoing issues with urinary continence and kidney function.

SC therapy offers a promising alternative for bladder reconstruction. Bladder tissue engineering has evolved significantly in recent decades, moving beyond simple cell-biomaterial combinations to more sophisticated approaches inspired by natural processes. These therapies have its roots in the late 1950s, with early attempts using simple plastic molds for bladder reconstruction. Unfortunately, these early efforts primarily resulted in fibrotic tissue, leading to complications and limited functional regeneration (Adamowicz et al., 2014; Bohne & Urwiller, 1956; Portilla Sanchez et al., 1958; Tsulukidze et al., 1964).

Over the years, two main approaches have emerged. The first involves using natural or synthetic scaffolds alone to support native cell ingrowth and tissue regeneration. The second combines autologous cells with biomaterials for a more directed approach. A diverse range of materials has been explored and clinically applied by researchers.

Research has shown that MSCs can be used to engineer bladder tissue by seeding these cells onto biodegradable scaffolds. These scaffolds, designed to mimic the structure of the bladder, provide a framework for the stem cells to grow and differentiate into bladder smooth muscle cells and urothelial cells. In animal models, bioengineered bladders have demonstrated the ability to integrate with host tissue, restore bladder capacity, and improve urinary function (Sivaraman et al., 2019).

One of the key studies in this area involved the use of BMSC to create bladder tissue in a rabbit model. The results showed that the bioengineered bladder not only integrated well with the host tissue but also exhibited normal bladder function, including compliance and capacity. This research has paved the way for further studies aimed at refining the techniques and materials used in bladder tissue engineering, with the ultimate goal of applying these findings to human patients (Kanno et al., 2014).

Nearly a decade ago, Anumanthan et al. (Anumanthan et al., 2008) reported the directed differentiation of BMSC into UCs using embryonic rat bladders stripped of their urothelium and seeded with mesenchymal stem cells. These constructs, implanted into the renal subcapsular space of athymic mice, resulted in the formation of a multilayered, urothelium-like structure after 42 days. More recently, Zhang et al. successfully differentiated adipose-derived stem cells (ADSCs) into urothelial-like cells by co-culturing them with the SV-HUC-1 human urothelial cancer cell line in a time-dependent manner (M. Zhang et al., 2013).

Another promising source of multipotent cells for genitourinary therapeutic applications is UDSCs. Bharadwaj et al. (Bharadwaj et al., 2013) isolated a small population of UDSCs from fresh urine samples of healthy individuals of various ages and demonstrated their ability to expand into a large cell population.

Moreover, UDSCs were efficiently differentiated into urothelial and other cell types simply by modifying the culture medium composition. While the exact origin of these cells remains unclear, it has been suggested that they may be of renal origin, as they express kidney-related markers like PAX8, NR3C2, and L1CAM (Qin et al., 2014; R. Wu et al., 2013).

Tsuji et al. (Tsuji et al., 1967) attempted a more advanced approach using gelatin sponges for bladder reconstruction in cancer patients. While the sponge offered a favorable microenvironment for tissue growth, it eventually degraded, leading to urinary incontinence and other complications (Adamowicz et al., 2014; Tsuji et al., 1967).

Subsequent trials treated gelatin sponges with nobecutane or rezifilm to enhance their properties. These modified sponges demonstrated improved bladder capacity, epithelial coverage, and smooth muscle regeneration without inflammation {Citation}. However, the lack of further clinical use suggests skepticism regarding the reliability of these results.

Taguchi et al. (H. Taguchi et al., 1977) confirmed these doubts, finding that gelatin sponges can disintegrate in vivo and, when combined with nobecutane, form non-degradable synthetic structures that can lead to calculus formation. These findings highlight the challenges associated with using gelatin sponges for bladder augmentation.

A partial success was achieved using formalin-preserved dog bladders as temporary scaffolds for bladder reconstruction in patients after cystectomy. While a pseudo-bladder formed quickly, it often led to decreasing capacity, incontinence, reflux, and ureteral issues (Tsuji et al., 1963).

In 2006, a clinical study used autologous cells combined with collagen or a collagen-PGA hybrid. Seeded with urothelial and smooth muscle cells, these constructs were implanted in young patients with bladder disease. After nearly five years, the collagen-PGA composite demonstrated superior bladder regeneration compared to collagen alone. An omental wrap further enhanced vascularization and cell survival. While morphological analysis showed a well-structured bladder wall, only a few patients experienced increased bladder compliance and capacity (Atala et al., 2006).

Most patients lacked improvement and developed fibrous tissue in their transplanted bladders. Although this study demonstrated feasibility and safety, more research is needed to achieve a functional bladder equivalent.

While pre-clinical studies have explored SC, effective bladder tissue engineering requires mastering SC techniques, designing functional scaffolds, and tailoring the microenvironment to meet specific stem cell needs

However, due to the rapid natural regeneration of urothelium after injury and the swift ingrowth of urothelial cells onto transplanted luminal biomaterial surfaces, strategies employing suitable biomaterials that mimic the basal membrane without urothelial cell coverage may be advantageous in regenerative medicine therapies for urothelial repair (Lavelle et al., 2002, p. 1; Salehi-Pourmehr et al., 2020)

Despite ongoing research, functional bladder restoration remains elusive. While therapeutic SC offer promise, limitations in our understanding of SC biology and their interactions with scaffolds, the immune system, and the bladder microenvironment hinder their full application.

Current SC approaches primarily rely on autologous cells and complex procedures involving isolation, differentiation, and transplantation. More natural approaches, like stimulating resident SC and modifying the bladder microenvironment, have received less attention.

A major challenge in early tissue engineering approaches for organ reconstruction is inadequate vascularization, particularly in larger grafts. Insufficient nutrient and oxygen supply in the graft center often led to incomplete regeneration and failure (Valentin et al., 2009).

A specific subset of hematopoietic SC (HSPCs, CD34+) has a unique ability to induce robust tissue vascularization (Iannaccone et al., 2018). In a study of partial cystectomy in nude rats, scaffolds seeded with various combinations of MSCs and CD34+ HSPCs demonstrated significant vascularization benefits. The combination of both SC types was particularly effective in promoting vascularization throughout the graft, including the central region.

The source of MSCs and HSPCs in this study included pediatric spina bifida patients and healthy donors. Interestingly, the spina bifida donor cells showed similar regeneration potential to those from healthy donors, suggesting that developmental defects may not negatively impact bladder functionality when using autologous SC.

Adequate and early vascularization should be a priority in tissue engineering research. Suboptimal oxygen and nutrient supply can severely impair tissue regeneration. Combining SC approaches with prevascularization strategies, such as generating preformed microvascular networks in grafts, may help address this fundamental challenge in large graft implantation (Laschke & Menger, 2016)

The complexity of the bladder requires a multidisciplinary approach and innovative thinking. Creating a multifactorial microenvironment that supports each cell type is essential. Further research is needed to fully understand the potential and optimal use of both transplanted and resident stem cells for functional bladder bioengineering.

KIDNEY REGENERATION

AKI and CKD are significant concerns in pediatric urology, particularly in children with CAKUT (Goligorsky et al., 2009; Tresa et al., 2017). These conditions can lead to end-stage renal disease (ESRD), requiring dialysis or kidney transplantation. SC therapy offers a potential solution for kidney regeneration by promoting the repair of damaged renal tissue and possibly reducing the progression to ESRD.

MSCs and iPSCs are the primary stem cell types being investigated for kidney regeneration. MSCs have demonstrated the ability to reduce inflammation, promote tissue repair, and improve kidney function in animal models of AKI. One of the key mechanisms by which MSCs exert their effects is through the secretion of bioactive molecules that promote the repair of damaged renal cells and modulate the immune response to reduce inflammation (Araoka et al., 2014; Bussolati & Camussi, 2015; Imberti et al., 2015; Marcheque et al., 2019; Tanigawa et al., 2018).

In a study involving a mouse model of AKI, MSCs were administered intravenously, resulting in a significant reduction in kidney injury markers and improved renal function compared to the control group. The MSC-treated mice also showed reduced inflammation and fibrosis in the kidneys, suggesting that MSCs could help prevent the progression to CKD {Citation}. iPSCs, with their ability to differentiate into any cell type, hold the potential to generate functional renal tissue for transplantation. Research is ongoing to develop protocols for differentiating iPSCs into renal progenitor cells that can form nephrons, the functional units of the kidney. While the clinical application of iPSC-derived renal tissue is still in the early stages, the potential for creating bioengineered kidneys for transplantation represents a promising avenue for addressing the shortage of donor organs for pediatric patients with ESRD.

Mulder et al. used UDSCs to generate kidney organoids, employing the protocol established by Takasato et al., using control iPSC lines and determined optimal experimental seeding densities, following the approach of van den Berg et al. (Mulder et al., 2020; Takasato et al., 2015; van den Berg et al., 2018)

Kidney organoids derived from UiPSCs closely resembled those obtained from control iPSC lines and those described in previous studies using the Takasato protocol. These findings collectively demonstrate that not only can pediatric patient-specific UiPSCs be generated within a timeframe comparable to prior reports, but also that kidney organoids can be successfully generated shortly thereafter.

This offers the potential for tool for modeling human kidney disease, such as renal hypodysplasia, in the context of actual patients. Previously, such modeling has only been achieved using fibroblast-derived iPSCs (Forbes et al., 2018; Tanigawa et al., 2018).

FUTURE DIRECTIONS AND CHALLENGES

The application of SC in pediatric urology is a rapidly evolving field with significant potential to transform the treatment of congenital and acquired urological conditions. However, several challenges must be addressed to realize the full potential of stem cell therapies:

Clinical Translation

While preclinical studies have provided valuable insights into the potential of SC therapies, the translation of these findings into clinical practice requires rigorous clinical trials. These trials must establish the safety and efficacy of stem cell treatments in pediatric patients, as well as determine the optimal delivery methods, dosages, and treatment protocols.

One of the challenges in clinical translation is the need for long-term follow-up studies to assess the durability of SC-based treatments. Pediatric patients, in particular, require careful monitoring to ensure that stem cell therapies do not interfere with growth and development or lead to unforeseen complications (Mulder et al., 2020).

Personalized Medicine

The development of personalized stem cell therapies, tailored to the individual needs of pediatric patients, represents an exciting area of research. iPSCs, in particular, offer the potential for creating patient-specific treatments that minimize the risk of immune rejection. By deriving iPSCs from a patient's own cells, researchers can create a source of pluripotent SC that can be used to generate any cell type needed for therapy (Imberti et al., 2015; Mandegar et al., 2016).

However, significant technical and logistical challenges remain in the development and implementation of personalized stem cell therapies. The process of creating iPSCs, differentiating them into the desired cell type, and ensuring their safety and efficacy is complex and time-consuming. Additionally, the cost of personalized SC therapies may be prohibitive, limiting their accessibility to patients.

Scaffold Development and Tissue Engineering

The success of stem cell-based tissue engineering relies on the development of suitable scaffolds that support cell growth, differentiation, and integration with host tissue. Scaffolds must be biocompatible, biodegradable, and provide the appropriate mechanical and biochemical cues to guide tissue formation. Advances in biomaterials and 3D printing technologies are likely to play a critical role in the future of SC-based tissue engineering in pediatric urology (Sack et al., 2016; Sivaraman et al., 2019; Yao et al., 2013).

Researchers are exploring a range of scaffold materials, including natural polymers like collagen and synthetic polymers like polylactic acid, to create scaffolds that mimic the structure and function of native tissues. The use of 3D printing allows for the precise fabrication of scaffolds with complex geometries, tailored to the specific needs of each patient. Combining SC with these advanced scaffolds holds the potential to create bioengineered tissues that can replace or repair damaged organs in pediatric patients.

Immune Modulation and Tolerance

Developing strategies to modulate the immune response and promote tolerance of stem cell-derived tissues is essential for the success of these therapies. While autologous stem cells are less likely to be rejected by the immune system, allogeneic SC pose a risk of immune rejection.

Researchers are exploring various approaches to induce immune tolerance, including the use of immunosuppressive drugs, genetic modification of SC, and the co-transplantation of stem cells with regulatory T cells that promote immune tolerance. Understanding the complex interactions between SC and the immune system is crucial for developing therapies that are both effective and safe for pediatric patients (Semedo et al., 2009; Stagg, 2007).

ETHICAL AND REGULATORY CONSIDERATIONS

SC offers the possibility of generating a limitless supply of human cells for various applications, including disease modeling, drug discovery, and toxicology testing.

The use of SC, particularly ESCs and iPSCs, raises ethical and regulatory challenges that must be carefully navigated. ESCs, derived from human embryos, pose significant ethical concerns related to the destruction of embryos, leading to debates about their use in research and therapy. iPSCs, while avoiding some of the ethical issues associated with ESCs, still require careful consideration of the genetic modifications used to reprogram adult cells.

Regulatory agencies, such as the U.S. FDA and the European Medicines Agency (EMA), have established guidelines for the development and approval of SC therapies. These guidelines require rigorous testing to ensure the safety, efficacy, and quality of SC products before they can be used in clinical practice. Researchers and clinicians must work closely with regulatory bodies to navigate these requirements and bring stem cell therapies to the clinic.

Animal research has played a crucial role in biomedical advancement, but ethical concerns have led to a push for alternative methods. While SC research currently relies on animal models, it holds the potential to reduce or eliminate certain types of animal testing. This outcome would be welcomed by animal advocates. The 3Rs (Replacement, Reduction, and Refinement) strategy and in vitro approaches have gained prominence.

SC offer a promising alternative to animal experimentation in predictive toxicology and the development of organoids, without needed the sacrifice of other being. Research on SC and their derived tissue-specific cells has focused on developing alternative methods and biomarkers for toxicity testing, including developmental, genotoxic, and tissue-specific toxicity.

SC possess unique properties like self-renewal, infinite proliferation, and multilineage differentiation, making human stem cell-based in vitro systems valuable for increasing the predictive power of toxicology. These systems provide better scientific information on human toxic risks, eliminating inter-species variability.

Personalized toxicological assessment using iPSCs is a particularly promising in vitro approach compared to immortalized cells with abnormal phenotypes or primary cells with limited quantity and batch-to-batch variation.

The availability and characterization of SC from large animals varies significantly between species. Establishing and comparing these cell lines with human analogues requires substantial effort, as does testing potential applications.

It's currently difficult to predict the extent to which embryonic SC-based testing will reduce or eliminate animal testing. Research in this area is still in its early stages, as scientists are learning to differentiate ESC into various cell types. The cost associated with using hESC lines, many of which are patented or patent-pending, may also limit research in this area. This lack of economic incentive could hinder the development of alternative testing methods that could replace animal models.

SC-based therapies pose significant safety challenges that cannot be adequately addressed with traditional methods, and interspecies differences are crucial to have a more accurate result and match what the reality of human biology entails.

Despite efforts by various groups to minimize animal use in scientific experiments, animal testing has increased due to technological advancements and medical research. This rise has raised concerns about ethical issues, high costs, complex procedures, and potential inaccuracies. SC technologies, such as organ-on-a-chip, organoids, and induced pluripotent SC models, offer promising alternatives to animal testing. While SC methodologies have shortcomings, they hold the potential to replace animal testing. Achieving this goal requires further research, with potential societal and technological benefits (Park et al., 2024).

Jurisdictions with significant biomedical research capabilities have enacted laws regulating various aspects of ESC research. These activities include human reproductive cloning, somatic cell nuclear transfer (SCNT), creation of human embryos for research, derivation of SC lines from existing embryos, and research using existing lines (*The Hinxton Group: World Stem Cell Policies*, s. f.).

Legal regimes vary widely. Many Asian countries, like China, South Korea, Japan, Singapore, and India, have permissive policies allowing most activities except human reproductive cloning. In contrast, some European and North American jurisdictions have restrictive policies due to religious or historical reasons. For example, Germany and Italy only permit research using existing human embryonic stem cell lines, while Austria and Poland have ESC research altogether (*The Hinxton Group: World Stem Cell Policies*, s. f.).

CONCLUSION

SC therapy represents a groundbreaking approach to addressing some of the most challenging conditions in pediatric urology. The potential of stem cells to regenerate damaged tissues, repair congenital anomalies, and restore normal function in the urinary tract offers hope for improving outcomes and quality of life for children with urological disorders.

While significant progress has been made in understanding the mechanisms of stem cell action and demonstrating their potential in preclinical models, much work remains to be done to translate these findings into clinical practice. The future of pediatric urology lies in the continued exploration of stem cell-based therapies, with the ultimate goal of developing safe, effective, and accessible treatments that can be integrated into standard care.

As research progresses, the pediatric urology community must remain at the forefront of these developments, embracing the potential of stem cells while addressing the challenges and ethical considerations that accompany this rapidly evolving field. By working together, researchers, clinicians, and regulatory bodies can ensure that stem cell therapies fulfill their promise to transform the treatment of pediatric urological conditions, offering new hope to patients and their families.

chapter provides a comprehensive introduction to stem cells, focusing on their classification, clinical translation, and the recommendations of the International Society for Stem Cell Research (ISSCR). It begins with a fundamental overview of stem cell biology, including the various types and their potential applications in the medical practice.

The argument emphasizes the importance of adhering to established standards and guidelines to ensure ethical practices and scientific rigor in stem cell research. Key topics include the clinical translation of stem cell therapies, the role of preclinical studies, highlighting both their importance and the ethical concerns associated with traditional animal testing, and the critical evaluation of current standards to guide future research. Aimed primarily at urologists and gynecologists, this chapter provides essential background information to bridge the gap between basic stem cell science and its practical implications, setting the stage for the subsequent content of the book and offering insights into emerging opportunities and challenges in the field.

REFERENCES

Adamowicz, J., Pokrywczyńska, M., & Drewa, T. (2014). Conditioned medium derived from mesenchymal stem cells culture as a intravesical therapy for cystitis interstitials. *Medical Hypotheses*, *82*(6), 670–673. DOI: 10.1016/j.mehy.2014.02.027 PMID: 24679668

Aitken, K. J., Schröder, A., Haddad, A., Sidler, M., Penna, F., Fernandez, N., Ahmed, T., Marino, V., Bechbache, M., Jiang, J.-X., Tolg, C., & Bägli, D. J. (2024). Epigenetic insights to pediatric uropathology: Celebrating the fundamental biology vision of Tony Khoury. *Journal of Pediatric Urology*, *20*, S43–S57. DOI: 10.1016/j.jpurol.2024.06.008 PMID: 38944627

Aldamanhori, R., & Chapple, C. R. (2017). Underactive bladder, detrusor underactivity, definition, symptoms, epidemiology, etiopathogenesis, and risk factors. *Current Opinion in Urology*, *27*(3), 293–299. DOI: 10.1097/MOU.0000000000000381 PMID: 28221218

Algarrahi, K., Affas, S., Sack, B. S., Yang, X., Costa, K., Seager, C., Estrada, C. R.Jr, & Mauney, J. R. (2018). Repair of injured urethras with silk fibroin scaffolds in a rabbit model of onlay urethroplasty. *The Journal of Surgical Research*, *229*, 192–199. DOI: 10.1016/j.jss.2018.04.006 PMID: 29936989

Anumanthan, G., Makari, J. H., Honea, L., Thomas, J. C., Wills, M. L., Bhowmick, N. A., Adams, M. C., Hayward, S. W., Matusik, R. J., Brock, J. W., & Pope, J. C. (2008). Directed differentiation of bone marrow derived mesenchymal stem cells into bladder urothelium. *The Journal of Urology*, *180*(4, Suppl), 1778–1783. DOI: 10.1016/j.juro.2008.04.076 PMID: 18721942

Araoka, T., Mae, S., Kurose, Y., Uesugi, M., Ohta, A., Yamanaka, S., & Osafune, K. (2014). Efficient and rapid induction of human iPSCs/ESCs into nephrogenic intermediate mesoderm using small molecule-based differentiation methods. *PLoS One*, *9*(1), e84881. DOI: 10.1371/journal.pone.0084881 PMID: 24454758

Atala, A., Bauer, S. B., Soker, S., Yoo, J. J., & Retik, A. B. (2006). Tissue-engineered autologous bladders for patients needing cystoplasty. *Lancet*, *367*(9518), 1241–1246. DOI: 10.1016/S0140-6736(06)68438-9 PMID: 16631879

Bharadwaj, S., Liu, G., Shi, Y., Wu, R., Yang, B., He, T., Fan, Y., Lu, X., Zhou, X., Liu, H., Atala, A., Rohozinski, J., & Zhang, Y. (2013). Multipotential differentiation of human urine-derived stem cells: Potential for therapeutic applications in urology. *Stem Cells (Dayton, Ohio)*, *31*(9), 1840–1856. DOI: 10.1002/stem.1424 PMID: 23666768

Bohne, A. W., & Urwiller, K. L. (1956). Experience with urinary bladder regeneration. *Postgraduate Seminar. American Urological Association. North Central Section*, 24-27.

Bouty, A., Ayers, K. L., Pask, A., Heloury, Y., & Sinclair, A. H. (2015). The Genetic and Environmental Factors Underlying Hypospadias. *Sexual Development: Genetics, Molecular Biology, Evolution, Endocrinology, Embryology, and Pathology of Sex Determination and Differentiation*, *9*(5), 239–259. DOI: 10.1159/000441988 PMID: 26613581

Bussolati, B., & Camussi, G. (2015). Therapeutic use of human renal progenitor cells for kidney regeneration. *Nature Reviews. Nephrology*, *11*(12), 695–706. DOI: 10.1038/nrneph.2015.126 PMID: 26241019

Chai, T. C., Russo, A., Yu, S., & Lu, M. (2016). Mucosal signaling in the bladder. *Autonomic Neuroscience*, *200*, 49–56. DOI: 10.1016/j.autneu.2015.08.009 PMID: 26422993

de Sèze, M., Ruffion, A., Denys, P., Joseph, P.-A., & Perrouin-Verbe, B. (2007). The neurogenic bladder in multiple sclerosis: Review of the literature and proposal of management guidelines. *Multiple Sclerosis*, *13*(7), 915–928. DOI: 10.1177/1352458506075651 PMID: 17881401

el-Kassaby, A., AbouShwareb, T., & Atala, A. (2008). Randomized comparative study between buccal mucosal and acellular bladder matrix grafts in complex anterior urethral strictures. *The Journal of Urology*, *179*(4), 1432–1436. DOI: 10.1016/j.juro.2007.11.101 PMID: 18295282

Forbes, T. A., Howden, S. E., Lawlor, K., Phipson, B., Maksimovic, J., Hale, L., Wilson, S., Quinlan, C., Ho, G., Holman, K., Bennetts, B., Crawford, J., Trnka, P., Oshlack, A., Patel, C., Mallett, A., Simons, C., & Little, M. H. (2018). Patient-iPSC-Derived Kidney Organoids Show Functional Validation of a Ciliopathic Renal Phenotype and Reveal Underlying Pathogenetic Mechanisms. *American Journal of Human Genetics*, *102*(5), 816–831. DOI: 10.1016/j.ajhg.2018.03.014 PMID: 29706353

Franck, D., Gil, E. S., Adam, R. M., Kaplan, D. L., Chung, Y. G., Estrada, C. R., & Mauney, J. R. (2013). Evaluation of silk biomaterials in combination with extracellular matrix coatings for bladder tissue engineering with primary and pluripotent cells. *PLoS One*, *8*(2), e56237. DOI: 10.1371/journal.pone.0056237 PMID: 23409160

Goligorsky, M. S., Kuo, M.-C., Patschan, D., & Verhaar, M. C. (2009). Review article: Endothelial progenitor cells in renal disease. *Nephrology (Carlton, Vic.)*, *14*(3), 291–297. DOI: 10.1111/j.1440-1797.2009.01112.x PMID: 19444962

Hanczar, M., Moazen, M., & Day, R. (2021). The Significance of Biomechanics and Scaffold Structure for Bladder Tissue Engineering. *International Journal of Molecular Sciences*, *22*(23), 12657. DOI: 10.3390/ijms222312657 PMID: 34884464

Horst, M., Madduri, S., Milleret, V., Sulser, T., Gobet, R., & Eberli, D. (2013). A bilayered hybrid microfibrous PLGA—Acellular matrix scaffold for hollow organ tissue engineering. *Biomaterials*, *34*(5), 1537–1545. DOI: 10.1016/j.biomaterials.2012.10.075 PMID: 23177021

Iannaccone, P. M., Galat, V., Bury, M. I., Ma, Y. C., & Sharma, A. K. (2018). The utility of stem cells in pediatric urinary bladder regeneration. *Pediatric Research*, *83*(1-2), 258–266. DOI: 10.1038/pr.2017.229 PMID: 28915233

Imberti, B., Tomasoni, S., Ciampi, O., Pezzotta, A., Derosas, M., Xinaris, C., Rizzo, P., Papadimou, E., Novelli, R., Benigni, A., Remuzzi, G., & Morigi, M. (2015). Renal progenitors derived from human iPSCs engraft and restore function in a mouse model of acute kidney injury. *Scientific Reports*, *5*(1), 8826. DOI: 10.1038/srep08826 PMID: 25744951

Jerman, U. D., Veranič, P., & Kreft, M. E. (2014). Amniotic membrane scaffolds enable the development of tissue-engineered urothelium with molecular and ultrastructural properties comparable to that of native urothelium. *Tissue Engineering. Part C, Methods*, *20*(4), 317–327. DOI: 10.1089/ten.tec.2013.0298 PMID: 23947657

Jiang, M. H., Li, G., Liu, J., Liu, L., Wu, B., Huang, W., He, W., Deng, C., Wang, D., Li, C., Lahn, B. T., Shi, C., & Xiang, A. P. (2015). Nestin(+) kidney resident mesenchymal stem cells for the treatment of acute kidney ischemia injury. *Biomaterials*, *50*, 56–66. DOI: 10.1016/j.biomaterials.2015.01.029 PMID: 25736496

Jost, S. P., Gosling, J. A., & Dixon, J. S. (1989). The morphology of normal human bladder urothelium. *Journal of Anatomy*, *167*, 103–115. PMID: 2630525

Kanno, Y., Mitsui, T., Sano, H., Kitta, T., Moriya, K., & Nonomura, K. (2014). Contribution of bone marrow-derived mesenchymal stem cells to the morphological changes in the bladder after partial outlet obstruction: A preliminary study. *International journal of urology : official journal of the Japanese Urological Association*, *21*(7), 714-718. DOI: 10.1111/iju.12406

Kapetanos, K., Light, A., Thakare, N., Mahbubani, K., Saeb-Parsy, K., & Saeb-Parsy, K. (2022). Bioengineering solutions for ureteric disorders: Clinical need, challenges and opportunities. *BJU International*, *130*(4), 408–419. DOI: 10.1111/bju.15741 PMID: 35388587

Kiss, A., Sulya, B., Szász, A. M., Romics, I., Kelemen, Z., Tóth, J., Merksz, M., Kemény, S., & Nyírády, P. (2011). Long-term psychological and sexual outcomes of severe penile hypospadias repair. *Journal of Sexual Medicine*, *8*(5), 1529–1539. DOI: 10.1111/j.1743-6109.2010.02120.x PMID: 21091883

Kollhoff, D. M., Cheng, E. Y., & Sharma, A. K. (2011). Urologic applications of engineered tissue. *Regenerative Medicine*, *6*(6), 757–765. DOI: 10.2217/rme.11.91 PMID: 22050527

Kropp, B. P. (1998). Small-intestinal submucosa for bladder augmentation: A review of preclinical studies. *World Journal of Urology*, *16*(4), 262–267. DOI: 10.1007/s003450050064 PMID: 9775425

Laschke, M. W., & Menger, M. D. (2016). Prevascularization in tissue engineering: Current concepts and future directions. *Biotechnology Advances*, *34*(2), 112–121. DOI: 10.1016/j.biotechadv.2015.12.004 PMID: 26674312

Lavelle, J., Meyers, S., Ramage, R., Bastacky, S., Doty, D., Apodaca, G., & Zeidel, M. L. (2002). Bladder permeability barrier: Recovery from selective injury of surface epithelial cells. *American Journal of Physiology. Renal Physiology*, *283*(2), F242–F253. DOI: 10.1152/ajprenal.00307.2001 PMID: 12110507

Liu, L., Guo, S., Shi, W., Liu, Q., Huo, F., Wu, Y., & Tian, W. (2021). Bone Marrow Mesenchymal Stem Cell-Derived Small Extracellular Vesicles Promote Periodontal Regeneration. *Tissue Engineering. Part A*, *27*(13-14), 962–976. DOI: 10.1089/ten.tea.2020.0141 PMID: 32962564

Lu, Y., Wang, L., Zhang, M., & Chen, Z. (2022). Mesenchymal Stem Cell-Derived Small Extracellular Vesicles: A Novel Approach for Kidney Disease Treatment. *International Journal of Nanomedicine*, *17*, 3603–3618. DOI: 10.2147/IJN.S372254 PMID: 35990308

Mahajan, P. V., Subramanian, S., Danke, A., & Kumar, A. (2016). Neurogenic Bladder Repair Using Autologous Mesenchymal Stem Cells. *Case Reports in Urology*, *2539320*, 1–3. Advance online publication. DOI: 10.1155/2016/2539320 PMID: 27656308

Mandegar, M. A., Huebsch, N., Frolov, E. B., Shin, E., Truong, A., Olvera, M. P., Chan, A. H., Miyaoka, Y., Holmes, K., Spencer, C. I., Judge, L. M., Gordon, D. E., Eskildsen, T. V., Villalta, J. E., Horlbeck, M. A., Gilbert, L. A., Krogan, N. J., Sheikh, S. P., Weissman, J. S., & Conklin, B. R. (2016). CRISPR Interference Efficiently Induces Specific and Reversible Gene Silencing in Human iPSCs. *Cell Stem Cell*, *18*(4), 541–553. DOI: 10.1016/j.stem.2016.01.022 PMID: 26971820

Marcheque, J., Bussolati, B., Csete, M., & Perin, L. (2019). Concise Reviews: Stem Cells and Kidney Regeneration: An Update. *Stem Cells Translational Medicine*, *8*(1), 82–92. DOI: 10.1002/sctm.18-0115 PMID: 30302937

Mauney, J. R., Cannon, G. M., Lovett, M. L., Gong, E. M., Di Vizio, D., Gomez, P.III, Kaplan, D. L., Adam, R. M., & Estrada, C. R.Jr. (2011). Evaluation of gel spun silk-based biomaterials in a murine model of bladder augmentation. *Biomaterials*, *32*(3), 808–818. DOI: 10.1016/j.biomaterials.2010.09.051 PMID: 20951426

Morizane, R., Lam, A. Q., Freedman, B. S., Kishi, S., Valerius, M. T., & Bonventre, J. V. (2015). Nephron organoids derived from human pluripotent stem cells model kidney development and injury. *Nature Biotechnology*, *33*(11), 1193–1200. DOI: 10.1038/nbt.3392 PMID: 26458176

Mouravas, V., Filippopoulos, A., & Sfoungaris, D. (2014). Urethral plate grafting improves the results of tubularized incised plate urethroplasty in primary hypospadias. *Journal of Pediatric Urology*, *10*(3), 463–468. DOI: 10.1016/j.jpurol.2013.11.012 PMID: 24360521

Mulder, J., Sharmin, S., Chow, T., Rodrigues, D. C., Hildebrandt, M. R., D'Cruz, R., Rogers, I., Ellis, J., & Rosenblum, N. D. (2020). Generation of infant- and pediatric-derived urinary induced pluripotent stem cells competent to form kidney organoids. *Pediatric Research*, *87*(4), 647–655. DOI: 10.1038/s41390-019-0618-y PMID: 31629364

Nicolaou, N., Renkema, K. Y., Bongers, E. M. H. F., Giles, R. H., & Knoers, N. V. A. M. (2015). Genetic, environmental, and epigenetic factors involved in CAKUT. *Nature Reviews. Nephrology*, *11*(12), 720–731. DOI: 10.1038/nrneph.2015.140 PMID: 26281895

Ochodnicky, P., Cruz, C. D., Yoshimura, N., & Cruz, F. (2012). Neurotrophins as regulators of urinary bladder function. *Nature Reviews. Urology*, *9*(11), 628–637. DOI: 10.1038/nrurol.2012.178 PMID: 23045265

Park, G., Rim, Y. A., Sohn, Y., Nam, Y., & Ju, J. H. (2024). Replacing Animal Testing with Stem Cell-Organoids: Advantages and Limitations. *Stem Cell Reviews and Reports*, *20*(6), 1375–1386. DOI: 10.1007/s12015-024-10723-5 PMID: 38639829

Portilla Sanchez, R., Blanco, F. L., Santamarina, A., Casals Roa, J., Mata, J., & Kaufman, A. (1958). Vesical regeneration in the human after total cystectomy and implantation of a plastic mould. *British Journal of Urology*, *30*(2), 180–188. DOI: 10.1111/j.1464-410X.1958.tb06231.x PMID: 13546597

Preynat-Seauve, O., & Krause, K.-H. (2011). Stem cell sources for regenerative medicine: The immunological point of view. *Seminars in Immunopathology*, *33*(6), 519–524. DOI: 10.1007/s00281-011-0271-y PMID: 21598094

Qi, X., Zhang, J., Yuan, H., Xu, Z., Li, Q., Niu, X., Hu, B., Wang, Y., & Li, X. (2016). Exosomes Secreted by Human-Induced Pluripotent Stem Cell-Derived Mesenchymal Stem Cells Repair Critical-Sized Bone Defects through Enhanced Angiogenesis and Osteogenesis in Osteoporotic Rats. *International Journal of Biological Sciences*, *12*(7), 836–849. DOI: 10.7150/ijbs.14809 PMID: 27313497

Qin, D., Long, T., Deng, J., & Zhang, Y. (2014). Urine-derived stem cells for potential use in bladder repair. *Stem Cell Research & Therapy*, *5*(3), 69. DOI: 10.1186/scrt458 PMID: 25157812

Rigaud, J., & Le Normand, L. (2004). Entérocystoplastie d'agrandissement. *Annales d'Urologie*, *38*(6), 298–310. DOI: 10.1016/j.anuro.2004.08.001 PMID: 15651484

Sack, B. S., Mauney, J. R., & Estrada, C. R.Jr. (2016). Silk Fibroin Scaffolds for Urologic Tissue Engineering. *Current Urology Reports*, *17*(2), 16. DOI: 10.1007/s11934-015-0567-x PMID: 26801192

Salehi-Pourmehr, H., Hajebrahimi, S., Rahbarghazi, R., Pashazadeh, F., Mahmoudi, J., Maasoumi, N., & Sadigh-Eteghad, S. (2020). Stem Cell Therapy for Neurogenic Bladder Dysfunction in Rodent Models: A Systematic Review. *International Neurourology Journal*, *24*(3), 241–257. DOI: 10.5213/inj.2040058.029 PMID: 33017895

Schäfer, F.-M., & Stehr, M. (2018). Tissue engineering in pediatric urology—A critical appraisal. *Innovative Surgical Sciences*, *3*(2), 107–118. DOI: 10.1515/iss-2018-0011 PMID: 31579774

Semedo, P., Correa-Costa, M., Antonio Cenedeze, M., Maria Avancini Costa Malheiros, D., Antonia dos Reis, M., Shimizu, M. H., Seguro, A. C., Pacheco-Silva, A., & Saraiva Camara, N. O. (2009). Mesenchymal stem cells attenuate renal fibrosis through immune modulation and remodeling properties in a rat remnant kidney model. *Stem Cells (Dayton, Ohio)*, *27*(12), 3063–3073. DOI: 10.1002/stem.214 PMID: 19750536

Sivaraman, S., Amoroso, N., Gu, X., Purves, J. T., Hughes, F. M.Jr, Wagner, W. R., & Nagatomi, J. (2019). Evaluation of Poly (Carbonate-Urethane) Urea (PCUU) Scaffolds for Urinary Bladder Tissue Engineering. *Annals of Biomedical Engineering*, *47*(3), 891–901. DOI: 10.1007/s10439-018-02182-0 PMID: 30542784

Smolar, J., Salemi, S., Horst, M., Sulser, T., & Eberli, D. (2016). Stem Cells in Functional Bladder Engineering. *Transfusion Medicine and Hemotherapy; Offizielles Organ der Deutschen Gesellschaft für Transfusionsmedizin und Immunhämatologie*, *43*(5), 328–335. DOI: 10.1159/000447977 PMID: 27781020

Stagg, J. (2007). Immune regulation by mesenchymal stem cells: Two sides to the coin. *Tissue Antigens*, *69*(1), 1–9. DOI: 10.1111/j.1399-0039.2006.00739.x PMID: 17212702

Taguchi, A., Kaku, Y., Ohmori, T., Sharmin, S., Ogawa, M., Sasaki, H., & Nishinakamura, R. (2014). Redefining the in vivo origin of metanephric nephron progenitors enables generation of complex kidney structures from pluripotent stem cells. *Cell Stem Cell*, *14*(1), 53–67. DOI: 10.1016/j.stem.2013.11.010 PMID: 24332837

Taguchi, H., Ishizuka, E., & Saito, K. (1977). Cystoplasty by regeneration of the bladder. *The Journal of Urology*, *118*(5), 752–756. DOI: 10.1016/S0022-5347(17)58181-1 PMID: 916093

Takasato, M., Er, P. X., Chiu, H. S., Maier, B., Baillie, G. J., Ferguson, C., Parton, R. G., Wolvetang, E. J., Roost, M. S., Chuva de Sousa Lopes, S. M., & Little, M. H. (2015). Kidney organoids from human iPS cells contain multiple lineages and model human nephrogenesis. *Nature*, *526*(7574), 564–568. DOI: 10.1038/nature15695 PMID: 26444236

Tanigawa, S., Islam, M., Sharmin, S., Naganuma, H., Yoshimura, Y., Haque, F., Era, T., Nakazato, H., Nakanishi, K., Sakuma, T., Yamamoto, T., Kurihara, H., Taguchi, A., & Nishinakamura, R. (2018). Organoids from Nephrotic Disease-Derived iPSCs Identify Impaired NEPHRIN Localization and Slit Diaphragm Formation in Kidney Podocytes. *Stem Cell Reports*, *11*(3), 727–740. DOI: 10.1016/j.stemcr.2018.08.003 PMID: 30174315

The Hinxton Group: World Stem Cell Policies. (s. f.). Recuperado 25 de agosto de 2024, de http://www.hinxtongroup.org/wp.html

Tresa, V., Yaseen, A., Lanewala, A. A., Hashmi, S., Khatri, S., Ali, I., & Mubarak, M. (2017). Etiology, clinical profile and short-term outcome of acute kidney injury in children at a tertiary care pediatric nephrology center in Pakistan. *Renal Failure*, *39*(1), 26–31. DOI: 10.1080/0886022X.2016.1244074 PMID: 27767356

Tsuji, I., Kuroda, K., Fujieda, J., Shiraishi, Y., Kassai, T., & Shida, H. (1963). A clinical and experimental study on cystoplasty not using the intestine. *The Journal of Urology*, *89*(2), 214–225. DOI: 10.1016/S0022-5347(17)64531-2 PMID: 13994538

Tsuji, I., Kuroda, K., Fujieda, J., Shiraishi, Y., Kunishima, K., & Orikasa, S. (1967). Clinical experiences of bladder reconstruction using preserved bladder and gelatin sponge bladder in the case of bladder cancer. *The Journal of Urology*, *98*(1), 91–92. DOI: 10.1016/S0022-5347(17)62828-3 PMID: 6067880

Tsulukidze, A., Murvanidze, D., Dvali, R., & Ivashchenko, G. (1964). FORMATION OF A BLADDER BY A PLASTIC SHELL AFTER TOTAL CYSTECTOMY. *British Journal of Urology*, *36*(1), 102–105. DOI: 10.1111/j.1464-410X.1964.tb09488.x PMID: 14126062

Valentin, J. E., Freytes, D. O., Grasman, J. M., Pesyna, C., Freund, J., Gilbert, T. W., & Badylak, S. F. (2009). Oxygen diffusivity of biologic and synthetic scaffold materials for tissue engineering. *Journal of Biomedical Materials Research. Part A*, *91*(4), 1010–1017. DOI: 10.1002/jbm.a.32328 PMID: 19097154

van den Berg, C. W., Ritsma, L., Avramut, M. C., Wiersma, L. E., van den Berg, B. M., Leuning, D. G., Lievers, E., Koning, M., Vanslambrouck, J. M., Koster, A. J., Howden, S. E., Takasato, M., Little, M. H., & Rabelink, T. J. (2018). Renal Subcapsular Transplantation of PSC-Derived Kidney Organoids Induces Neo-vasculogenesis and Significant Glomerular and Tubular Maturation In Vivo. *Stem Cell Reports*, *10*(3), 751–765. DOI: 10.1016/j.stemcr.2018.01.041 PMID: 29503086

Wang, D., Li, M., Huang, W., Lu, M., Hu, C., Li, K., Qiu, J., & Gao, X. (2015). Repair of urethral defects with polylactid acid fibrous membrane seeded with adipose-derived stem cells in a rabbit model. *Connective Tissue Research*, *56*(6), 434–439. DOI: 10.3109/03008207.2015.1035376 PMID: 25943462

Wang, Y., & Liao, L. (2014). Histologic and functional outcomes of small intestine submucosa-regenerated bladder tissue. *BMC Urology*, *14*(1), 69. DOI: 10.1186/1471-2490-14-69 PMID: 25148849

Wu, R., Liu, G., Fan, Y., Rohozinski, J., Lu, X., Rodriguez, G., Farney, A., Atala, A., & Zhang, Y. (2013). 249 HUMAN URINE-DERIVED STEM CELLS ORIGINATE FROM PARIETAL STEM CELLS. *The Journal of Urology*, *189*(4S). Advance online publication. DOI: 10.1016/j.juro.2013.02.1631

Wu, X., Wang, Y., Zhu, C., Tong, X., Yang, M., Yang, L., Liu, Z., Huang, W., Wu, F., Zong, H., Li, H., & He, H. (2016). Preclinical animal study and human clinical trial data of co-electrospun poly(L-lactide-co-caprolactone) and fibrinogen mesh for anterior pelvic floor reconstruction. *International Journal of Nanomedicine*, *11*, 389–397. DOI: 10.2147/IJN.S88803 PMID: 26893556

Yao, C., Hedrick, M., Pareek, G., Renzulli, J., Haleblian, G., & Webster, T. J. (2013). Nanostructured polyurethane-poly-lactic-co-glycolic acid scaffolds increase bladder tissue regeneration: An in vivo study. *International Journal of Nanomedicine*, *8*, 3285–3296. DOI: 10.2147/IJN.S44901 PMID: 24039415

Zhang, M., Peng, Y., Zhou, Z., Zhou, J., Wang, Z., & Lu, M. (2013). Differentiation of human adipose-derived stem cells co-cultured with urothelium cell line toward a urothelium-like phenotype in a nude murine model. *Urology*, *81*(2), 465.e15–465.e22. DOI: 10.1016/j.urology.2012.10.030 PMID: 23374843

Zhang, S., Li, J., & Li, C. XuminXie, null, He, J., Ling, F., & Liu, G. (2022). Isolation and identification of a mesenchymal stem/stromal cell-like population from pediatric urethral tissue. *In Vitro Cellular & Developmental Biology. Animal, 58*(6), 503-511. DOI: 10.1007/s11626-022-00697-4

Zhang, S., Li, J., Li, C., Xie, X., He, J., Ling, F., Li, B., Wu, H., Li, Z., Zhen, J., & Liu, G. (2024). CD73-positive pediatric urethral mesenchymal stem-like cell-derived small extracellular vesicles stimulate angiogenesis. *Regenerative Therapy*, *25*, 77–84. DOI: 10.1016/j.reth.2023.12.002 PMID: 38111468

Zhang, S., Li, J., Li, C., Xie, X., He, J., Ling, F., Li, B., Wu, H., Li, Z., & Zheng, J. (2022). CD73-Positive Small Extracellular Vesicles Derived From Umbilical Cord Mesenchymal Stem Cells Promote the Proliferation and Migration of Pediatric Urethral Smooth Muscle Cells Through Adenosine Pathway. *Frontiers in Bioengineering and Biotechnology*, *10*, 895998. DOI: 10.3389/fbioe.2022.895998 PMID: 35573239

Zou, J., Yang, W., Cui, W., Li, C., Ma, C., Ji, X., Hong, J., Qu, Z., Chen, J., Liu, A., & Wu, H. (2023). Therapeutic potential and mechanisms of mesenchymal stem cell-derived exosomes as bioactive materials in tendon-bone healing. *Journal of Nanobiotechnology*, *21*(1), 14. DOI: 10.1186/s12951-023-01778-6 PMID: 36642728

Conclusion

The journey of SC research, particularly in the realms of urology and gynecology, has been marked by significant advancements and promising potential. This book has explored the diverse applications of regenerative treatments, addressing a wide range of conditions, including urinary incontinence, fertility issues, andrological concerns, organ regeneration, tissue engineering, and gene editing.

SC have shown promise in treating urinary incontinence, both stress and urge types, by regenerating damaged tissues and improving bladder function and having an important impact in the quality of life. Further studies are needed to optimize treatment protocols and assess long-term outcomes. Also, their uses have been investigated and proved to be effective for treating erectile dysfunction, male infertility, and benign prostatic hyperplasia. Ongoing research is focused on understanding the mechanisms of action and developing safe and effective therapies.

SC-based therapies offer potential solutions for infertility, including the creation of functional gametes and the regeneration of reproductive tissues. Advancements in this area could revolutionize assisted reproductive technologies. However, ethical concerns must be taken in account.

While the complete regeneration of organs remains a challenging goal, SC have shown potential in creating functional tissues, such as bladder, kidney, and uterine linings. Continued research is essential to overcome technical hurdles and translate laboratory findings into clinical applications.

These therapies can be used to engineer functional tissues and organs, providing a potential source of replacement tissues for patients with various urological and gynaecological conditions. Advances in biomaterials and bioprinting technologies are accelerating progress in this field.

The advent of gene editing tools, such as CRISPR-Cas9, has opened new possibilities for correcting genetic defects associated with urological and gynaecological diseases, improving and modelling the SC for the final use. Precise gene editing could potentially cure genetic disorders and prevent disease development.

Assessing the long-term durability of SC-based therapies is essential to understand their sustainability and effectiveness over time. Identifying and addressing potential long-term side effects associated with SC treatments is crucial for patient safety.

Despite the immense potential of SC therapy, several ethical considerations and challenges must be addressed. The use of embryonic SC raises ethical concerns related to the destruction of embryos. Induced pluripotent SC offer an alternative approach that avoids these ethical dilemmas, as well as the use of autologous mesenchymal cells, that can be edited for producing SC. While iPSCs offer an alternative without the ethical concerns associated with ESC, their generation and use may also raise ethical questions, such as the potential for misuse or genetic manipulation.

Ensuring the safety and efficacy of generative therapies is crucial. Rigorous clinical trials are necessary to evaluate the risks and benefits of these treatments. Concerns about cancerogenesis or induced disorders in the immune system appear as a very important topic of debate, to find a way to avoid these possible complications, associated with cellular manipulation. The body's immune system may recognize transplanted SC as foreign and mount an immune response, leading to their rejection. Strategies to suppress or manipulate the immune response or engineer immune-tolerant cells are necessary. Establishing clear regulatory guidelines is essential to promote responsible research and clinical translation of SC therapies.

The cost of SC therapies can be high, limiting their accessibility to a wider population. In addition, developing the necessary infrastructure, including specialized laboratories and trained personnel, for widespread implementation of SC therapies can be challenging. Efforts must be made to ensure equitable access to SC therapies, particularly in underserved populations, who cannot access due to economic reasons, an ethical context must be ensured when experimenting on vulnerable populations, including animals. Navigating the regulatory landscape for stem cell therapies can be complex, as regulations vary between countries and may be evolving.

The use of animals in research raises ethical questions about the moral justification of causing harm to sentient beings. There is a growing emphasis on developing in vitro models, such as cell cultures and organ-on-a-chip systems, to reduce the reliance on animals. So, the use of SC could reduce in the future the use of animals, profiting the autologous tissues for culture, improving as well the efficacy of the research and the final reports. Computational modelling can provide insights into biological processes without the need for animal experiments.

Addressing these challenges will be essential for the successful translation of SC research into clinically relevant treatments. The future of SC therapy in urology and gynaecology is bright. Continued research and development are essential to unlock the full potential of these versatile cells. By addressing ethical concerns, overcoming technical challenges, and establishing robust regulatory frameworks, we can pave the way for a future where stem cell-based therapies revolutionize the treatment of a wide range of urological and gynaecological conditions.

Glossary

α-SMA: alpha-smooth muscle actin.
ADM: Acellular dermal matrix.
AdMSCs: Autologous adipose-derived MSCs.
ADSC: adipose derived stem cells.
AD-SVF: adipose-derived stromal vascular fraction.
AFSCs: amniotic fluid stem cells.
AGE: advanced glycation.
AKI: acute kidney injury.
AMDC: adult muscle-derived cells.
AMP: adenosine monophosphate.
ATP: adenosine triphosphate. Ca: Calcium.
AUA: American Urological Association.
ASC: Adipose Stem Cells.
BADM: bovine acellular dermal matrix.
BLSF: bilayer silk fibroin.
BMSC: Bone marrow-derived mesenchymal stem cells.
BOO: Bladder outlet obstruction.
BUN: blood urea nitrogen.
cAMP: cyclic adenosine monophosphate.
CAKUT: Congenital anomalies of the kidney and urinary tract.
Cas9: associated protein-9 nucleases.
cGMP: cyclic guanosine monophosphate.
cGPT: cyclic guanosine triphosphate.
CKD: chronic kidney disease.
CNI: cavernous nerve injury.
CS/UC-MSC: collagen scaffold/umbilical cord mesenchymal SC.
CSM: Chitosan Microspheres.
CRISPR: clustered regularly interspaced short palindromic repeats.
CYP-IC: cyclophosphamide-induced cystitis.

DMED: Diabetes Mellitus Erectile Dysfunction
DNA: deoxyribonucleic acid.
DSBs: double-strand breaks.
DUA: Detrusor underactivity.
ECM: Extracellular Matrix.
ED: Erectile Dysfunction
Edu: 5-ethynyl-2′-deoxyuridine.
EHS: erectile hardness score.
EMA: European Medicines Agency.
EMRO: embryo research oversight.
eNOS: Endothelial Nitric Oxide Synthase
EPCs: Endothelial progenitor cells.
ESCs: embryonic SCs.
ESRD: end-stage renal disease.
EUA: European Urology Association.
FDA: Food and Drug Administration.
FET: frozen-thawed embryo transfers.
FP: fertility preservation.
GAG: glycosaminoglycan.
GFT: green fluorescent protein.
GMP: guanosine monophosphate.
GPT: guanosine triphosphate.
FD: Fabry disease.
FUS: female urethral stricture.
h: human.
HADM: human acellular dermal matrix.
hAFSCs: human amniotic fluid.
HCL-IC: hydrochloric acid-induced cystitis.
HDR: homology-directed repair.
hESC; Human embryonic stem cells.
HGF: Hepatocyte growth factor.
HUCBs: Human umbilical cord blood stem cell.
hESC: Human Embryonic Stem Cells
HSPC: hematopoietic stem and progenitor cell.
hPSC: human pluripotent stem cell.
ICI: intra cavernosa injection.
IC/BPS: Interstitial cystitis/bladder pain syndrome.
ICS: International Continence Society.
ICM: inner cell mass.
IIEF-15: International Index of Erectile Function-15.

ISD: intrinsic sphincter deficiency.
iPSCs: induced pluripotent stem cells.
ISSCR: International Society for Stem Cell Research.
KC: ketamine-induced cystitis.
LESW: low-energy shock waves.
LPS: lipopolysaccharide.
LS: Lichen sclerosus.
m: mouse.
MDSCs: Muscle Derived Stem Cells.
Men-MSCs: menstrual blood-derived MSCs
MFAT: Microfragmented adipose tissue.
Mfn2: mitofusin 2.
M-MSCs: multipotent stem cells.
MS: multiple sclerosis.
MSCs: mesenchymal stem cells.
mRNA: messenger RNA.
MTs: micro tissues.
MV: microvesicles.
NGF: nerve growth factor.
NHEJ: non-homologous end joining.
NOA: nonobstructive azoospermia.
NOS: nitric oxide synthase.
OAB: Overactive Bladder.
OSC: ovarian stem cell.
OUI: Overflow Urinary Incontinence.
PADM: porcine acellular dermal matrix.
PD: Peyronie's disease.
PaD: Parkinson's disease.
PDE5: Phosphodiesterase type 5 inhibitors.
PDSCs: placenta-derived stem cell.
PEDF: pigment epithelium-derived factor.
PET–MRI: Positron emission tomography–magnetic resonance imaging.
PFD: Pelvic floor dysfunction.
PGCs: Primordial germ cells.
PGC-LC: primordial germ cell-like cell.
PKD: polycystic kidney disease.
PLGA: poly lactic-co-glycolic acid.
POP: pelvic organ prolapse.
PRP: plasma rich platelets.
PS: protamine sulphate.

PSC: pluripotent stem cell.
PTNS: Percutaneous Tibial Nerve Stimulation.
PU-MSCs: pediatric urethra.
PUSMCs: pediatric urethral smooth muscle cells.
PUV: Posterior urethral valves.
RNA: Ribonucleic acid.
RP: Radical Prostatectomy.
RPs: renal progenitors.
RR: relative risk.
RVFs: rectovaginal fistulas.
SCI: spinal cord injury.
SC: Stem cells
SCT: Stem Cell Therapy.
SDF-1: stromal cell derivedfactor 1.
sEVs: secrete small extracellular vesicles.
SIS: small intestinal submucosa.
SkMSCs: skeletal muscle-derived SC.
SSC: spermatogonial stem cell.
SSEA: stage-specificantigen.
SVF: stromal vascular fractions.
SUI: Stress urinary incontinence.
STIs: Sexually transmitted infections.
TA: Transit-amplifying.
TGF-β: Transforming growth factor-beta
TGFβ-1: transforming growth factor beta-1.
TNCs: Total nucleated cells.
UCs: for urothelial cells.
UCSCs: umbilical cord stem cell.
UDSCs: Urine-derived stem cells.
UI: Urinary incontinence.
VD: vaginal dilatation.
VEGF: vascular endothelial growth factor.
VECs: vascular endothelial cells.
VSELs: Very Small Embryonic-Like Stem Cells.
VUR: Vesicoureteral Reflux
WJSC. Wharton's gelatine-derived Stem Cells.

Compilation of References

Abdelaal, N. E., Tanga, B. M., Abdelgawad, M., Allam, S., Fathi, M., Saadeldin, I. M., Bang, S., & Cho, J. (2021). Cellular Therapy via Spermatogonial Stem Cells for Treating Impaired Spermatogenesis, Non-Obstructive Azoospermia. *Cells*, *10*(7), 1779. Advance online publication. DOI: 10.3390/cells10071779 PMID: 34359947

ACOG Committee on Ethics. (2006). ACOG Committee Opinion No. 347, November 2006: Using preimplantation embryos for research. *Obstetrics and Gynecology*, *108*(5), 1305–1317. DOI: 10.1097/00006250-200611000-00057 PMID: 17077265

Adamowicz, J., Pokrywczyńska, M., & Drewa, T. (2014). Conditioned medium derived from mesenchymal stem cells culture as a intravesical therapy for cystitis interstitials. *Medical Hypotheses*, *82*(6), 670–673. DOI: 10.1016/j.mehy.2014.02.027 PMID: 24679668

Aggarwal, S., Grange, C., Iampietro, C., Camussi, G., & Bussolati, B. (2016). Human CD133+ Renal Progenitor Cells Induce Erythropoietin Production and Limit Fibrosis After Acute Tubular Injury. *Scientific Reports*, *6*(1), 37270. DOI: 10.1038/srep37270 PMID: 27853265

Aghamir, S. M. K. (2020). Introduction. In Aghamir, S. M. K. (Ed.), *Stem Cells in Urology* (pp. 1–6). Springer International Publishing., DOI: 10.1007/978-3-030-41476-4_1

Ahmadian, S., Mahdipour, M., Pazhang, M., Sheshpari, S., Mobarak, H., Bedate, A. M., Rahbarghazi, R., & Nouri, M. (2020). Effectiveness of Stem Cell Therapy in the Treatment of Ovarian Disorders and Female Infertility: A Systematic Review. *Current Stem Cell Research & Therapy*, *15*(2), 173–186. DOI: 10.2174/1574888X14666191119122159 PMID: 31746298

Aitken, K. J., Schröder, A., Haddad, A., Sidler, M., Penna, F., Fernandez, N., Ahmed, T., Marino, V., Bechbache, M., Jiang, J.-X., Tolg, C., & Bägli, D. J. (2024). Epigenetic insights to pediatric uropathology: Celebrating the fundamental biology vision of Tony Khoury. *Journal of Pediatric Urology*, *20*, S43–S57. DOI: 10.1016/j.jpurol.2024.06.008 PMID: 38944627

Al Demour, S., Adwan, S., Jafar, H., Rahmeh, R., Alhawari, H., & Awidi, A. (2021). Safety and Efficacy of 2 Intracavernous Injections of Allogeneic Wharton's Jelly-Derived Mesenchymal Stem Cells in Diabetic Patients with Erectile Dysfunction: Phase 1/2 Clinical Trial. *Urologia Internationalis*, *105*(11-12), 935–943. DOI: 10.1159/000517364 PMID: 34384079

Al Demour, S., Jafar, H., Adwan, S., AlSharif, A., Alhawari, H., Alrabadi, A., Zayed, A., Jaradat, A., & Awidi, A. (2018). Safety and Potential Therapeutic Effect of Two Intracavernous Autologous Bone Marrow Derived Mesenchymal Stem Cells injections in Diabetic Patients with Erectile Dysfunction: An Open Label Phase I Clinical Trial. *Urologia Internationalis*, *101*(3), 358–365. DOI: 10.1159/000492120 PMID: 30173210

Alamein, M. A., Stephens, S., Liu, Q., Skabo, S., & Warnke, P. H. (2013). Mass production of nanofibrous extracellular matrix with controlled 3D morphology for large-scale soft tissue regeneration. *Tissue Engineering. Part C, Methods*, *19*(6), 458–472. DOI: 10.1089/ten.tec.2012.0417 PMID: 23102268

Albersen, M., Fandel, T. M., Lin, G., Wang, G., Banie, L., Lin, C. S., & Lue, T. F. (2010). Injections of adipose tissue-derived stem cells and stem cell lysate improve recovery of erectile function in a rat model of cavernous nerve injury. *Journal of Sexual Medicine*, *7*(10), 3331–3340. DOI: 10.1111/j.1743-6109.2010.01875.x PMID: 20561166

Aldamanhori, R., & Chapple, C. R. (2017). Underactive bladder, detrusor underactivity, definition, symptoms, epidemiology, etiopathogenesis, and risk factors. *Current Opinion in Urology*, *27*(3), 293–299. DOI: 10.1097/MOU.0000000000000381 PMID: 28221218

Aleixandre-Benavent, R., Simon, C., & Fauser, B. C. J. M. (2015). Trends in clinical reproductive medicine research: 10 years of growth. *Fertility and Sterility*, *104*(1), 131–137.e5. DOI: 10.1016/j.fertnstert.2015.03.025 PMID: 25936232

Algarrahi, K., Affas, S., Sack, B. S., Yang, X., Costa, K., Seager, C., Estrada, C. R.Jr, & Mauney, J. R. (2018). Repair of injured urethras with silk fibroin scaffolds in a rabbit model of onlay urethroplasty. *The Journal of Surgical Research*, *229*, 192–199. DOI: 10.1016/j.jss.2018.04.006 PMID: 29936989

Al-Husseiny, F., Sobh, M. A., Ashour, R. H., Foud, S., Medhat, T., El-Gilany, A.-H., Elghannam, D., Abdel-Ghaffar, H., Saad, M.-A., & Sobh, M. (2016). Amniotic Fluid-Derived Mesenchymal Stem Cells Cut Short the Acuteness of Cisplatin-Induced Nephrotoxicity in Sprague-Dawley Rats. *International Journal of Stem Cells*, *9*(1), 70–78. DOI: 10.15283/ijsc.2016.9.1.70 PMID: 27426088

Alwaal, A., Hussein, A. A., Lin, C. S., & Lue, T. F. (2015). Prospects of stem cell treatment in benign urological diseases. *Korean Journal of Urology*, *56*(4), 257–265. DOI: 10.4111/kju.2015.56.4.257 PMID: 25874038

Anders, C., Niewoehner, O., Duerst, A., & Jinek, M. (2014). Structural basis of PAM-dependent target DNA recognition by the Cas9 endonuclease. *Nature*, *513*(7519), 569–573. DOI: 10.1038/nature13579 PMID: 25079318

Anumanthan, G., Makari, J. H., Honea, L., Thomas, J. C., Wills, M. L., Bhowmick, N. A., Adams, M. C., Hayward, S. W., Matusik, R. J., Brock, J. W., & Pope, J. C. (2008). Directed differentiation of bone marrow derived mesenchymal stem cells into bladder urothelium. *The Journal of Urology*, *180*(4, Suppl), 1778–1783. DOI: 10.1016/j.juro.2008.04.076 PMID: 18721942

Anway, M. D., & Skinner, M. K. (2008). Epigenetic programming of the germ line: Effects of endocrine disruptors on the development of transgenerational disease. *Reproductive Biomedicine Online*, *16*(1), 23–25. DOI: 10.1016/S1472-6483(10)60553-6 PMID: 18252044

Apodaca, G., Kiss, S., Ruiz, W., Meyers, S., Zeidel, M., & Birder, L. (2003). Disruption of bladder epithelium barrier function after spinal cord injury. *American Journal of Physiology. Renal Physiology*, *284*(5), F966–F976. DOI: 10.1152/ajprenal.00359.2002 PMID: 12527557

Araoka, T., Mae, S., Kurose, Y., Uesugi, M., Ohta, A., Yamanaka, S., & Osafune, K. (2014). Efficient and rapid induction of human iPSCs/ESCs into nephrogenic intermediate mesoderm using small molecule-based differentiation methods. *PLoS One*, *9*(1), e84881. DOI: 10.1371/journal.pone.0084881 PMID: 24454758

Arcolino, F. O., Zia, S., Held, K., Papadimitriou, E., Theunis, K., Bussolati, B., Raaijmakers, A., Allegaert, K., Voet, T., Deprest, J., Vriens, J., Toelen, J., van den Heuvel, L., & Levtchenko, E. (2016). Urine of Preterm Neonates as a Novel Source of Kidney Progenitor Cells. *Journal of the American Society of Nephrology*, *27*(9), 2762–2770. DOI: 10.1681/ASN.2015060664 PMID: 26940093

Arjmand, B., Safavi, M., Heidari, R., Aghayan, H., Bazargani, S. T., Dehghani, S., & Pourmand, G. (2017). Concomitant transurethral and transvaginal-periurethral injection of autologous adipose derived stem cells for treatment of female stress urinary incontinence: A phase one clinical trial. *Acta Medica Iranica*, •••, 368–374.

Ashour, R. H., Saad, M.-A., Sobh, M.-A., Al-Husseiny, F., Abouelkheir, M., Awad, A., Elghannam, D., Abdel-Ghaffar, H., & Sobh, M. (2016). Comparative study of allogenic and xenogeneic mesenchymal stem cells on cisplatin-induced acute kidney injury in Sprague-Dawley rats. *Stem Cell Research & Therapy*, *7*(1), 126. DOI: 10.1186/s13287-016-0386-0 PMID: 27585525

ASRM. (2020). *Ethics in embryo research: A position statement by the ASRM Ethics in Embryo Research Task Force and the ASRM Ethics Committee (2020)* [2020]. https://www.asrm.org/practice-guidance/ethics-opinions/ethics-in-embryo-research-a-position-statement-by-the-asrm-ethics-in-embryo-research-task-force-and-the-asrm-ethics-committee-2020/

Atala, A., Bauer, S. B., Soker, S., Yoo, J. J., & Retik, A. B. (2006). Tissue-engineered autologous bladders for patients needing cystoplasty. *Lancet*, *367*(9518), 1241–1246. DOI: 10.1016/S0140-6736(06)68438-9 PMID: 16631879

Avarbock, M. R., Brinster, C. J., & Brinster, R. L. (1996). Reconstitution of spermatogenesis from frozen spermatogonial stem cells. *Nature Medicine*, *2*(6), 693–696. DOI: 10.1038/nm0696-693 PMID: 8640563

Avior, Y., Lezmi, E., Yanuka, D., & Benvenisty, N. (2017). Modeling Developmental and Tumorigenic Aspects of Trilateral Retinoblastoma via Human Embryonic Stem Cells. *Stem Cell Reports*, *8*(5), 1354–1365. DOI: 10.1016/j.stemcr.2017.03.005 PMID: 28392220

Baban, B., Liu, J. Y., Payne, S., Abebe, W., Yu, J. C., & Mozaffari, M. S. (2016). Status of stem cells in diabetic nephropathy: Predictive and preventive potentials. *The EPMA Journal*, *7*(1), 21. DOI: 10.1186/s13167-016-0070-6 PMID: 27729946

Bachmann, G. A., & Nevadunsky, N. S. (2000). Diagnosis and treatment of atrophic vaginitis. *American Family Physician*, *61*(10), 3090–3096. PMID: 10839558

Badra, S., Andersson, K. E., Dean, A., Mourad, S., & Williams, J. K. (2013). Long-term structural and functional effects of autologous muscle precursor cell therapy in a nonhuman primate model of urinary sphincter deficiency. *The Journal of Urology*, *190*(5), 1938–1945. DOI: 10.1016/j.juro.2013.04.052 PMID: 23618586

Badylak, S. F., Freytes, D. O., & Gilbert, T. W. (2009). Extracellular matrix as a biological scaffold material: Structure and function. *Acta Biomaterialia*, *5*(1), 1–13. DOI: 10.1016/j.actbio.2008.09.013 PMID: 18938117

Bahk, J. Y., Jung, J. H., Han, H., Min, S. K., & Lee, Y. S. (2010). Treatment of diabetic impotence with umbilical cord blood stem cell intracavernosal transplant: Preliminary report of 7 cases. *Experimental and Clinical Transplantation*, *8*(2). PMID: 20565373

Barakat, B., Franke, K., Schakaki, S., Hijazi, S., Hasselhof, V., & Vögeli, T. A. (2020a). Stem cell applications in regenerative medicine for stress urinary incontinence: A review of effectiveness based on clinical trials. *Arab Journal of Urology*, *18*(3), 194–205. DOI: 10.1080/2090598X.2020.1750864 PMID: 33029431

Barnes, C. P., Pemble, C. W.IV, Brand, D. D., Simpson, D. G., & Bowlin, G. L. (2007). Cross-linking electrospun type II collagen tissue engineering scaffolds with carbodiimide in ethanol. *Tissue Engineering*, *13*(7), 1593–1605. DOI: 10.1089/ten.2006.0292 PMID: 17523878

Batchelder, C. A., Lee, C. C. I., Matsell, D. G., Yoder, M. C., & Tarantal, A. F. (2009). Renal ontogeny in the rhesus monkey (Macaca mulatta) and directed differentiation of human embryonic stem cells towards kidney precursors. *Differentiation. Differentiation*, *78*(1), 45–56. DOI: 10.1016/j.diff.2009.05.001 PMID: 19500897

Baulier, E., Favreau, F., Le Corf, A., Jayle, C., Schneider, F., Goujon, J.-M., Feraud, O., Bennaceur-Griscelli, A., Hauet, T., & Turhan, A. G. (2014). Amniotic fluid-derived mesenchymal stem cells prevent fibrosis and preserve renal function in a preclinical porcine model of kidney transplantation. *Stem Cells Translational Medicine*, *3*(7), 809–820. DOI: 10.5966/sctm.2013-0186 PMID: 24797827

Behr, L., Hekmati, M., Lucchini, A., Houcinet, K., Faussat, A.-M., Borenstein, N., Noel, L.-H., Lelievre-Pegorier, M., & Laborde, K. (2009). Evaluation of the effect of autologous mesenchymal stem cell injection in a large-animal model of bilateral kidney ischaemia reperfusion injury. *Cell Proliferation*, *42*(3), 284–297. DOI: 10.1111/j.1365-2184.2009.00591.x PMID: 19438896

Bergman, A., Karram, M. M., & Bhatia, N. N. (1990). Changes in urethral cytology following estrogen administration. *Gynecologic and Obstetric Investigation*, *29*(3), 211–213. DOI: 10.1159/000293384 PMID: 2358196

Bharadwaj, S., Liu, G., Shi, Y., Wu, R., Yang, B., He, T., Fan, Y., Lu, X., Zhou, X., Liu, H., Atala, A., Rohozinski, J., & Zhang, Y. (2013). Multipotential differentiation of human urine-derived stem cells: Potential for therapeutic applications in urology. *Stem Cells (Dayton, Ohio)*, *31*(9), 1840–1856. DOI: 10.1002/stem.1424 PMID: 23666768

Bhartiya, D., Hinduja, I., Patel, H., & Bhilawadikar, R. (2014). Making gametes from pluripotent stem cells—A promising role for very small embryonic-like stem cells. *Reproductive biology and endocrinology. RB&E, 12*(1). Advance online publication. DOI: 10.1186/1477-7827-12-114 PMID: 25421462

Bhartiya, D., Shaikh, A., Anand, S., Patel, H., Kapoor, S., Sriraman, K., Parte, S., & Unni, S. (2016). Endogenous, very small embryonic-like stem cells: Critical review, therapeutic potential and a look ahead. *Human Reproduction Update, 23*(1), 1–36. DOI: 10.1093/humupd/dmw030 PMID: 27614362

Bhartiya, D., Sriraman, K., Parte, S., & Patel, H. (2013). Ovarian stem cells: Absence of evidence is not evidence of absence. *Journal of Ovarian Research, 6*(1), 65. Advance online publication. DOI: 10.1186/1757-2215-6-65 PMID: 24044496

Bieri, M., Said, E., Antonini, G., Dickerson, D., Tuma, J., Bartlett, C. E., Patel, A. N., & Gershman, A. (2020). Phase I and registry study of autologous bone marrow concentrate evaluated in PDE5 inhibitor refractory erectile dysfunction. *Journal of Translational Medicine, 18*(1), 24. DOI: 10.1186/s12967-019-02195-w PMID: 31937310

Bilsel, Y., & Abci, I. (2012). The search for ideal hernia repair; mesh materials and types. *International Journal of Surgery, 10*(6), 317–321. DOI: 10.1016/j.ijsu.2012.05.002 PMID: 22588090

Birder, L., & Andersson, K. E. (2018). Animal Modelling of Interstitial Cystitis/Bladder Pain Syndrome. *International Neurourology Journal, 22*(Suppl 1), S3–S9. DOI: 10.5213/inj.1835062.531 PMID: 29385788

Bivalacqua, T. J., Deng, W., Kendirci, M., Usta, M. F., Robinson, C., Taylor, B. K., Murthy, S. N., Champion, H. C., Hellstrom, W. J. G., & Kadowitz, P. J. (2007). Mesenchymal stem cells alone or ex vivo gene modified with endothelial nitric oxide synthase reverse age-associated erectile dysfunction. *American Journal of Physiology. Heart and Circulatory Physiology, 292*(3), H1278–H1290. Advance online publication. DOI: 10.1152/ajpheart.00685.2006 PMID: 17071732

Blaganje, M., & Lukanović, A. (2012). Intrasphincteric autologous myoblast injections with electrical stimulation for stress urinary incontinence. *International Journal of Gynaecology and Obstetrics: the Official Organ of the International Federation of Gynaecology and Obstetrics, 117*(2), 164–167. DOI: 10.1016/j.ijgo.2011.11.029 PMID: 22342057

Blaivas, J. G., & Olsson, C. A. (1988). Stress incontinence: Classification and surgical approach. *Journal of Urology, 139*(4 I), 727-731. DOI: 10.1016/S0022-5347(17)42611-5

Bochinski, D., Lin, G. T., Nunes, L., Carrion, R., Rahman, N., Lin, C. S., & Lue, T. F. (2004). The effect of neural embryonic stem cell therapy in a rat model of cavernosal nerve injury. *BJU International*, *94*(6), 904–909. DOI: 10.1111/j.1464-410X.2003.05057.x PMID: 15476533

Boffito, M., Sirianni, P., Di Rienzo, A. M., & Chiono, V. (2015). Thermosensitive block copolymer hydrogels based on poly(ε-caprolactone) and polyethylene glycol for biomedical applications: State of the art and future perspectives. *Journal of Biomedical Materials Research. Part A*, *103*(3), 1276–1290. DOI: 10.1002/jbm.a.35253 PMID: 24912941

Bohne, A. W., & Urwiller, K. L. (1956). Experience with urinary bladder regeneration *Postgraduate Seminar. American Urological Association. North Central Section*, 24-27.

Borowiak, M., Maehr, R., Chen, S., Chen, A. E., Tang, W., Fox, J. L., Schreiber, S. L., & Melton, D. A. (2009). Small molecules efficiently direct endodermal differentiation of mouse and human embryonic stem cells. *Cell Stem Cell*, *4*(4), 348–358. DOI: 10.1016/j.stem.2009.01.014 PMID: 19341624

Bouty, A., Ayers, K. L., Pask, A., Heloury, Y., & Sinclair, A. H. (2015). The Genetic and Environmental Factors Underlying Hypospadias. *Sexual Development: Genetics, Molecular Biology, Evolution, Endocrinology, Embryology, and Pathology of Sex Determination and Differentiation*, *9*(5), 239–259. DOI: 10.1159/000441988 PMID: 26613581

Brinster, R. L., & Zimmermann, J. W. (1994). Spermatogenesis following male germ-cell transplantation. *Proceedings of the National Academy of Sciences of the United States of America*, *91*(24), 11298–11302. DOI: 10.1073/pnas.91.24.11298 PMID: 7972053

Bruce, S. J., Rea, R. W., Steptoe, A. L., Busslinger, M., Bertram, J. F., & Perkins, A. C. (2007). In vitro differentiation of murine embryonic stem cells toward a renal lineage. *Differentiation. Differentiation*, *75*(5), 337–349. DOI: 10.1111/j.1432-0436.2006.00149.x PMID: 17286599

Bruno, S., Chiabotto, G., & Camussi, G. (2014). Concise review: Different mesenchymal stromal/stem cell populations reside in the adult kidney. *Stem Cells Translational Medicine*, *3*(12), 1451–1455. DOI: 10.5966/sctm.2014-0142 PMID: 25355731

Bruno, S., Grange, C., Deregibus, M. C., Calogero, R. A., Saviozzi, S., Collino, F., Morando, L., Busca, A., Falda, M., Bussolati, B., Tetta, C., & Camussi, G. (2009). Mesenchymal stem cell-derived microvesicles protect against acute tubular injury. *Journal of the American Society of Nephrology*, *20*(5), 1053–1067. DOI: 10.1681/ASN.2008070798 PMID: 19389847

Bugos, O., Bhide, M., & Zilka, N. (2009). Beyond the rat models of human neurodegenerative disorders. *Cellular and Molecular Neurobiology*, *29*(6-7), 859–869. DOI: 10.1007/s10571-009-9367-5 PMID: 19263215

Burkhard, F. C., Bosch, J. L. H. R., Cruz, F., Lemack, G. E., Nambiar, A. K., Thiruchelvam, N., & Tubaro, A. (2018). EAU Guidelines on Urinary Incontinence in Adults. *European Association of Urology*, 99.

Bussolati, B., & Camussi, G. (2015). Therapeutic use of human renal progenitor cells for kidney regeneration. *Nature Reviews. Nephrology*, *11*(12), 695–706. DOI: 10.1038/nrneph.2015.126 PMID: 26241019

Cai, H., Xia, X., Wang, L., Liu, Y., He, Z., Guo, Q., & Xu, C. (2013). In vitro and in vivo differentiation of induced pluripotent stem cells into male germ cells. *Biochemical and Biophysical Research Communications*, *433*(3), 286–291. DOI: 10.1016/j.bbrc.2013.02.107 PMID: 23524261

Campeau, L., Soler, R., Sittadjody, S., Pareta, R., Nomiya, M., Zarifpour, M., Opara, E. C., Yoo, J. J., & Andersson, K.-E. (2014). Effects of allogeneic bone marrow derived mesenchymal stromal cell therapy on voiding function in a rat model of Parkinson disease. *The Journal of Urology*, *191*(3), 850–859. DOI: 10.1016/j.juro.2013.08.026 PMID: 23973520

Cannon, T. W., Lee, J. Y., Somogyi, G., Pruchnic, R., Smith, C. P., Huard, J., & Chancellor, M. B. (2003). Improved sphincter contractility after allogenic muscle-derived progenitor cell injection into the denervated rat urethra. *Urology*, *62*(5), 958–963. DOI: 10.1016/S0090-4295(03)00679-4 PMID: 14624934

Cao, B., Wang, Y., Wen, D., Liu, W., Wang, J., Fan, G., Ruan, L., Song, B., Cai, Y., Wei, M., Li, X., Xia, J., Chen, N., Xiang, J., Yu, T., Bai, T., Xie, X., Zhang, L., Li, C., & Wang, C. (2020). A trial of lopinavir-ritonavir in adults hospitalized with severe covid-19. *The New England Journal of Medicine*, *382*(19), 1787–1799. DOI: 10.1056/NEJMoa2001282 PMID: 32187464

Cao, Y., Su, Q., Zhang, B., Shen, F., & Li, S. (2021). Efficacy of stem cells therapy for Crohn's fistula: A meta-analysis and systematic review. *Stem Cell Research & Therapy*, *12*(1), 32. DOI: 10.1186/s13287-020-02095-7 PMID: 33413661

Carr, L. K., Robert, M., Kultgen, P. L., Herschorn, S., Birch, C., Murphy, M., & Chancellor, M. B. (2013). Autologous muscle derived cell therapy for stress urinary incontinence: A prospective, dose ranging study. *The Journal of Urology*, *189*(2), 595–601. DOI: 10.1016/j.juro.2012.09.028 PMID: 23260547

Carr, L. K., Steele, D., Steele, S., Wagner, D., Pruchnic, R., Jankowski, R., Erickson, J., Huard, J., & Chancellor, M. B. (2008). 1-Year Follow-Up of Autologous Muscle-Derived Stem Cell Injection Pilot Study To Treat Stress Urinary Incontinence. *International Urogynecology Journal*, *19*(6), 881–883. DOI: 10.1007/s00192-007-0553-z PMID: 18204978

Cary, R. B., Peterson, S. R., Wang, J., Bear, D. G., Bradbury, E. M., & Chen, D. J. (1997). DNA looping by Ku and the DNA-dependent protein kinase. *Proceedings of the National Academy of Sciences of the United States of America*, *94*(9), 4267–4272. DOI: 10.1073/pnas.94.9.4267 PMID: 9113978

Casarotti, G. A., Chiodera, P., & Tremolada, C. (2018). Menopause: New frontiers in the treatment of urogenital atrophy. *European Review for Medical and Pharmacological Sciences*, *22*(2), 567–574. DOI: 10.26355/eurrev_201801_14211 PMID: 29424920

Casarotti, G., & Tremolada, C. (2020). A new treatment of genito-urinary post-menopausal atrophy with autologous micro-fragmented fat tissue: A thirty-six months follow up case series. *European Review for Medical and Pharmacological Sciences*, *24*(13), 7420–7426. DOI: 10.26355/eurrev_202007_21910 PMID: 32706081

Cassim, M. I., & Mohamed, T. (2019). *Case Report A Novel Therapy for the Treatment of Malefactor Infertility Due to Non-obstructive Azoospermia: A Case Report*. *6*(1), 129-131.

Castiglione, F., Hedlund, P., Van Der Aa, F., Bivalacqua, T. J., Rigatti, P., Van Poppel, H., Montorsi, F., De Ridder, D., & Albersen, M. (2013). Intratunical injection of human adipose tissue-derived stem cells prevents fibrosis and is associated with improved erectile function in a rat model of Peyronie's disease. *European Urology*, *63*(3), 551–560. DOI: 10.1016/j.eururo.2012.09.034 PMID: 23040209

Cengiz, T., Kaya, E., Oral, D. Y., Ozakca, I., Bayatli, N., Karabay, A. Z., Ensari, T. A., Karahan, T., Yilmaz, E., & Gur, S. (2017). Intracavernous Injection of Human Umbilical Cord Blood Mononuclear Cells Improves Erectile Dysfunction in Streptozotocin-Induced Diabetic Rats. *Journal of Sexual Medicine*, *14*(1), 50–58. DOI: 10.1016/j.jsxm.2016.11.314 PMID: 28065360

Cervelló, I., Medrano, J. V., & Simón, C. (2016). Regenerative Medicine and Tissue Engineering in Reproductive Medicine: Future Clinical Applications in Human Infertility. *Translating Regenerative Medicine to the Clinic*, 139-151. DOI: 10.1016/B978-0-12-800548-4.00010-3

Cervelló, I., & Simón, C. (2009). Somatic stem cells in the endometrium. *Reproductive Sciences (Thousand Oaks, Calif.)*, *16*(2), 200–205. DOI: 10.1177/1933719108329955 PMID: 19208788

Chade, A. R., Zhu, X., Lavi, R., Krier, J. D., Pislaru, S., Simari, R. D., Napoli, C., Lerman, A., & Lerman, L. O. (2009). Endothelial progenitor cells restore renal function in chronic experimental renovascular disease. *Circulation*, *119*(4), 547–557. DOI: 10.1161/CIRCULATIONAHA.108.788653 PMID: 19153272

Chai, T. C., Russo, A., Yu, S., & Lu, M. (2016). Mucosal signaling in the bladder. *Autonomic Neuroscience*, *200*, 49–56. DOI: 10.1016/j.autneu.2015.08.009 PMID: 26422993

Chancellor, M. B., Yokoyama, T., Tirney, S., Mattes, C. E., Ozawa, H., Yoshimura, N., De Groat, W. C., & Huard, J. (2000). Preliminary results of myoblast injection into the urethra and bladder wall: A possible method for the treatment of stress urinary incontinence and impaired detrusor contractility. *Neurourology and Urodynamics*, *19*(3), 279–287. DOI: 10.1002/(SICI)1520-6777(2000)19:3<279::AID-NAU9>3.0.CO;2-M PMID: 10797585

Chen, F., Zhang, H., Wang, Z., Ding, W., Zeng, Q., Liu, W., Huang, C., He, S., & Wei, A. (2017). Adipose-Derived Stem Cell-Derived Exosomes Ameliorate Erectile Dysfunction in a Rat Model of Type 2 Diabetes. *Journal of Sexual Medicine*, *14*(9), 1084–1094. DOI: 10.1016/j.jsxm.2017.07.005 PMID: 28781215

Chen, J., Park, H.-C., Addabbo, F., Ni, J., Pelger, E., Li, H., Plotkin, M., & Goligorsky, M. S. (2008). Kidney-derived mesenchymal stem cells contribute to vasculogenesis, angiogenesis and endothelial repair. *Kidney International*, *74*(7), 879–889. DOI: 10.1038/ki.2008.304 PMID: 18596729

Chen, S., Zhang, H. Y., Zhang, N., Li, W. H., Shan, H., Liu, K., & Yang, Y. (2012). Treatment for chronic ischaemia-induced bladder detrusor dysfunction using bone marrow mesenchymal stem cells: An experimental study. *International Journal of Molecular Medicine*, *29*(3), 416–422. DOI: 10.3892/ijmm.2011.846 PMID: 22109789

Chen, X., Yang, Q., Zheng, T., Bian, J., Sun, X., Shi, Y., Liang, X., Gao, G., Liu, G., & Deng, C. (2016). Neurotrophic effect of adipose tissue-derived stem cells on erectile function recovery by pigment epithelium-derived factor secretion in a rat model of cavernous nerve injury. *Stem Cells International*, *2016*(1), 5161248. Advance online publication. DOI: 10.1155/2016/5161248 PMID: 26783403

Chen, Y. T., Chiang, H. J., Chen, C. H., Sung, P. H., Lee, F. Y., Tsai, T. H., Lo Chang, C., Chen, H. H., Sun, C. K., Leu, S., Chang, H. W., Yang, C. C., & Yip, H. K. (2014a). Melatonin treatment further improves adipose-derived mesenchymal stem cell therapy for acute interstitial cystitis in rat. *Journal of Pineal Research*, *57*(3), 248–261. DOI: 10.1111/jpi.12164 PMID: 25132326

Chiang, B. J., Liao, C. H., Mao, S. H., & Chien, C. T. (2021). Adipose-derived stem cells and their derived microvesicles ameliorate detrusor overactivity secondary to bilateral partial iliac arterial occlusion-induced bladder ischemia. *International Journal of Molecular Sciences*, *22*(13), 7000. Advance online publication. DOI: 10.3390/ijms22137000 PMID: 34210091

Chinnadurai, R., Rajan, D., Qayed, M., Arafat, D., Garcia, M., Liu, Y., Kugathasan, S., Anderson, L. J., Gibson, G., & Galipeau, J. (2018). Potency Analysis of Mesenchymal Stromal Cells Using a Combinatorial Assay Matrix Approach. *Cell Reports*, *22*(9), 2504–2517. DOI: 10.1016/j.celrep.2018.02.013 PMID: 29490284

Choi, J. Y., Kim, T. H., Yang, J. D., Suh, J. S., & Kwon, T. G. (2016a). Adipose-derived regenerative cell injection therapy for postprostatectomy incontinence: A phase i clinical study. *Yonsei Medical Journal*, *57*(5), 1152–1158. DOI: 10.3349/ymj.2016.57.5.1152 PMID: 27401646

Chow, D., Nunalee, M. L., Lim, D. W., Simnick, A. J., & Chilkoti, A. (2008). Peptide-based Biopolymers in Biomedicine and Biotechnology. *Materials Science & Engineering. R, Reports: A Review Journal, 62*(4), 125-155. DOI: 10.1016/j.mser.2008.04.004

Cho, Y.-S., Ko, I.-G., Kim, S.-E., Lee, S.-M., Shin, M.-S., Kim, C.-J., Kim, S.-H., Jin, J.-J., & Kim, K.-H. (2014). Oral mucosa stem cells alleviates spinal cord injury-induced neurogenic bladder symptoms in rats. *Journal of Biomedical Science*, *21*(1), 43. DOI: 10.1186/1423-0127-21-43 PMID: 24884998

Clark, A. T., Bodnar, M. S., Fox, M., Rodriquez, R. T., Abeyta, M. J., Firpo, M. T., & Reijo Pera, R. A. (2004). Spontaneous differentiation of germ cells from human embryonic stem cells in vitro. *Human Molecular Genetics*, *13*(7), 727–739. DOI: 10.1093/hmg/ddh088 PMID: 14962983

Coburn, J. M., Gibson, M., Monagle, S., Patterson, Z., & Elisseeff, J. H. (2012). Bioinspired nanofibers support chondrogenesis for articular cartilage repair. *Proceedings of the National Academy of Sciences of the United States of America, 109*(25), 10012–10017. DOI: 10.1073/pnas.1121605109 PMID: 22665791

Cocuzza, M., Alvarenga, C., & Pagani, R. (2013). The epidemiology and etiology of azoospermia. *Clinics (Sao Paulo, Brazil), 68 Suppl 1*(Suppl 1), 15-26. DOI: 10.6061/clinics/2013(Sup01)03

Cong, L., Ran, F. A., Cox, D., Lin, S., Barretto, R., Habib, N., Hsu, P. D., Wu, X., Jiang, W., Marraffini, L. A., & Zhang, F. (2013). Multiplex genome engineering using CRISPR/Cas systems. *Science, 339*(6121), 819–823. DOI: 10.1126/science.1231143 PMID: 23287718

Coresh, J., Selvin, E., Stevens, L. A., Manzi, J., Kusek, J. W., Eggers, P., Van Lente, F., & Levey, A. S. (2007). Prevalence of chronic kidney disease in the United States. *Journal of the American Medical Association, 298*(17), 2038–2047. DOI: 10.1001/jama.298.17.2038 PMID: 17986697

Craggs, M. D., Balasubramaniam, A. V., Chung, E. A. L., & Emmanuel, A. V. (2006). Aberrant reflexes and function of the pelvic organs following spinal cord injury in man. *Autonomic Neuroscience, 126-127*, 355–370. DOI: 10.1016/j.autneu.2006.03.010 PMID: 16731049

Crapo, P. M., Gilbert, T. W., & Badylak, S. F. (2011). An overview of tissue and whole organ decellularization processes. *Biomaterials, 32*(12), 3233–3243. DOI: 10.1016/j.biomaterials.2011.01.057 PMID: 21296410

Cui, F.-Z., Li, Y., & Ge, J. (2007). Self-assembly of mineralized collagen composites. *Materials Science and Engineering R Reports, 57*(1), 1–27. DOI: 10.1016/j.mser.2007.04.001

Culha, M. G., Erkan, E., Cay, T., & Yücetaş, U. (2019). The Effect of Platelet-Rich Plasma on Peyronie's Disease in Rat Model. *Urologia Internationalis, 102*(2), 218–223. DOI: 10.1159/000492755 PMID: 30317233

Cyranoski, D. (2013). Stem cells: Egg engineers. *Nature, 500*(7463), 392–394. DOI: 10.1038/500392a PMID: 23969442

D'Ancona, C., Haylen, B., Oelke, M., Abranches-Monteiro, L., Arnold, E., Goldman, H., Hamid, R., Homma, Y., Marcelissen, T., Rademakers, K., Schizas, A., Singla, A., Soto, I., Tse, V., de Wachter, S., & Herschorn, S. (2019). The International Continence Society (ICS) report on the terminology for adult male lower urinary tract and pelvic floor symptoms and dysfunction. *Neurourology and Urodynamics*, *38*(2), 433–477. DOI: 10.1002/nau.23897 PMID: 30681183

Da Sacco, S., Perin, L., & Sedrakyan, S. (2018). Amniotic fluid cells: Current progress and emerging challenges in renal regeneration. *Pediatric Nephrology (Berlin, Germany)*, *33*(6), 935–945. DOI: 10.1007/s00467-017-3711-7 PMID: 28620747

Dawkins, M. (2012). *Animal Suffering: The Science of Animal Welfare*. Springer Science & Business Media.

De Coppi, P., Bartsch, G.Jr, Siddiqui, M. M., Xu, T., Santos, C. C., Perin, L., Mostoslavsky, G., Serre, A. C., Snyder, E. Y., Yoo, J. J., Furth, M. E., Soker, S., & Atala, A. (2007). Isolation of amniotic stem cell lines with potential for therapy. *Nature Biotechnology*, *25*(1), 100–106. DOI: 10.1038/nbt1274 PMID: 17206138

de Sèze, M., Ruffion, A., Denys, P., Joseph, P.-A., & Perrouin-Verbe, B. (2007). The neurogenic bladder in multiple sclerosis: Review of the literature and proposal of management guidelines. *Multiple Sclerosis*, *13*(7), 915–928. DOI: 10.1177/1352458506075651 PMID: 17881401

de Weerd, L., Weum, S., & Norderval, S. (2015). Novel treatment for recalcitrant rectovaginal fistulas: Fat injection. *International Urogynecology Journal*, *26*(1), 139–144. DOI: 10.1007/s00192-014-2497-4 PMID: 25199495

de Wolf, C., van de Bovenkamp, M., & Hoefnagel, M. (2017). Regulatory perspective on in vitro potency assays for human mesenchymal stromal cells used in immunotherapy. *Cytotherapy*, *19*(7), 784–797. DOI: 10.1016/j.jcyt.2017.03.076 PMID: 28457740

Definition of reproductive medicine—NCI Dictionary of Cancer Terms—NCI. (s. f.). Recuperado 19 de agosto de 2023, de https://www.cancer.gov/publications/dictionaries/cancer-terms/def/reproductive-medicine

DePaul, M. A., Palmer, M., Lang, B. T., Cutrone, R., Tran, A. P., Madalena, K. M., Bogaerts, A., Hamilton, J. A., Deans, R. J., Mays, R. W., Busch, S. A., & Silver, J. (2015). Intravenous multipotent adult progenitor cell treatment decreases inflammation leading to functional recovery following spinal cord injury. *Scientific Reports*, *5*(1), 16795. DOI: 10.1038/srep16795 PMID: 26582249

Diallo, A. B., Diallo, T. M. O., Bah, I., Bah, M. D., & Barry, M. (2015). Vesicovaginal Fistulas: Anatomical Clinical and Surgical Aspects in the Conakry University Hospital Center. *Open Journal of Urology*, *5*(12), 12. Advance online publication. DOI: 10.4236/oju.2015.512036

Dissanayake, D. (2018). In Vitro Spermatogenesis; Past, Present, and Future. In *Spermatozoa—Facts and Perspectives*. InTech., DOI: 10.5772/intechopen.73505

Dissaranan, C., Cruz, M. A., Kiedrowski, M. J., Balog, B. M., Gill, B. C., Penn, M. S., Goldman, H. B., & Damaser, M. S. (2014). Rat mesenchymal stem cell secretome promotes elastogenesis and facilitates recovery from simulated childbirth injury. *Cell Transplantation*, *23*(11), 1395–1406. DOI: 10.3727/096368913X670921 PMID: 23866688

Drake, M. J., Williams, J., & Bijos, D. A. (2014). Voiding dysfunction due to detrusor underactivity: An overview. *Nature Reviews. Urology*, *11*(8), 454–464. DOI: 10.1038/nrurol.2014.156 PMID: 25002201

Durruthy, J. D., Ramathal, C., Sukhwani, M., Fang, F., Cui, J., Orwig, K. E., & Reijo Pera, R. A. (2014). Fate of induced pluripotent stem cells following transplantation to murine seminiferous tubules. *Human Molecular Genetics*, *23*(12), 3071–3084. DOI: 10.1093/hmg/ddu012 PMID: 24449759

Easley, C. A.IV, Phillips, B. T., McGuire, M. M., Barringer, J. M., Valli, H., Hermann, B. P., Simerly, C. R., Rajkovic, A., Miki, T., Orwig, K. E., & Schatten, G. P. (2012). Direct differentiation of human pluripotent stem cells into haploid spermatogenic cells. *Cell Reports*, *2*(3), 440–446. DOI: 10.1016/j.celrep.2012.07.015 PMID: 22921399

EAU Guidelines on Urethral Strictures—DEFINITION EPIDEMIOLOGY AETIOLOGY AND PREVENTION - Uroweb. (s. f.). Uroweb - European Association of Urology. Recuperado 11 de agosto de 2024, de https://uroweb.org/guidelines/urethral-strictures/chapter/definition-epidemiology-aetiology-and-prevention

Eberli, D., Andersson, K. E., Yoo, J. J., & Atala, A. (2009). A canine model of irreversible urethral sphincter insufficiency. *BJU International*, *103*(2), 248–253. DOI: 10.1111/j.1464-410X.2008.08001.x PMID: 18782310

Eggan, K., Jurga, S., Gosden, R., Min, I. M., & Wagers, A. J. (2006). Ovulated oocytes in adult mice derive from non-circulating germ cells. *Nature*, *441*(7097), 1109–1114. DOI: 10.1038/nature04929 PMID: 16799565

Eguizabal, C., Montserrat, N., Veiga, A., & Belmonte, J. I. (2013). Dedifferentiation, transdifferentiation, and reprogramming: Future directions in regenerative medicine. *Seminars in Reproductive Medicine*, *31*(1), 82–94. DOI: 10.1055/s-0032-1331802 PMID: 23329641

Eirin, A., & Lerman, L. O. (2014). Mesenchymal stem cell treatment for chronic renal failure. *Stem Cell Research & Therapy*, *5*(4), 83. DOI: 10.1186/scrt472 PMID: 25158205

Eirin, A., Zhu, X.-Y., Jonnada, S., Lerman, A., van Wijnen, A. J., & Lerman, L. O. (2018). Mesenchymal Stem Cell-Derived Extracellular Vesicles Improve the Renal Microvasculature in Metabolic Renovascular Disease in Swine. *Cell Transplantation*, *27*(7), 1080–1095. DOI: 10.1177/0963689718780942 PMID: 29954220

El-Ansary, M., Saadi, G., & Abd El-Hamid, S. M. (2012). Mesenchymal stem cells are a rescue approach for recovery of deteriorating kidney function. *Nephrology (Carlton, Vic.)*, *17*(7), 650–657. DOI: 10.1111/j.1440-1797.2012.01622.x PMID: 22640266

Elbadawi, A., Yalla, S. V., & Resnick, N. M. (1993). Structural basis of geriatric voiding dysfunction. III. Detrusor overactivity. *Journal of Urology, 150*(5 II), 1668-1680. DOI: 10.1016/S0022-5347(17)35868-8

el-Kassaby, A., AbouShwareb, T., & Atala, A. (2008). Randomized comparative study between buccal mucosal and acellular bladder matrix grafts in complex anterior urethral strictures. *The Journal of Urology*, *179*(4), 1432–1436. DOI: 10.1016/j.juro.2007.11.101 PMID: 18295282

Elser, D. M. (2012). Stress urinary incontinence and overactive bladder syndrome: Current options and new targets for management. *Postgraduate Medicine*, *124*(3), 42–49. DOI: 10.3810/pgm.2012.05.2547 PMID: 22691898

Epifanova, M. V., Gvasalia, B. R., Durashov, M. A., & Artemenko, S. A. (2020). Platelet-Rich Plasma Therapy for Male Sexual Dysfunction: Myth or Reality? En *Sexual Medicine Reviews* (Vol. 8, Número 1, pp. 106-113). Elsevier B.V. DOI: 10.1016/j.sxmr.2019.02.002

Erceg Ivkošić, I., Fureš, R., Ćosić, V., Mikelin, N., Bulić, L., Dobranić, D., Brlek, P., & Primorac, D. (2023). Unlocking the Potential of Mesenchymal Stem Cells in Gynecology: Where Are We Now? *Journal of Personalized Medicine*, *13*(8), 1253. DOI: 10.3390/jpm13081253 PMID: 37623503

ESHRE. (s. f.). *Documents of the Task Force Ethics Law*. ESHRE Ethical Considerations. Recuperado 6 de julio de 2025, de https://www.eshre.eu/Specialty-groups/Special-Interest-Groups/Ethics-and-Law/Documents-of-the-Task-Force-Ethics-Law

Eshtiaghi, P., & Sadownik, L. A. (2019). Fact or Fiction? Adipose-Derived Stem Cells and Platelet-Rich Plasma for the Treatment of Vulvar Lichen Sclerosus. *Journal of Lower Genital Tract Disease*, *23*(1), 65–70. DOI: 10.1097/LGT.0000000000000440 PMID: 30252710

Ezquer, F. E., Ezquer, M. E., Parrau, D. B., Carpio, D., Yañez, A. J., & Conget, P. A. (2008). Systemic administration of multipotent mesenchymal stromal cells reverts hyperglycemia and prevents nephropathy in type 1 diabetic mice. *Biology of Blood and Marrow Transplantation*, *14*(6), 631–640. DOI: 10.1016/j.bbmt.2008.01.006 PMID: 18489988

Fandel, T. M., Trivedi, A., Nicholas, C. R., Zhang, H., Chen, J., Martinez, A. F., Noble-Haeusslein, L. J., & Kriegstein, A. R. (2016). Transplanted Human Stem Cell-Derived Interneuron Precursors Mitigate Mouse Bladder Dysfunction and Central Neuropathic Pain after Spinal Cord Injury. *Cell Stem Cell*, *19*(4), 544–557. DOI: 10.1016/j.stem.2016.08.020 PMID: 27666009

Fang, F., Li, Z., Zhao, Q., Li, H., & Xiong, C. (2018). Human induced pluripotent stem cells and male infertility: An overview of current progress and perspectives. *Human Reproduction (Oxford, England)*, *33*(2), 188–195. DOI: 10.1093/humrep/dex369 PMID: 29315416

Feng, H., Liu, Q., Deng, Z., Li, H., Zhang, H., Song, J., Liu, X., Liu, J., Wen, B., & Wang, T. (2022). Human umbilical cord mesenchymal stem cells ameliorate erectile dysfunction in rats with diabetes mellitus through the attenuation of ferroptosis. *Stem Cell Research & Therapy*, *13*(1), 450. Advance online publication. DOI: 10.1186/s13287-022-03147-w PMID: 36064453

Feng, J., Lu, C., Dai, Q., Sheng, J., & Xu, M. (2018). SIRT3 Facilitates Amniotic Fluid Stem Cells to Repair Diabetic Nephropathy Through Protecting Mitochondrial Homeostasis by Modulation of Mitophagy. *Cellular Physiology and Biochemistry: International Journal of Experimental Cellular Physiology. Biochemical Pharmacology*, *46*(4), 1508–1524. DOI: 10.1159/000489194 PMID: 29689547

Feng, P., Li, P., & Tan, J. (2019). Human Menstrual Blood-Derived Stromal Cells Promote Recovery of Premature Ovarian Insufficiency Via Regulating the ECM-Dependent FAK/AKT Signaling. *Stem Cell Reviews and Reports*, *15*(2), 241–255. DOI: 10.1007/s12015-018-9867-0 PMID: 30560467

Ferlin, A., Raicu, F., Gatta, V., Zuccarello, D., Palka, G., & Foresta, C. (2007). Male infertility: Role of genetic background. *Reproductive Biomedicine Online*, *14*(6), 734–745. DOI: 10.1016/S1472-6483(10)60677-3 PMID: 17579990

Fiala, R., Vidlar, A., Vrtal, R., Belej, K., & Student, V. (2007). Porcine small intestinal submucosa graft for repair of anterior urethral strictures. *European Urology*, *51*(6), 1702–1708. DOI: 10.1016/j.eururo.2007.01.099 PMID: 17306922

Fode, M., Nadler, N., Lund, L., & Azawi, N. (2023). Feasibility of minimally invasive, same-day injection of autologous adipose-derived stem cells in the treatment of erectile dysfunction. *Scandinavian Journal of Urology*, *57*(1-6), 110–114. DOI: 10.1080/21681805.2022.2162117 PMID: 36586416

Forbes, C. M., Flannigan, R., & Schlegel, P. N. (2017). Spermatogonial stem cell transplantation and male infertility: Current status and future directions. *Arab Journal of Urology*, *16*(1), 171–180. DOI: 10.1016/j.aju.2017.11.015 PMID: 29713548

Forbes, T. A., Howden, S. E., Lawlor, K., Phipson, B., Maksimovic, J., Hale, L., Wilson, S., Quinlan, C., Ho, G., Holman, K., Bennetts, B., Crawford, J., Trnka, P., Oshlack, A., Patel, C., Mallett, A., Simons, C., & Little, M. H. (2018). Patient-iPSC-Derived Kidney Organoids Show Functional Validation of a Ciliopathic Renal Phenotype and Reveal Underlying Pathogenetic Mechanisms. *American Journal of Human Genetics*, *102*(5), 816–831. DOI: 10.1016/j.ajhg.2018.03.014 PMID: 29706353

Franck, D., Gil, E. S., Adam, R. M., Kaplan, D. L., Chung, Y. G., Estrada, C. R., & Mauney, J. R. (2013). Evaluation of silk biomaterials in combination with extracellular matrix coatings for bladder tissue engineering with primary and pluripotent cells. *PLoS One*, *8*(2), e56237. DOI: 10.1371/journal.pone.0056237 PMID: 23409160

Freedman, B. S., Brooks, C. R., Lam, A. Q., Fu, H., Morizane, R., Agrawal, V., Saad, A. F., Li, M. K., Hughes, M. R., Werff, R. V., Peters, D. T., Lu, J., Baccei, A., Siedlecki, A. M., Valerius, M. T., Musunuru, K., McNagny, K. M., Steinman, T. I., Zhou, J., & Bonventre, J. V. (2015). Modelling kidney disease with CRISPR-mutant kidney organoids derived from human pluripotent epiblast spheroids. *Nature Communications*, *6*(1), 8715. DOI: 10.1038/ncomms9715 PMID: 26493500

Freedman, B. S., Lam, A. Q., Sundsbak, J. L., Iatrino, R., Su, X., Koon, S. J., Wu, M., Daheron, L., Harris, P. C., Zhou, J., & Bonventre, J. V. (2013). Reduced ciliary polycystin-2 in induced pluripotent stem cells from polycystic kidney disease patients with PKD1 mutations. *Journal of the American Society of Nephrology*, *24*(10), 1571–1586. DOI: 10.1681/ASN.2012111089 PMID: 24009235

Furuta, A., Jankowski, R. J., Honda, M., Pruchnic, R., Yoshimura, N., & Chancellor, M. B. (2007). State of the art of where we are at using stem cells for stress urinary incontinence. *Neurourology and Urodynamics*, *26*(7), 966–971. DOI: 10.1002/nau.20448 PMID: 17580339

Furuta, A., Yamamoto, T., Igarashi, T., Suzuki, Y., Egawa, S., & Yoshimura, N. (2018). Bladder wall injection of mesenchymal stem cells ameliorates bladder inflammation, overactivity, and nociception in a chemically induced interstitial cystitis-like rat model. *International Urogynecology Journal*, *29*(11), 1615–1622. DOI: 10.1007/s00192-018-3592-8 PMID: 29508044

Fu, W.-J., Zhang, X., Zhang, B.-H., Zhang, P., Hong, B.-F., Gao, J.-P., Meng, B., Kun, H., & Cui, F.-Z. (2009). Biodegradable urethral stents seeded with autologous urethral epithelial cells in the treatment of post-traumatic urethral stricture: A feasibility study in a rabbit model. *BJU International*, *104*(2), 263–268. DOI: 10.1111/j.1464-410X.2009.08366.x PMID: 19220257

Gallo, F., Ninotta, G., Schenone, M., Cortese, P., & Giberti, C. (2019). Advances in stem cell therapy for male stress urinary incontinence. En *Expert Opinion on Biological Therapy* (Vol. 19, Número 4, pp. 293-300). Taylor and Francis Ltd. DOI: 10.1080/14712598.2019.1578343

Garber, M. G., & Carlos, N. D. (2015). Intracavernous administration of adipose stem cells: A new technique of treating erectile dysfunction in diabetic patient, preliminary report of 6 cases. *MOJ Cell Science & Report*, *2*(1). Advance online publication. DOI: 10.15406/mojcsr.2015.02.00018

Garcia-Arranz, M., Alonso-Gregorio, S., Fontana-Portella, P., Bravo, E., Diez Sebastian, J., Fernandez-Santos, M. E., & Garcia-Olmo, D. (2020). Two phase I/II clinical trials for the treatment of urinary incontinence with autologous mesenchymal stem cells. *Stem Cells Translational Medicine*, *9*(12), 1500–1508. DOI: 10.1002/sctm.19-0431 PMID: 32864818

García-Arranz, M., Herreros, M. D., González-Gómez, C., de la Quintana, P., Guadalajara, H., Georgiev-Hristov, T., Trébol, J., & Garcia-Olmo, D. (2016). Treatment of Crohn's-Related Rectovaginal Fistula With Allogeneic Expanded-Adipose Derived Stem Cells: A Phase I-IIa Clinical Trial. *Stem Cells Translational Medicine*, *5*(11), 1441–1446. DOI: 10.5966/sctm.2015-0356 PMID: 27412883

Garcia, M. M., Fandel, T. M., Lin, G., Shindel, A. W., Banie, L., Lin, C. S., & Lue, T. F. (2010). Treatment of erectile dysfunction in the obese Type 2 diabetic ZDF rat with adipose tissue-derived stem cells. *Journal of Sexual Medicine*, *7*(1 PART 1), 89–98. DOI: 10.1111/j.1743-6109.2009.01541.x PMID: 20104670

García-Perdomo, H. A., & Jurado-Penagos, A. (2022). Application of regenerative medicine and 3d bioprinting in urology. *Actas Urológicas Españolas (English Ed.)*, *46*(6), 323–328. DOI: 10.1016/j.acuroe.2022.03.006 PMID: 35660078

Gargett, C. E. (2004). Stem cells in gynaecology. *Australian and New Zealand Journal of Obstetrics and Gynaecology*, *44*(5), 380–386. DOI: 10.1111/j.1479-828X.2004.00290.x PMID: 15387855

Gargett, C. E. (2010). Stem cells in human reproduction. *Reproduction (Cambridge, England)*, *140*(1), 1–2. DOI: 10.1530/REP-10-0225 PMID: 20584995

Gargett, C. E., & Ye, L. (2012). Endometrial reconstruction from stem cells. *Fertility and Sterility*, *98*(1), 11–20. DOI: 10.1016/j.fertnstert.2012.05.004 PMID: 22657248

Gauthier-Fisher, A., Kauffman, A., & Librach, C. L. (2020). Potential use of stem cells for fertility preservation. *Andrology*, *8*(4), 862–878. DOI: 10.1111/andr.12713 PMID: 31560823

Gene Editing. (s. f.). CRISPR Therapeutics. Recuperado 5 de julio de 2025, de https://crisprtx.com/gene-editing

Geng, X.-D., Zheng, W., Wu, C.-M., Wang, S.-Q., Hong, Q., Cai, G.-Y., Chen, X.-M., & Wu, D. (2016). Embryonic Stem Cells-loaded Gelatin Microcryogels Slow Progression of Chronic Kidney Disease. *Chinese Medical Journal*, *129*(4), 392–398. DOI: 10.4103/0366-6999.176088 PMID: 26879011

Gerullis, H., Eimer, C., Georgas, E., Homburger, M., El-Baz, A. G., Wishahi, M., Borós, M., Ecke, T. H., & Otto, T. (2012a). Muscle-derived cells for treatment of iatrogenic sphincter damage and urinary incontinence in men. *TheScientificWorldJournal*, *2012*, 1–6. Advance online publication. DOI: 10.1100/2012/898535 PMID: 22919359

Gil-Sanchis, C., Cervelló, I., Khurana, S., Faus, A., Verfaillie, C., & Simón, C. (2015). Contribution of different bone marrow-derived cell types in endometrial regeneration using an irradiated murine model. *Fertility and Sterility*, *103*(6), 1596–1605.e1. DOI: 10.1016/j.fertnstert.2015.02.030 PMID: 25813284

Gluckman, E., Broxmeyer, H. E., Auerbach, A. D., Friedman, H. S., Douglas, G. W., Devergie, A., Esperou, H., Thierry, D., Socie, G., Lehn, P., Cooper, S., English, D., Kurtzberg, J., Bard, J., & Boyse, E. A. (1989). Hematopoietic Reconstitution in a Patient with Fanconi's Anemia by Means of Umbilical-Cord Blood from an HLA-Identical Sibling. *The New England Journal of Medicine*, *321*(17), 1174–1178. DOI: 10.1056/NEJM198910263211707 PMID: 2571931

Gluckman, E., Rocha, V., Boyer-Chammard, A., Locatelli, F., Arcese, W., Pasquini, R., Ortega, J., Souillet, G., Ferreira, E., Laporte, J.-P., Fernandez, M., & Chastang, C. (1997). Outcome of Cord-Blood Transplantation from Related and Unrelated Donors. *The New England Journal of Medicine*, *337*(6), 373–381. DOI: 10.1056/NEJM199708073370602 PMID: 9241126

Gokce, A., Abd Elmageed, Z. Y., Lasker, G. F., Bouljihad, M., Braun, S. E., Kim, H., Kadowitz, P. J., Abdel-Mageed, A. B., Sikka, S. C., & Hellstrom, W. J. (2015). Intratunical Injection of Genetically Modified Adipose Tissue-Derived Stem Cells with Human Interferon α-2b for Treatment of Erectile Dysfunction in a Rat Model of Tunica Albugineal Fibrosis. *Journal of Sexual Medicine*, *12*(7), 1533–1544. DOI: 10.1111/jsm.12916 PMID: 26062100

Gokce, A., Abd Elmageed, Z. Y., Lasker, G. F., Bouljihad, M., Kim, H., Trost, L. W., Kadowitz, P. J., Abdel-Mageed, A. B., Sikka, S. C., & Hellstrom, W. J. (2014a). Adipose tissue-derived stem cell therapy for prevention and treatment of erectile dysfunction in a rat model of Peyronie's disease. *Andrology*, *2*(2), 244–251. DOI: 10.1111/j.2047-2927.2013.00181.x PMID: 24574095

Goldman, H. B., Sievert, K. D., & Damaser, M. S. (2012). Will we ever use stem cells for the treatment of SUI? ICI-RS 2011. *Neurourology and Urodynamics*, *31*(3), 386–389. DOI: 10.1002/nau.22217 PMID: 22431263

Goligorsky, M. S., Kuo, M.-C., Patschan, D., & Verhaar, M. C. (2009). Review article: Endothelial progenitor cells in renal disease. *Nephrology (Carlton, Vic.)*, *14*(3), 291–297. DOI: 10.1111/j.1440-1797.2009.01112.x PMID: 19444962

Gomes, S. A., Hare, J. M., & Rangel, E. B. (2018). Kidney-Derived c-Kit+ Cells Possess Regenerative Potential. *Stem Cells Translational Medicine*, *7*(4), 317–324. DOI: 10.1002/sctm.17-0232 PMID: 29575816

González Enguita, C., Garranzo García-Ibarrola, M., Tufet, I., Jaumont, J. J., Garde García, H., González López, R., Quintana Franco, L. M., Torres Zambrano, G. M., & García-Arranz, M. (2024). Cell Therapy in the Treatment of Female Stress Urinary Incontinence: Current Status and Future Proposals. *Life (Chicago, Ill.)*, *14*(7), 7. Advance online publication. DOI: 10.3390/life14070861 PMID: 39063615

Gotoh, M., Yamamoto, T., Kato, M., Majima, T., Toriyama, K., Kamei, Y., Matsukawa, Y., Hirakawa, A., & Funahashi, Y. (2014). Regenerative treatment of male stress urinary incontinence by periurethral injection of autologous adipose-derived regenerative cells: 1-year outcomes in 11 patients. *International Journal of Urology*, *21*(3), 294–300. DOI: 10.1111/iju.12266 PMID: 24033774

Guidelines for Stem Cell Research and Clinical Translation. - Search Results—PubMed. (s. f.). Recuperado 28 de marzo de 2021, de https://pubmed.ncbi.nlm.nih.gov/?term=Guidelines+for+Stem+Cell+Research+and+Clinical+Translation

Guidelines—International Society for Stem Cell Research. (2021). https://www.isscr.org/guidelines

Gul, M., Hildorf, S., Dong, L., Thorup, J., Hoffmann, E. R., Jensen, C. F. S., Sønksen, J., Cortes, D., Fedder, J., Andersen, C. Y., & Goossens, E. (2020). Review of injection techniques for spermatogonial stem cell transplantation. En *Human Reproduction Update* (Vol. 26, Número 3, pp. 368-391). Oxford University Press. DOI: 10.1093/humupd/dmaa003

Gunetti, M., Tomasi, S., Giammò, A., Boido, M., Rustichelli, D., Mareschi, K., Errichiello, E., Parola, M., Ferrero, I., Fagioli, F., Vercelli, A., & Carone, R. (2012). Myogenic Potential of Whole Bone Marrow Mesenchymal Stem Cells In Vitro and In Vivo for Usage in Urinary Incontinence. *PLoS One*, *7*(9), e45538. Advance online publication. DOI: 10.1371/journal.pone.0045538 PMID: 23029081

Gupta, S., Lodha, P., Karthick, M. S., & Tandulwadkar, S. (2018). Role of Autologous Bone Marrow-Derived Stem Cell Therapy for Follicular Recruitment in Premature Ovarian Insufficiency: Review of Literature and a Case Report of World's First Baby with Ovarian Autologous Stem Cell Therapy in a Perimenopausal Woman of Age 45 Year. *Journal of Human Reproductive Sciences*, *11*(2), 125–130. DOI: 10.4103/jhrs.JHRS_57_18 PMID: 30158807

Haahr, M. K., Harken Jensen, C., Toyserkani, N. M., Andersen, D. C., Damkier, P., Sørensen, J. A., Sheikh, S. P., & Lund, L. (2018). A 12-Month Follow-up After a Single Intracavernous Injection of Autologous Adipose-Derived Regenerative Cells in Patients with Erectile Dysfunction Following Radical Prostatectomy: An Open-Label Phase I Clinical Trial. *Urology*, *121*, 203.e6–203.e13. DOI: 10.1016/j.urology.2018.06.018 PMID: 29958973

Haahr, M. K., Jensen, C. H., Toyserkani, N. M., Andersen, D. C., Damkier, P., Sørensen, J. A., Lund, L., & Sheikh, S. P. (2016). Safety and Potential Effect of a Single Intracavernous Injection of Autologous Adipose-Derived Regenerative Cells in Patients with Erectile Dysfunction Following Radical Prostatectomy: An Open-Label Phase I Clinical Trial. *EBioMedicine*, *5*, 204–210. DOI: 10.1016/j.ebiom.2016.01.024 PMID: 27077129

Hafezparast, M., Kaur, G. P., Zdzienicka, M., Athwal, R. S., Lehmann, A. R., & Jeggo, P. A. (1993). Subchromosomal localization of a gene (XRCC5) involved in double strand break repair to the region 2q34-36. *Somatic Cell and Molecular Genetics*, *19*(5), 413–421. DOI: 10.1007/BF01233246 PMID: 8291019

Hakim, L., De Ridder, D., & Van der Aa, F. (2015). Slings for urinary incontinence and the application of cell-based therapy. In *Advanced Drug Delivery Reviews* (Vol. 82, pp. 22–30). Elsevier B.V., DOI: 10.1016/j.addr.2014.11.006

Hanczar, M., Moazen, M., & Day, R. (2021). The Significance of Biomechanics and Scaffold Structure for Bladder Tissue Engineering. *International Journal of Molecular Sciences*, *22*(23), 12657. DOI: 10.3390/ijms222312657 PMID: 34884464

Han, D., & Gouma, P.-I. (2006). Electrospun bioscaffolds that mimic the topology of extracellular matrix. *Nanomedicine; Nanotechnology, Biology, and Medicine*, *2*(1), 37–41. DOI: 10.1016/j.nano.2006.01.002 PMID: 17292114

Harari-Steinberg, O., Metsuyanim, S., Omer, D., Gnatek, Y., Gershon, R., Pri-Chen, S., Ozdemir, D. D., Lerenthal, Y., Noiman, T., Ben-Hur, H., Vaknin, Z., Schneider, D. F., Aronow, B. J., Goldstein, R. S., Hohenstein, P., & Dekel, B. (2013). Identification of human nephron progenitors capable of generation of kidney structures and functional repair of chronic renal disease. *EMBO Molecular Medicine*, *5*(10), 1556–1568. DOI: 10.1002/emmm.201201584 PMID: 23996934

Hauser, P. V., De Fazio, R., Bruno, S., Sdei, S., Grange, C., Bussolati, B., Benedetto, C., & Camussi, G. (2010). Stem cells derived from human amniotic fluid contribute to acute kidney injury recovery. *American Journal of Pathology*, *177*(4), 2011–2021. DOI: 10.2353/ajpath.2010.091245 PMID: 20724594

Hayashi, K., Ogushi, S., Kurimoto, K., Shimamoto, S., Ohta, H., & Saitou, M. (2012). Offspring from oocytes derived from in vitro primordial germ cell-like cells in mice. *Science*, *338*(6109), 971–975. DOI: 10.1126/science.1226889 PMID: 23042295

Hayashi, K., Ohta, H., Kurimoto, K., Aramaki, S., & Saitou, M. (2011). Reconstitution of the mouse germ cell specification pathway in culture by pluripotent stem cells. *Cell*, *146*(4), 519–532. DOI: 10.1016/j.cell.2011.06.052 PMID: 21820164

Hayashi, K., & Saitou, M. (2013). Generation of eggs from mouse embryonic stem cells and induced pluripotent stem cells. *Nature Protocols*, *8*(8), 1513–1524. DOI: 10.1038/nprot.2013.090 PMID: 23845963

Hayashi, Y., Saitou, M., & Yamanaka, S. (2012). Germline development from human pluripotent stem cells toward disease modeling of infertility. *Fertility and Sterility*, *97*(6), 1250–1259. DOI: 10.1016/j.fertnstert.2012.04.037 PMID: 22656305

Heinemann, C., & Reid, G. (2005). Vaginal microbial diversity among postmenopausal women with and without hormone replacement therapy. *Canadian Journal of Microbiology*, *51*(9), 777–781. DOI: 10.1139/w05-070 PMID: 16391657

Hellstrom, W. J. G., Kendirci, M., Matern, R., Cockerham, Y., Myers, L., Sikka, S. C., Venable, D., Honig, S., McCullough, A., Hakim, L. S., Nehra, A., Templeton, L. E., & Pryor, J. L. (2006). Single-Blind, Multicenter, Placebo Controlled, Parallel Study to Assess the Safety and Efficacy of Intralesional Interferon α-2b for Minimally Invasive Treatment for Peyronie's Disease. *The Journal of Urology*, *176*(1), 394–398. DOI: 10.1016/S0022-5347(06)00517-9 PMID: 16753449

Hendry, C. E., Vanslambrouck, J. M., Ineson, J., Suhaimi, N., Takasato, M., Rae, F., & Little, M. H. (2013). Direct transcriptional reprogramming of adult cells to embryonic nephron progenitors. *Journal of the American Society of Nephrology*, *24*(9), 1424–1434. DOI: 10.1681/ASN.2012121143 PMID: 23766537

Hermann, B. P., Sukhwani, M., Winkler, F., Pascarella, J. N., Peters, K. A., Sheng, Y., Valli, H., Rodriguez, M., Ezzelarab, M., Dargo, G., Peterson, K., Masterson, K., Ramsey, C., Ward, T., Lienesch, M., Volk, A., Cooper, D. K., Thomson, A. W., Kiss, J. E., & Orwig, K. E. (2012). Spermatogonial stem cell transplantation into rhesus testes regenerates spermatogenesis producing functional sperm. *Cell Stem Cell*, *11*(5), 715–726. DOI: 10.1016/j.stem.2012.07.017 PMID: 23122294

Hernandez, A.B. (s. f.). *STEM CELLS THERAPY MANUAL*. ADSCC.

Herrera, M. B., Bussolati, B., Bruno, S., Morando, L., Mauriello-Romanazzi, G., Sanavio, F., Stamenkovic, I., Biancone, L., & Camussi, G. (2007). Exogenous mesenchymal stem cells localize to the kidney by means of CD44 following acute tubular injury. *Kidney International*, *72*(4), 430–441. DOI: 10.1038/sj.ki.5002334 PMID: 17507906

HFEA. (2019). *HFEA: UK fertility regulator*. Human Fertilisation & Embryology Authority (HFEA). https://www.hfea.gov.uk/

Hickson, L. J., Eirin, A., & Lerman, L. O. (2016). Challenges and opportunities for stem cell therapy in patients with chronic kidney disease. *Kidney International*, *89*(4), 767–778. DOI: 10.1016/j.kint.2015.11.023 PMID: 26924058

Hikabe, O., Hamazaki, N., Nagamatsu, G., Obata, Y., Hirao, Y., Hamada, N., Shimamoto, S., Imamura, T., Nakashima, K., Saitou, M., & Hayashi, K. (2016). Reconstitution in vitro of the entire cycle of the mouse female germ line. *Nature*, *539*(7628), 299–303. DOI: 10.1038/nature20104 PMID: 27750280

Hillary, C. J., Roman, S., MacNeil, S., Aicher, W. K., Stenzl, A., & Chapple, C. R. (2020). Regenerative medicine and injection therapies in stress urinary incontinence. En *Nature Reviews Urology* (Vol. 17, Número 3, pp. 151-161). Nature Research. DOI: 10.1038/s41585-019-0273-4

Hirose, Y., Yamamoto, T., Nakashima, M., Funahashi, Y., Matsukawa, Y., Yamaguchi, M., Kawabata, S., & Gotoh, M. (2016). Injection of Dental Pulp Stem Cells Promotes Healing of Damaged Bladder Tissue in a Rat Model of Chemically Induced Cystitis. *Cell Transplantation*, *25*(3), 425–436. DOI: 10.3727/096368915X689523 PMID: 26395427

Hoeijmakers, J. H. (2001). Genome maintenance mechanisms for preventing cancer. *Nature*, *411*(6835), 366–374. DOI: 10.1038/35077232 PMID: 11357144

Horan, R. L., Bramono, D. S., Stanley, J. R. L., Simmons, Q., Chen, J., Boepple, H. E., & Altman, G. H. (2009). Biological and biomechanical assessment of a long-term bioresorbable silk-derived surgical mesh in an abdominal body wall defect model. *Hernia*, *13*(2), 189–199. DOI: 10.1007/s10029-008-0459-9 PMID: 19198755

Horst, M., Madduri, S., Milleret, V., Sulser, T., Gobet, R., & Eberli, D. (2013). A bilayered hybrid microfibrous PLGA—Acellular matrix scaffold for hollow organ tissue engineering. *Biomaterials*, *34*(5), 1537–1545. DOI: 10.1016/j.biomaterials.2012.10.075 PMID: 23177021

Hou, Q., Ge, M., Zhang, C., Tian, D., Wang, L., Tian, H., Wang, W., & Zhang, W. (2017). Adipose tissue-derived stem cell therapy for erectile dysfunction in rats: A systematic review and meta-analysis. *International Urology and Nephrology*, *49*(7), 1127–1137. DOI: 10.1007/s11255-017-1590-2 PMID: 28417342

Huang, Y. C., Ning, H., Shindel, A. W., Fandel, T. M., Lin, G., Harraz, A. M., Lue, T. F., & Lin, C. S. (2010). The effect of intracavernous injection of adipose tissue-derived stem cells on hyperlipidemia-associated erectile dysfunction in a rat model. *Journal of Sexual Medicine*, *7*(4 PART 1), 1391–1400. DOI: 10.1111/j.1743-6109.2009.01697.x PMID: 20141586

Huang, Y. C., Shindel, A. W., Ning, H., Lin, G., Harraz, A. M., Wang, G., Garcia, M., Lue, T. F., & Lin, C. S. (2010). Adipose Derived Stem Cells Ameliorate Hyperlipidemia Associated Detrusor Overactivity in a Rat Model. *The Journal of Urology*, *183*(3), 1232–1240. DOI: 10.1016/j.juro.2009.11.012 PMID: 20096880

Huard, J., Yokoyama, T., Pruchnic, R., Qu, Z., Li, Y., Lee, J. Y., Somogyi, G. T., de Groat, W. C., & Chancellor, M. B. (2002). Muscle-derived cell-mediated ex vivo gene therapy for urological dysfunction. *Gene Therapy*, *9*(23), 1617–1626. DOI: 10.1038/sj.gt.3301816 PMID: 12424614

Hübner, K., Fuhrmann, G., Christenson, L. K., Kehler, J., Reinbold, R., De La Fuente, R., Wood, J., Strauss, J. F.III, Boiani, M., & Schöler, H. R. (2003). Derivation of oocytes from mouse embryonic stem cells. *Science*, *300*(5623), 1251–1256. DOI: 10.1126/science.1083452 PMID: 12730498

Hudson, N., Culley, L., Rapport, F., Johnson, M., & Bharadwaj, A. (2009). «Public» perceptions of gamete donation: A research review. *Public Understanding of Science (Bristol, England)*, *18*(1), 61–77. DOI: 10.1177/0963662507078396 PMID: 19579535

Hu, J., Schokrpur, S., Archang, M., Hermann, K., Sharrow, A. C., Khanna, P., Novak, J., Signoretti, S., Bhatt, R. S., Knudsen, B. S., Xu, H., & Wu, L. (2018). A Non-integrating Lentiviral Approach Overcomes Cas9-Induced Immune Rejection to Establish an Immunocompetent Metastatic Renal Cancer Model. *Molecular Therapy. Methods & Clinical Development*, *9*, 203–210. DOI: 10.1016/j.omtm.2018.02.009 PMID: 29766028

Hu, Y., Liao, L. M., Ju, Y. H., Fu, G., Zhang, H. Y., & Wu, H. X. (2012). Intravenously transplanted bone marrow stromal cells promote recovery of lower urinary tract function in rats with complete spinal cord injury. *Spinal Cord*, *50*(3), 202–207. DOI: 10.1038/sc.2011.128 PMID: 22124344

Iannaccone, P. M., Galat, V., Bury, M. I., Ma, Y. C., & Sharma, A. K. (2018). The utility of stem cells in pediatric urinary bladder regeneration. *Pediatric Research*, *83*(1-2), 258–266. DOI: 10.1038/pr.2017.229 PMID: 28915233

Ibtisham, F., & Honaramooz, A. (2020). Spermatogonial Stem Cells for In Vitro Spermatogenesis and In Vivo Restoration of Fertility. *Cells*, *9*(3), 745. Advance online publication. DOI: 10.3390/cells9030745 PMID: 32197440

Ichim, T. E., Warbington, T., Cristea, O., Chin, J. L., & Patel, A. N. (2013). Intracavernous administration of bone marrow mononuclear cells: A new method of treating erectile dysfunction? En *Journal of Translational Medicine* (Vol. 11, Número 1). J Transl Med. DOI: 10.1186/1479-5876-11-139

Imberti, B., Tomasoni, S., Ciampi, O., Pezzotta, A., Derosas, M., Xinaris, C., Rizzo, P., Papadimou, E., Novelli, R., Benigni, A., Remuzzi, G., & Morigi, M. (2015). Renal progenitors derived from human iPSCs engraft and restore function in a mouse model of acute kidney injury. *Scientific Reports*, *5*(1), 8826. DOI: 10.1038/srep08826 PMID: 25744951

Ishikawa, F., Drake, C. J., Yang, S., Fleming, P. A., Minamiguchi, H., Visconti, R. P., Crosby, C. V., Argraves, W. S., Harada, M., Key, L. L.Jr, Livingston, A. G., Wingard, J. R., & Ogawa, M. (2003). Transplanted human cord blood cells give rise to hepatocytes in engrafted mice. *Annals of the New York Academy of Sciences, 996*(1), 174–185. DOI: 10.1111/j.1749-6632.2003.tb03245.x PMID: 12799295

Izadyar, F., Den Ouden, K., Stout, T. A. E., Stout, J., Coret, J., Lankveld, D. P. K., Spoormakers, T. J. P., Colenbrander, B., Oldenbroek, J. K., Van der Ploeg, K. D., Woelders, H., Kal, H. B., & De Rooij, D. G. (2003). Autologous and homologous transplantation of bovine spermatogonial stem cells. *Reproduction (Cambridge, England), 126*(6), 765–774. DOI: 10.1530/rep.0.1260765 PMID: 14748695

Jack, G. S., Almeida, F. G., Zhang, R., Alfonso, Z. C., Zuk, P. A., & Rodríguez, L. V. (2005). Processed lipoaspirate cells for tissue engineering of the lower urinary tract: Implications for the treatment of stress urinary incontinence and bladder reconstruction. *The Journal of Urology, 174*(5), 2041–2045. DOI: 10.1097/01.ju.0000176489.96993.84 PMID: 16217390

Jack, G. S., Zhang, R., Lee, M., Xu, Y., Wu, B. M., & Rodríguez, L. V. (2009). Urinary bladder smooth muscle engineered from adipose stem cells and a three dimensional synthetic composite. *Biomaterials, 30*(19), 3259–3270. DOI: 10.1016/j.biomaterials.2009.02.035 PMID: 19345408

Jackson, S., James, M., & Abrams, P. (2002). The effect of oestradiol on vaginal collagen metabolism in postmenopausal women with genuine stress incontinence. *BJOG, 109*(3), 339–344. DOI: 10.1111/j.1471-0528.2002.01052.x PMID: 11950190

Jackson, W. M., Nesti, L. J., & Tuan, R. S. (2012). Mesenchymal stem cell therapy for attenuation of scar formation during wound healing. *Stem Cell Research & Therapy, 3*(3), 20. Advance online publication. DOI: 10.1186/scrt111 PMID: 22668751

Jahn, P., Beutner, K., & Langer, G. (2012). Types of indwelling urinary catheters for long-term bladder drainage in adults. *Cochrane Database of Systematic Reviews, 10*(10), CD004997. Advance online publication. DOI: 10.1002/14651858.CD004997.pub3 PMID: 23076911

James, M. T., Hemmelgarn, B. R., & Tonelli, M. (2010). Early recognition and prevention of chronic kidney disease. *Lancet, 375*(9722), 1296–1309. DOI: 10.1016/S0140-6736(09)62004-3 PMID: 20382326

Jeggo, P. A., Pearl, L. H., & Carr, A. M. (2016). DNA repair, genome stability and cancer: A historical perspective. *Nature Reviews. Cancer, 16*(1), 35–42. DOI: 10.1038/nrc.2015.4 PMID: 26667849

Jeon, S. H., Shrestha, K. R., Kim, R. Y., Jung, A. R., Park, Y. H., Kwon, O., Kim, G. E., Kim, S. H., Kim, K. H., & Lee, J. Y. (2016). Combination therapy using human adipose-derived stem cells on the cavernous nerve and low-energy shockwaves on the corpus cavernosum in a rat model of post-prostatectomy erectile dysfunction. *Urology*, *88*, 226.e1–226.e9. DOI: 10.1016/j.urology.2015.10.021 PMID: 26522972

Jerman, U. D., Veranič, P., & Kreft, M. E. (2014). Amniotic membrane scaffolds enable the development of tissue-engineered urothelium with molecular and ultrastructural properties comparable to that of native urothelium. *Tissue Engineering. Part C, Methods*, *20*(4), 317–327. DOI: 10.1089/ten.tec.2013.0298 PMID: 23947657

Jiang, M. H., Li, G., Liu, J., Liu, L., Wu, B., Huang, W., He, W., Deng, C., Wang, D., Li, C., Lahn, B. T., Shi, C., & Xiang, A. P. (2015). Nestin(+) kidney resident mesenchymal stem cells for the treatment of acute kidney ischemia injury. *Biomaterials*, *50*, 56–66. DOI: 10.1016/j.biomaterials.2015.01.029 PMID: 25736496

Jiao, J., Milwid, J. M., Yarmush, M. L., & Parekkadan, B. (2011). A mesenchymal stem cell potency assay. *Methods in Molecular Biology (Clifton, N.J.)*, *677*, 221–231. DOI: 10.1007/978-1-60761-869-0_16 PMID: 20941614

Jinek, M., Chylinski, K., Fonfara, I., Hauer, M., Doudna, J. A., & Charpentier, E. (2012). A programmable dual-RNA-guided DNA endonuclease in adaptive bacterial immunity. *Science*, *337*(6096), 816–821. DOI: 10.1126/science.1225829 PMID: 22745249

Jin, M., Chen, Y., Zhou, Y., Mei, Y., Liu, W., Pan, C., & Hua, X. (2016). Transplantation of bone marrow-derived mesenchymal stem cells expressing elastin alleviates pelvic floor dysfunction. *Stem Cell Research & Therapy*, *7*(1), 51. DOI: 10.1186/s13287-016-0308-1 PMID: 27048404

Jin, Y., Bouyer, J., Shumsky, J. S., Haas, C., & Fischer, I. (2016). Transplantation of neural progenitor cells in chronic spinal cord injury. *Neuroscience*, *320*, 69–82. DOI: 10.1016/j.neuroscience.2016.01.066 PMID: 26852702

Jin, Y., Neuhuber, B., Singh, A., Bouyer, J., Lepore, A., Bonner, J., Himes, T., Campanelli, J. T., & Fischer, I. (2011). Transplantation of human glial restricted progenitors and derived astrocytes into a contusion model of spinal cord injury. *Journal of Neurotrauma*, *28*(4), 579–594. DOI: 10.1089/neu.2010.1626 PMID: 21222572

Jin, Y., Xu, L., Zhao, Y., Wang, M., Jin, X., & Zhang, H. (2017). Endogenous Stem Cells Were Recruited by Defocused Low-Energy Shock Wave in Treating Diabetic Bladder Dysfunction. *Stem Cell Reviews and Reports*, *13*(2), 287–298. DOI: 10.1007/s12015-016-9705-1 PMID: 27921202

Jin, Z., Ding, Y., Xue, R., Jia, Z., Huang, Z., Ding, Y., Gu, C., & Yang, J. (2017). Involvement of interstitial cells of Cajal in bladder dysfunction in mice with experimental autoimmune encephalomyelitis. *International Urology and Nephrology*, *49*(8), 1353–1359. DOI: 10.1007/s11255-017-1597-8 PMID: 28425078

Johnson, E. K., Finlayson, C., Rowell, E. E., Gosiengfiao, Y., Pavone, M. E., Lockart, B., Orwig, K. E., Brannigan, R. E., & Woodruff, T. K. (2017). Fertility Preservation for Pediatric Patients: Current State and Future Possibilities. *The Journal of Urology*, *198*(1), 186–194. DOI: 10.1016/j.juro.2016.09.159 PMID: 28189577

Johnson, J., Bagley, J., Skaznik-Wikiel, M., Lee, H. J., Adams, G. B., Niikura, Y., Tschudy, K. S., Tilly, J. C., Cortes, M. L., Forkert, R., Spitzer, T., Iacomini, J., Scadden, D. T., & Tilly, J. L. (2005). Oocyte generation in adult mammalian ovaries by putative germ cells in bone marrow and peripheral blood. *Cell*, *122*(2), 303–315. DOI: 10.1016/j.cell.2005.06.031 PMID: 16051153

Johnson, J., Canning, J., Kaneko, T., Pru, J. K., & Tilly, J. L. (2004). Germline stem cells and follicular renewal in the postnatal mammalian ovary. *Nature*, *428*(6979), 145–150. DOI: 10.1038/nature02316 PMID: 15014492

Jost, S. P., Gosling, J. A., & Dixon, J. S. (1989). The morphology of normal human bladder urothelium. *Journal of Anatomy*, *167*, 103–115. PMID: 2630525

Jung, D., Xiong, J., Ye, M., Qin, X., Li, L., Cheng, S., Luo, M., Peng, J., Dong, J., Tang, F., Shen, W., Matzuk, M. M., & Kee, K. (2017). In vitro differentiation of human embryonic stem cells into ovarian follicle-like cells. *Nature Communications*, *8*(1), 15680. Advance online publication. DOI: 10.1038/ncomms15680 PMID: 28604658

Kanatsu-Shinohara, M., Ogonuki, N., Inoue, K., Miki, H., Ogura, A., Toyokuni, S., & Shinohara, T. (2003). Long-term proliferation in culture and germline transmission of mouse male germline stem cells. *Biology of Reproduction*, *69*(2), 612–616. DOI: 10.1095/biolreprod.103.017012 PMID: 12700182

Kanematsu, A., Yamamoto, S., Iwai-Kanai, E., Kanatani, I., Imamura, M., Adam, R. M., Tabata, Y., & Ogawa, O. (2005). Induction of smooth muscle cell-like phenotype in marrow-derived cells among regenerating urinary bladder smooth muscle cells. *American Journal of Pathology*, *166*(2), 565–573. DOI: 10.1016/S0002-9440(10)62278-X PMID: 15681839

Kang, M., & Han, Y.-M. (2014). Differentiation of human pluripotent stem cells into nephron progenitor cells in a serum and feeder free system. *PLoS One*, *9*(4), e94888. DOI: 10.1371/journal.pone.0094888 PMID: 24728509

Kanno, Y., Mitsui, T., Sano, H., Kitta, T., Moriya, K., & Nonomura, K. (2014). Contribution of bone marrow-derived mesenchymal stem cells to the morphological changes in the bladder after partial outlet obstruction: A preliminary study. *International journal of urology : official journal of the Japanese Urological Association, 21*(7), 714-718. DOI: 10.1111/iju.12406

Kapetanos, K., Light, A., Thakare, N., Mahbubani, K., Saeb-Parsy, K., & Saeb-Parsy, K. (2022). Bioengineering solutions for ureteric disorders: Clinical need, challenges and opportunities. *BJU International, 130*(4), 408–419. DOI: 10.1111/bju.15741 PMID: 35388587

Karon, M., & Chatterjee, S. (2019). Sacrocolpopexy: Patient Outcomes Support the Use of Non-Crosslinked Acellular Dermal Matrix as an Alternative to the Synthetic Polypropylene Mesh. *Journal of Gynecologic Surgery, 35*(6), 337–344. DOI: 10.1089/gyn.2019.0010 PMID: 31871389

Kaufman, M. R. (2020). Contemporary application of autologous muscle-derived cells for urinary sphincter regeneration. *World Journal of Urology, 38*(9), 2095–2099. DOI: 10.1007/s00345-019-03018-9 PMID: 31834471

Kee, K., Angeles, V. T., Flores, M., Nguyen, H. N., & Reijo Pera, R. A. (2009). Human DAZL, DAZ and BOULE genes modulate primordial germ-cell and haploid gamete formation. *Nature, 462*(7270), 222–225. DOI: 10.1038/nature08562 PMID: 19865085

Kendirci, M., Trost, L., Bakondi, B., Whitney, M. J., Hellstrom, W. J. G., & Spees, J. L. (2010). Transplantation of nonhematopoietic adult bone marrow stem/progenitor cells isolated by p75 nerve growth factor receptor into the penis rescues erectile function in a rat model of cavernous nerve injury. *The Journal of Urology, 184*(4), 1560–1566. DOI: 10.1016/j.juro.2010.05.088 PMID: 20728109

Kharazi, U., & Badalzadeh, R. (2020). A review on the stem cell therapy and an introduction to exosomes as a new tool in reproductive medicine. *Reproductive Biology, 20*(4), 447–459. DOI: 10.1016/j.repbio.2020.07.002 PMID: 32900639

Kim, J. H., Lee, H. J., & Song, Y. S. (2014). Treatment of bladder dysfunction using stem cell or tissue engineering technique. En *Korean Journal of Urology* (Vol. 55, Número 4, pp. 228-238). Korean Urological Association. DOI: 10.4111/kju.2014.55.4.228

Kim, J. H., Lee, S. R., Song, Y. S., & Lee, H. J. (2013). Stem cell therapy in bladder dysfunction: Where are we? And where do we have to go? En *BioMed Research International* (Vol. 2013). DOI: 10.1155/2013/930713

Kim, Y. K., Refaeli, I., Brooks, C. R., Jing, P., Gulieva, R. E., Hughes, M. R., Cruz, N. M., Liu, Y., Churchill, A. J., Wang, Y., Fu, H., Pippin, J. W., Lin, L. Y., Shankland, S. J., Vogl, A. W., McNagny, K. M., & Freedman, B. S. (2017). Gene-Edited Human Kidney Organoids Reveal Mechanisms of Disease in Podocyte Development. *Stem Cells (Dayton, Ohio), 35*(12), 2366-2378. DOI: 10.1002/stem.2707

Kim, A., Han, J.-Y., Ryu, C.-M., Yu, H. Y., Lee, S., Kim, Y., Jeong, S. U., Cho, Y. M., Shin, D.-M., & Choo, M.-S. (2017). Histopathological characteristics of interstitial cystitis/bladder pain syndrome without Hunner lesion. *Histopathology, 71*(3), 415–424. DOI: 10.1111/his.13235 PMID: 28394416

Kim, A., Yu, H. Y., Heo, J., Song, M., Shin, J. H., Lim, J., Yoon, S. J., Kim, Y., Lee, S., Kim, S. W., Oh, W., Choi, S. J., Shin, D. M., & Choo, M. S. (2016). Mesenchymal stem cells protect against the tissue fibrosis of ketamine-induced cystitis in rat bladder. *Scientific Reports, 6*(1), 30881. Advance online publication. DOI: 10.1038/srep30881 PMID: 27481042

Kim, A., Yu, H. Y., Lim, J., Ryu, C. M., Kim, Y. H., Heo, J., Han, J. Y., Lee, S., Bae, Y. S., Kim, J. Y., Bae, D. J., Kim, S. Y., Noh, B. J., Hong, K. S., Han, J. Y., Lee, S. W., Song, M., Chung, H. M., Kim, J. K., & Choo, M. S. (2017a). Improved efficacy and in vivo cellular properties of human embryonic stem cell derivative in a preclinical model of bladder pain syndrome. *Scientific Reports, 7*(1), 8872. Advance online publication. DOI: 10.1038/s41598-017-09330-x PMID: 28827631

Kim, D., & Dressler, G. R. (2005). Nephrogenic factors promote differentiation of mouse embryonic stem cells into renal epithelia. *Journal of the American Society of Nephrology, 16*(12), 3527–3534. DOI: 10.1681/ASN.2005050544 PMID: 16267156

Kim, J. H., Shim, S. R., Doo, S. W., Yang, W. J., Yoo, B. W., Kim, J. M., Ko, Y. M., Song, E. S., Lim, I. S., Lee, H. J., & Song, Y. S. (2015). Bladder recovery by stem cell based cell therapy in the bladder dysfunction induced by spinal cord injury: Systematic review and meta-analysis. *PLoS One, 10*(3), e0113491. DOI: 10.1371/journal.pone.0113491 PMID: 25781610

Kim, J.-S., Lee, J.-H., Kwon, O., Cho, J.-H., Choi, J.-Y., Park, S.-H., Kim, C.-D., Kim, Y.-J., & Kim, Y.-L. (2017). Rapid deterioration of preexisting renal insufficiency after autologous mesenchymal stem cell therapy. *Kidney Research and Clinical Practice, 36*(2), 200–204. DOI: 10.23876/j.krcp.2017.36.2.200 PMID: 28680828

Kim, S. J., Cho, Y. S., Park, J. M., Na, Y. G., & Kim, K. H. (2020). Stem Cell Therapy for Neurogenic Bladder After Spinal Cord Injury: Clinically Possible? *International Neurourology Journal, 24*(Suppl 1), S3–S10. DOI: 10.5213/inj.2040150.075 PMID: 32482052

Kinebuchi, Y., Aizawa, N., Imamura, T., Ishizuka, O., Igawa, Y., & Nishizawa, O. (2010). Autologous bone-marrow-derived mesenchymal stem cell transplantation into injured rat urethral sphincter. *International Journal of Urology*, *17*(4), 359–368. DOI: 10.1111/j.1442-2042.2010.02471.x PMID: 20202003

Kiss, A., Sulya, B., Szász, A. M., Romics, I., Kelemen, Z., Tóth, J., Merksz, M., Kemény, S., & Nyírády, P. (2011). Long-term psychological and sexual outcomes of severe penile hypospadias repair. *Journal of Sexual Medicine*, *8*(5), 1529–1539. DOI: 10.1111/j.1743-6109.2010.02120.x PMID: 21091883

Knoll, J., Amend, B., Harland, N., Isser, S., Bézière, N., Kraushaar, U., Stenzl, A., & Aicher, W. K. (2024). Cell Therapy by Mesenchymal Stromal Cells Versus Myoblasts in a Pig Model of Urinary Incontinence. *Tissue Engineering. Part A*, *30*(1-2), 14–30. DOI: 10.1089/ten.tea.2023.0103 PMID: 37933911

Kobashi, K. C., Albo, M. E., Dmochowski, R. R., Ginsberg, D. A., Goldman, H. B., Gomelsky, A., Kraus, S. R., Sandhu, J. S., Shepler, T., Treadwell, J. R., Vasavada, S., & Lemack, G. E. (2017). Surgical Treatment of Female Stress Urinary Incontinence: AUA/SUFU Guideline. *The Journal of Urology*, *198*(4), 875–883. DOI: 10.1016/j.juro.2017.06.061 PMID: 28625508

Koga, S., & Horiguchi, Y. (2022). Efficacy of a cultured conditioned medium of exfoliated deciduous dental pulp stem cells in erectile dysfunction patients. *Journal of Cellular and Molecular Medicine*, *26*(1), 195–201. DOI: 10.1111/jcmm.17072 PMID: 34845823

Kollhoff, D. M., Cheng, E. Y., & Sharma, A. K. (2011). Urologic applications of engineered tissue. *Regenerative Medicine*, *6*(6), 757–765. DOI: 10.2217/rme.11.91 PMID: 22050527

Kong, K. Y., Ren, J., Kraus, M., Finklestein, S. P., & Brown, R. H.Jr. (2004). Human Umbilical Cord Blood Cells Differentiate into Muscle in sjl Muscular Dystrophy Mice. *Stem Cells (Dayton, Ohio)*, *22*(6), 981–993. DOI: 10.1634/stemcells.22-6-981 PMID: 15536189

Krapf, J. M., Mitchell, L., Holton, M. A., & Goldstein, A. T. (2020). Vulvar Lichen Sclerosus: Current Perspectives. *International Journal of Women's Health*, *12*, 11–20. DOI: 10.2147/IJWH.S191200 PMID: 32021489

Kravvas, G., Muneer, A., Watchorn, R. E., Castiglione, F., Haider, A., Freeman, A., Hadway, P., Alnajjar, H., Lynch, M., & Bunker, C. B. (2022). Male genital lichen sclerosus, microincontinence and occlusion: Mapping the disease across the prepuce. *Clinical and Experimental Dermatology*, *47*(6), 1124–1130. DOI: 10.1111/ced.15127 PMID: 35150005

Kropp, B. P. (1998). Small-intestinal submucosa for bladder augmentation: A review of preclinical studies. *World Journal of Urology*, *16*(4), 262–267. DOI: 10.1007/s003450050064 PMID: 9775425

Kubota, H., & Brinster, R. L. (2006). Technology insight: In vitro culture of spermatogonial stem cells and their potential therapeutic uses. *Nature Clinical Practice. Endocrinology & Metabolism*, *2*(2), 99–108. DOI: 10.1038/ncpendmet0098 PMID: 16932264

Kucia, M., Halasa, M., Wysoczynski, M., Baskiewicz-Masiuk, M., Moldenhawer, S., Zuba-Surma, E., Czajka, R., Wojakowski, W., Machalinski, B., & Ratajczak, M. Z. (2007). Morphological and molecular characterization of novel population of CXCR4+ SSEA-4+ Oct-4+ very small embryonic-like cells purified from human cord blood—Preliminary report. *Leukemia*, *21*(2), 297–303. DOI: 10.1038/sj.leu.2404470 PMID: 17136117

Kucia, M., Reca, R., Campbell, F. R., Zuba-Surma, E., Majka, M., Ratajczak, J., & Ratajczak, M. Z. (2006). A population of very small embryonic-like (VSEL) CXCR4+ SSEA-1+Oct-4+ stem cells identified in adult bone marrow. *Leukemia*, *20*(5), 857–869. DOI: 10.1038/sj.leu.2404171 PMID: 16498386

Kuismanen, K., Sartoneva, R., Haimi, S., Mannerström, B., Tomás, E., Miettinen, S., & Nieminen, K. (2014). Autologous Adipose Stem Cells in Treatment of Female Stress Urinary Incontinence: Results of a Pilot Study. *Stem Cells Translational Medicine*, *3*(8), 936–941. DOI: 10.5966/sctm.2013-0197 PMID: 24985079

Kuppe, C., & Kramann, R. (2016). Role of mesenchymal stem cells in kidney injury and fibrosis. *Current Opinion in Nephrology and Hypertension*, *25*(4), 372–377. DOI: 10.1097/MNH.0000000000000230 PMID: 27191350

Kurniawati, E. M., Santoso, B., Rantam, F. A., Santoso, B. I., Widjiati, , Hadi, T. H. S., Hardianto, G., & Paraton, H. (2022). Does freeze dried amnion useful as scaffold for mesenchymal stem cell for repair of vesicovaginal fistula? An overview of proliferative and remodeling phase in the wound healing process. *European Journal of Obstetrics, Gynecology, and Reproductive Biology*, *274*, 113–116. DOI: 10.1016/j.ejogrb.2022.05.017 PMID: 35640439

Kwon, D., Minnery, B., Kim, Y., Jang, H. K., De Miguel, F., Yoshimura, N., & Chancellor, M. B. (2005). Neurologic recovery and improved detrusor contractility using muscle-derived cells in rat model of unilateral pelvic nerve transection. *Urology*, *65*(6), 1249–1253. DOI: 10.1016/j.urology.2005.01.037 PMID: 15922415

Lacham-Kaplan, O., Chy, H., & Trounson, A. (2006). Testicular cell conditioned medium supports differentiation of embryonic stem cells into ovarian structures containing oocytes. *Stem Cells (Dayton, Ohio)*, *24*(2), 266–273. DOI: 10.1634/stemcells.2005-0204 PMID: 16109761

Lahlil, R., Scrofani, M., Barbet, R., Tancredi, C., Aries, A., & Hénon, P. (2018). VSELs Maintain their Pluripotency and Competence to Differentiate after Enhanced Ex Vivo Expansion. *Stem Cell Reviews and Reports*, *14*(4), 510–524. DOI: 10.1007/s12015-018-9821-1 PMID: 29736843

Lam, A. Q., Freedman, B. S., Morizane, R., Lerou, P. H., Valerius, M. T., & Bonventre, J. V. (2014). Rapid and efficient differentiation of human pluripotent stem cells into intermediate mesoderm that forms tubules expressing kidney proximal tubular markers. *Journal of the American Society of Nephrology*, *25*(6), 1211–1225. DOI: 10.1681/ASN.2013080831 PMID: 24357672

Laschke, M. W., & Menger, M. D. (2016). Prevascularization in tissue engineering: Current concepts and future directions. *Biotechnology Advances*, *34*(2), 112–121. DOI: 10.1016/j.biotechadv.2015.12.004 PMID: 26674312

Latini, J. M., McAninch, J. W., Brandes, S. B., Chung, J. Y., & Rosenstein, D. (2014). SIU/ICUD Consultation On Urethral Strictures: Epidemiology, etiology, anatomy, and nomenclature of urethral stenoses, strictures, and pelvic fracture urethral disruption injuries. *Urology*, *83*(3, Suppl), S1–S7. DOI: 10.1016/j.urology.2013.09.009 PMID: 24210733

Lau, R. W., Wang, B., & Ricardo, S. D. (2018). Gene editing of stem cells for kidney disease modelling and therapeutic intervention. *Nephrology (Carlton, Vic.)*, *23*(11), 981–990. DOI: 10.1111/nep.13410 PMID: 29851168

Lavelle, J., Meyers, S., Ramage, R., Bastacky, S., Doty, D., Apodaca, G., & Zeidel, M. L. (2002). Bladder permeability barrier: Recovery from selective injury of surface epithelial cells. *American Journal of Physiology. Renal Physiology*, *283*(2), F242–F253. DOI: 10.1152/ajprenal.00307.2001 PMID: 12110507

Lazzeri, E., Romagnani, P., & Lasagni, L. (2015). Stem cell therapy for kidney disease. *Expert Opinion on Biological Therapy*, *15*(10), 1455–1468. DOI: 10.1517/14712598.2015.1067300 PMID: 26160420

Lazzeri, E., Ronconi, E., Angelotti, M. L., Peired, A., Mazzinghi, B., Becherucci, F., Conti, S., Sansavini, G., Sisti, A., Ravaglia, F., Lombardi, D., Provenzano, A., Manonelles, A., Cruzado, J. M., Giglio, S., Roperto, R. M., Materassi, M., Lasagni, L., & Romagnani, P. (2015). Human Urine-Derived Renal Progenitors for Personalized Modeling of Genetic Kidney Disorders. *Journal of the American Society of Nephrology*, *26*(8), 1961–1974. DOI: 10.1681/ASN.2014010057 PMID: 25568173

Le Page, M. (s. f.). What is CRISPR? *NewScientist*. https://www.newscientist.com/definition/what-is-crispr/#:~:text=A%20technology%20that%20can%20be,of%20DNA%20inside%20a%20cell

Lee, C. N., Jang, J. B., Kim, J. Y., Koh, C., Baek, J. Y., & Lee, K. J. (2010). Human cord blood stem cell therapy for treatment of stress urinary incontinence. *Journal of Korean Medical Science*, *25*(6), 813–816. DOI: 10.3346/jkms.2010.25.6.813 PMID: 20514298

Lee, H. J., An, J., Doo, S. W., Kim, J. H., Choi, S. S., Lee, S.-R., Park, S. W., Song, Y. S., & Kim, S. U. (2015). Improvement in Spinal Cord Injury-Induced Bladder Fibrosis Using Mesenchymal Stem Cell Transplantation Into the Bladder Wall. *Cell Transplantation*, *24*(7), 1253–1263. DOI: 10.3727/096368914X682125 PMID: 24912020

Lee, H. J., Selesniemi, K., Niikura, Y., Niikura, T., Klein, R., Dombkowski, D. M., & Tilly, J. L. (2007). Bone marrow transplantation generates immature oocytes and rescues long-term fertility in a preclinical mouse model of chemotherapy-induced premature ovarian failure. *Journal of Clinical Oncology*, *25*(22), 3198–3204. DOI: 10.1200/JCO.2006.10.3028 PMID: 17664466

Lee, H. J., Won, J. H., Doo, S. H., Kim, J. H., Song, K. Y., Lee, S. J., Lim, I., Chang, K. T., Song, Y. S., & Kim, S. U. (2012). Inhibition of collagen deposit in obstructed rat bladder outlet by transplantation of superparamagnetic iron oxide-labeled human mesenchymal stem cells as monitored by molecular magnetic resonance imaging (MRI). *Cell Transplantation*, *21*(5), 959–970. DOI: 10.3727/096368911X627516 PMID: 22449414

Lee, J. Y., Qu-Petersen, Z., Cao, B., Kimura, S., Jankowski, R., Cummins, J., Usas, A., Gates, C., Robbins, P., Wernig, A., & Huard, J. (2000). Clonal isolation of muscle-derived cells capable of enhancing muscle regeneration and bone healing. *The Journal of Cell Biology*, *150*(5), 1085–1099. DOI: 10.1083/jcb.150.5.1085 PMID: 10973997

Lee, P.-Y., Chien, Y., Chiou, G.-Y., Lin, C.-H., Chiou, C.-H., & Tarng, D.-C. (2012). Induced pluripotent stem cells without c-Myc attenuate acute kidney injury via down-regulating the signaling of oxidative stress and inflammation in ischemia-reperfusion rats. *Cell Transplantation*, *21*(12), 2569–2585. DOI: 10.3727/096368912X636902 PMID: 22507855

Lee, R. H., Seo, M. J., Reger, R. L., Spees, J. L., Pulin, A. A., Olson, S. D., & Prockop, D. J. (2006). Multipotent stromal cells from human marrow home to and promote repair of pancreatic islets and renal glomeruli in diabetic NOD/scid mice. *Proceedings of the National Academy of Sciences of the United States of America*, *103*(46), 17438–17443. DOI: 10.1073/pnas.0608249103 PMID: 17088535

Lee, S. W., Ryu, C. M., Shin, J. H., Choi, D., Kim, A., Yu, H. Y., Han, J. Y., Lee, H. Y., Lim, J., Kim, Y. H., Heo, J., Lee, S., Ju, H., Kim, S., Hong, K. S., Han, J. Y., Song, M., Chung, H. M., Kim, J. K., & Choo, M. S. (2018). The therapeutic effect of human embryonic stem cell-derived multipotent mesenchymal stem cells on chemical-induced cystitis in rats. *International Neurourology Journal*, *22*(Suppl 1), S34–S45. DOI: 10.5213/inj.1836014.007 PMID: 29385783

Lee, S.-J., Ryu, M.-O., Seo, M.-S., Park, S.-B., Ahn, J.-O., Han, S.-M., Kang, K.-S., Bhang, D.-H., & Youn, H.-Y. (2017). Mesenchymal Stem Cells Contribute to Improvement of Renal Function in a Canine Kidney Injury Model. *In Vivo (Athens, Greece)*, *31*(6), 1115–1124. DOI: 10.21873/invivo.11177 PMID: 29102933

Lee, S.-R., Lee, S.-H., Moon, J.-Y., Park, J.-Y., Lee, D., Lim, S. J., Jeong, K.-H., Park, J.-K., Lee, T.-W., & Ihm, C.-G. (2010). Repeated administration of bone marrow-derived mesenchymal stem cells improved the protective effects on a remnant kidney model. *Renal Failure*, *32*(7), 840–848. DOI: 10.3109/0886022X.2010.494803 PMID: 20662698

Lee, Y., & Kang, E. (2019). Stem cells and reproduction. *BMB Reports*, *52*(8), 482–489. DOI: 10.5483/BMBRep.2019.52.8.141 PMID: 31234956

Lei, L., & Spradling, A. C. (2013). Female mice lack adult germ-line stem cells but sustain oogenesis using stable primordial follicles. *Proceedings of the National Academy of Sciences of the United States of America*, *110*(21), 8585–8590. DOI: 10.1073/pnas.1306189110 PMID: 23630252

Levanovich, P. E., Diokno, A., Hasenau, D. L., Lajiness, M., Pruchnic, R., & Chancellor, M. B. (2015). Intradetrusor injection of adult muscle-derived cells for the treatment of underactive bladder: Pilot study. *International Urology and Nephrology*, *47*(3), 465–467. DOI: 10.1007/s11255-015-0924-1 PMID: 25663053

Levey, A. S., & Coresh, J. (2012). Chronic kidney disease. *Lancet, 379*(9811), 165–180. DOI: 10.1016/S0140-6736(11)60178-5 PMID: 21840587

Levy, J. A., Marchand, M., Iorio, L., Cassini, W., & Zahalsky, M. P. (2016). Determining the feasibility of managing erectile dysfunction in humans with placental-derived stem cells. *The Journal of the American Osteopathic Association, 116*(1), e1–e5. DOI: 10.7556/jaoa.2016.007 PMID: 26745574

Liang, C. C., Lee, T. H., & Chang, S. D. (2016). Effect of umbilical cord blood stem cells transplantation on bladder dysfunction induced by cerebral ischemia in rats. *Taiwanese Journal of Obstetrics & Gynecology, 55*(5), 672–679. DOI: 10.1016/j.tjog.2015.11.002 PMID: 27751414

Liang, C. C., Shaw, S. W. S., Huang, Y. H., Lin, Y. H., & Lee, T. H. (2018). Improvement in bladder dysfunction after bladder transplantation of amniotic fluid stem cells in diabetic rats. *Scientific Reports, 8*(1), 2105. Advance online publication. DOI: 10.1038/s41598-018-20512-z PMID: 29391467

Liang, C. C., Shaw, S. W. S., Lin, Y. H., & Lee, T. H. (2018). Amniotic fluid stem cells ameliorate bladder dysfunction induced by chronic bladder ischemia in rat. *Neurourology and Urodynamics, 37*(1), 123–131. DOI: 10.1002/nau.23316 PMID: 28605059

Liang, C. C., Steven Shaw, S. W., Huang, Y. H., Lin, Y. H., & Lee, T. H. (2017). Bladder Transplantation of Amniotic Fluid Stem Cell may Ameliorate Bladder Dysfunction After Focal Cerebral Ischemia in Rat. *Stem Cells Translational Medicine, 6*(4), 1227–1236. DOI: 10.1002/sctm.16-0212 PMID: 28186672

Liang, R., Knight, K., Barone, W., Powers, R. W., Nolfi, A., Palcsey, S., Abramowitch, S., & Moalli, P. A. (2017). Extracellular matrix regenerative graft attenuates the negative impact of polypropylene prolapse mesh on vagina in rhesus macaque. *American Journal of Obstetrics and Gynecology, 216*(2), 153.e1–153.e9. DOI: 10.1016/j.ajog.2016.09.073 PMID: 27615441

Liang, R., Knight, K., Easley, D., Palcsey, S., Abramowitch, S., & Moalli, P. A. (2017). Towards rebuilding vaginal support utilizing an extracellular matrix bioscaffold. *Acta Biomaterialia, 57*, 324–333. DOI: 10.1016/j.actbio.2017.05.015 PMID: 28487243

Liang, R., Zong, W., Palcsey, S., Abramowitch, S., & Moalli, P. A. (2015). Impact of prolapse meshes on the metabolism of vaginal extracellular matrix in rhesus macaque. *American Journal of Obstetrics and Gynecology, 212*(2), 174.e1–174.e7. DOI: 10.1016/j.ajog.2014.08.008 PMID: 25128444

Lieber, M. R., Ma, Y., Pannicke, U., & Schwarz, K. (2004). The mechanism of vertebrate nonhomologous DNA end joining and its role in V(D)J recombination. *DNA Repair*, *3*(8-9), 817–826. DOI: 10.1016/j.dnarep.2004.03.015 PMID: 15279766

Li, G., Yu, C., Yu, P., Peng, Q., Wang, Q., Ren, S., Li, H., Li, M., Li, P., & He, R. (2022). Periurethral and Intravenous Injections of Adipose-Derived Stem Cells to Promote Local Tissue Recovery in a Rat Model of Stress Urinary Incontinence. *Urology*, *167*, 82–89. DOI: 10.1016/j.urology.2022.05.018 PMID: 35654272

Li, J., Luo, H., Dong, X., Liu, Q., Wu, C., Zhang, T., Hu, X., Zhang, Y., Song, B., & Li, L. (2017). Therapeutic effect of urine-derived stem cells for protamine/lipopolysaccharide-induced interstitial cystitis in a rat model. *Stem Cell Research & Therapy*, *8*(1), 107. Advance online publication. DOI: 10.1186/s13287-017-0547-9 PMID: 28482861

Li, L., & Xie, T. (2005). Stem cell niche: Structure and function. *Annual Review of Cell and Developmental Biology*, *21*(1), 605–631. DOI: 10.1146/annurev.cellbio.21.012704.131525 PMID: 16212509

Li, M., Mondrinos, M. J., Gandhi, M. R., Ko, F. K., Weiss, A. S., & Lelkes, P. I. (2005). Electrospun protein fibers as matrices for tissue engineering. *Biomaterials*, *26*(30), 5999–6008. DOI: 10.1016/j.biomaterials.2005.03.030 PMID: 15894371

Lin, G., Banie, L., Ning, H., Bella, A. J., Lin, C. S., & Lue, T. F. (2009). Potential of adipose-derived stem cells for treatment of erectile dysfunction. *The journal of sexual medicine, 6 Suppl 3*(Suppl 3), 320-327. DOI: 10.1111/j.1743-6109.2008.01190.x

Lin, A. S., Carrier, S., Morgan, D. M., & Lue, T. F. (1998). Effect of simulated birth trauma on the urinary continence mechanism in the rat. *Urology*, *52*(1), 143–151. DOI: 10.1016/S0090-4295(98)00136-8 PMID: 9671889

Lin, H., Dhanani, N., Tseng, H., Souza, G. R., Wang, G., Cao, Y., Ko, T. C., Jiang, H., & Wang, R. (2016). Nanoparticle Improved Stem Cell Therapy for Erectile Dysfunction in a Rat Model of Cavernous Nerve Injury. *The Journal of Urology*, *195*(3), 788–795. DOI: 10.1016/j.juro.2015.10.129 PMID: 26519654

Lin, H.-K., Madihally, S. V., Palmer, B., Frimberger, D., Fung, K.-M., & Kropp, B. P. (2015). Biomatrices for bladder reconstruction. *Advanced Drug Delivery Reviews*, *82-83*, 47–63. DOI: 10.1016/j.addr.2014.11.020 PMID: 25477305

Lin, M., Lu, Y., & Chen, J. (2022). Tissue-engineered repair material for pelvic floor dysfunction. *Frontiers in Bioengineering and Biotechnology*, *10*(September), 1–12. DOI: 10.3389/fbioe.2022.968482 PMID: 36147522

Lin, S. A., Kolle, G., Grimmond, S. M., Zhou, Q., Doust, E., Little, M. H., Aronow, B., Ricardo, S. D., Pera, M. F., Bertram, J. F., & Laslett, A. L. (2010). Subfractionation of differentiating human embryonic stem cell populations allows the isolation of a mesodermal population enriched for intermediate mesoderm and putative renal progenitors. *Stem Cells and Development, 19*(10), 1637–1648. DOI: 10.1089/scd.2010.0017 PMID: 20143954

Li, P., Hu, H., Yang, S., Tian, R., Zhang, Z., Zhang, W., Ma, M., Zhu, Y., Guo, X., Huang, Y., He, Z., & Li, Z. (2013). Differentiation of induced pluripotent stem cells into male germ cells in vitro through embryoid body formation and retinoic acid or testosterone induction. *BioMed Research International, 2013*, 1–9. Advance online publication. DOI: 10.1155/2013/608728 PMID: 23509752

Liu, G., Sun, X., Bian, J., Wu, R., Guan, X., Ouyang, B., Huang, Y., Xiao, H., Luo, D., Atala, A., Zhang, Y., & Deng, C. (2013). Correction of Diabetic Erectile Dysfunction with Adipose Derived Stem Cells Modified with the Vascular Endothelial Growth Factor Gene in a Rodent Diabetic Model. *PLoS One, 8*(8), e72790. Advance online publication. DOI: 10.1371/journal.pone.0072790 PMID: 24023647

Liu, L., Guo, S., Shi, W., Liu, Q., Huo, F., Wu, Y., & Tian, W. (2021). Bone Marrow Mesenchymal Stem Cell-Derived Small Extracellular Vesicles Promote Periodontal Regeneration. *Tissue Engineering. Part A, 27*(13-14), 962–976. DOI: 10.1089/ten.tea.2020.0141 PMID: 32962564

Liu, M., Lu, Y., Sun, F., Li, Y., Wu, J., & Zou, Q. (2025). The Nerve-Induced Adipose Stem Cells Promote Nerve Repair in Stress Urinary Incontinence by Regulating Schwann Cell Repair Phenotype Conversion Through Activation of the Notch Pathway. *Molecular Neurobiology, 62*(6), 7330–7344. DOI: 10.1007/s12035-025-04704-z PMID: 39881114

Liu, T., Peng, Y., Jia, C., Fang, X., Li, J., & Zhong, W. (2015). Hepatocyte growth factor-modified adipose tissue-derived stem cells improve erectile function in streptozotocin-induced diabetic rats. *Growth Factors (Chur, Switzerland), 33*(4), 282–289. DOI: 10.3109/08977194.2015.1077825 PMID: 26339935

Liu, X., Holzwarth, J. M., & Ma, P. X. (2012). Functionalized synthetic biodegradable polymer scaffolds for tissue engineering. *Macromolecular Bioscience, 12*(7), 911–919. DOI: 10.1002/mabi.201100466 PMID: 22396193

Liu, X., Smith, L. A., Hu, J., & Ma, P. X. (2009). Biomimetic nanofibrous gelatin/apatite composite scaffolds for bone tissue engineering. *Biomaterials, 30*(12), 2252–2258. DOI: 10.1016/j.biomaterials.2008.12.068 PMID: 19152974

Li, X., Wang, Y., Sharif-Afshar, A.-R., Uwamariya, C., Yi, A., Ishii, K., Hayward, S. W., Matusik, R. J., & Bhowmick, N. A. (2009). Urothelial transdifferentiation to prostate epithelia is mediated by paracrine TGF-beta signaling. *Differentiation. Differentiation, 77*(1), 95–102. DOI: 10.1016/j.diff.2008.09.012 PMID: 19281768

Li, X., Xie, J., Yuan, X., & Xia, Y. (2008). Coating electrospun poly(epsilon-caprolactone) fibers with gelatin and calcium phosphate and their use as biomimetic scaffolds for bone tissue engineering. *Langmuir, 24*(24), 14145–14150. DOI: 10.1021/la802984a PMID: 19053657

Li, Y., Pan, H., & Huard, J. (2010). Isolating stem cells from soft musculoskeletal tissues. *Journal of Visualized Experiments, 41*(41). Advance online publication. DOI: 10.3791/2011 PMID: 20644509

Li, Z., & Tan, B. H. (2014). Towards the development of polycaprolactone based amphiphilic block copolymers: Molecular design, self-assembly and biomedical applications. *Materials Science and Engineering C, 45*, 620–634. DOI: 10.1016/j.msec.2014.06.003 PMID: 25491872

Lokeshwar, S. D., Patel, P., Shah, S. M., & Ramasamy, R. (2020). A Systematic Review of Human Trials Using Stem Cell Therapy for Erectile Dysfunction. En *Sexual Medicine Reviews* (Vol. 8, Número 1, pp. 122-130). Elsevier B.V. DOI: 10.1016/j.sxmr.2019.08.003

Lorenzi, B., Pessina, F., Lorenzoni, P., Urbani, S., Vernillo, R., Sgaragli, G., Gerli, R., Mazzanti, B., Bosi, A., Saccardi, R., & Lorenzi, M. (2008). Treatment of experimental injury of anal sphincters with primary surgical repair and injection of bone marrow-derived mesenchymal stem cells. *Diseases of the Colon and Rectum, 51*(4), 411–420. DOI: 10.1007/s10350-007-9153-8 PMID: 18224375

Lu, H., Hoshiba, T., Kawazoe, N., & Chen, G. (2012). Comparison of decellularization techniques for preparation of extracellular matrix scaffolds derived from three-dimensional cell culture. *Journal of Biomedical Materials Research. Part A, 100*(9), 2507–2516. DOI: 10.1002/jbm.a.34150 PMID: 22623317

Lu, J., Xin, Z., Zhang, Q., Cui, D., Xiao, Y., Zhuo, J., Sun, F., Xia, S., & Shao, Y. (2016). Beneficial effect of PEDF-transfected ADSCs on erectile dysfunction in a streptozotocin-diabetic rat model. *Cell and Tissue Research, 366*(3), 623–637. DOI: 10.1007/s00441-016-2494-7 PMID: 27655233

Lu, L., Peter, S. J., Lyman, M. D., Lai, H. L., Leite, S. M., Tamada, J. A., Uyama, S., Vacanti, J. P., Langer, R., & Mikos, A. G. (2000). In vitro and in vivo degradation of porous poly(DL-lactic-co-glycolic acid) foams. *Biomaterials, 21*(18), 1837–1845. DOI: 10.1016/S0142-9612(00)00047-8 PMID: 10919687

Luo, C., Peng, Y., Zhou, X., Fan, J., Chen, W., Zhang, H., & Wei, A. (2022). NLRP3 downregulation enhances engraftment and functionality of adipose-derived stem cells to alleviate erectile dysfunction in diabetic rats. *Frontiers in Endocrinology*, *13*, 913296. Advance online publication. DOI: 10.3389/fendo.2022.913296 PMID: 35937790

Luo, Y., Xu, X., An, X., Sun, X., Wang, S., & Zhu, D. (2016). Targeted Inhibition of the miR-199a/214 Cluster by CRISPR Interference Augments the Tumor Tropism of Human Induced Pluripotent Stem Cell-Derived Neural Stem Cells under Hypoxic Condition. *Stem Cells International*, *3598542*(1), 3598542. Advance online publication. DOI: 10.1155/2016/3598542 PMID: 27965712

Lu, S. H., Cannon, T. W., Chermanski, C., Pruchnic, R., Somogyi, G., Sacks, M., De Groat, W. C., Huard, J., & Chancellor, M. B. (2003). Muscle-derived stem cells seeded into acellular scaffolds develop calcium-dependent contractile activity that is modulated by nicotinic receptors. *Urology*, *61*(6), 1285–1291. DOI: 10.1016/S0090-4295(03)00020-7 PMID: 12809930

Lu, Y., Wang, L., Zhang, M., & Chen, Z. (2022). Mesenchymal Stem Cell-Derived Small Extracellular Vesicles: A Novel Approach for Kidney Disease Treatment. *International Journal of Nanomedicine*, *17*, 3603–3618. DOI: 10.2147/IJN.S372254 PMID: 35990308

Mae, S.-I., Shono, A., Shiota, F., Yasuno, T., Kajiwara, M., Gotoda-Nishimura, N., Arai, S., Sato-Otubo, A., Toyoda, T., Takahashi, K., Nakayama, N., Cowan, C. A., Aoi, T., Ogawa, S., McMahon, A. P., Yamanaka, S., & Osafune, K. (2013). Monitoring and robust induction of nephrogenic intermediate mesoderm from human pluripotent stem cells. *Nature Communications*, *4*(1), 1367. DOI: 10.1038/ncomms2378 PMID: 23340407

Ma, H., Liu, M., Li, Y., Wang, W., Yang, K., Lu, L., He, M., Deng, T., Li, M., & Wu, D. (2020). Intrauterine transplantation of autologous menstrual blood stem cells increases endometrial thickness and pregnancy potential in patients with refractory intrauterine adhesion. *Journal of Obstetrics and Gynaecology Research*, *46*(11), 2347–2355. DOI: 10.1111/jog.14449 PMID: 32856391

Mahajan, P. V., Subramanian, S., Danke, A., & Kumar, A. (2016). Neurogenic Bladder Repair Using Autologous Mesenchymal Stem Cells. *Case Reports in Urology*, *2539320*, 1–3. Advance online publication. DOI: 10.1155/2016/2539320 PMID: 27656308

Makhlough, A., Shekarchian, S., Moghadasali, R., Einollahi, B., Dastgheib, M., Janbabaee, G., Hosseini, S. E., Falah, N., Abbasi, F., Baharvand, H., & Aghdami, N. (2018). Bone marrow-mesenchymal stromal cell infusion in patients with chronic kidney disease: A safety study with 18 months of follow-up. *Cytotherapy*, *20*(5), 660–669. DOI: 10.1016/j.jcyt.2018.02.368 PMID: 29580865

Makhlough, A., Shekarchian, S., Moghadasali, R., Einollahi, B., Hosseini, S. E., Jaroughi, N., Bolurieh, T., Baharvand, H., & Aghdami, N. (2017). Safety and tolerability of autologous bone marrow mesenchymal stromal cells in ADPKD patients. *Stem Cell Research & Therapy*, *8*(1), 116. DOI: 10.1186/s13287-017-0557-7 PMID: 28535817

Mali, P., Yang, L., Esvelt, K. M., Aach, J., Guell, M., DiCarlo, J. E., Norville, J. E., & Church, G. M. (2013). RNA-guided human genome engineering via Cas9. *Science*, *339*(6121), 823–826. DOI: 10.1126/science.1232033 PMID: 23287722

Mandegar, M. A., Huebsch, N., Frolov, E. B., Shin, E., Truong, A., Olvera, M. P., Chan, A. H., Miyaoka, Y., Holmes, K., Spencer, C. I., Judge, L. M., Gordon, D. E., Eskildsen, T. V., Villalta, J. E., Horlbeck, M. A., Gilbert, L. A., Krogan, N. J., Sheikh, S. P., Weissman, J. S., & Conklin, B. R. (2016). CRISPR Interference Efficiently Induces Specific and Reversible Gene Silencing in Human iPSCs. *Cell Stem Cell*, *18*(4), 541–553. DOI: 10.1016/j.stem.2016.01.022 PMID: 26971820

Mangera, A., Bullock, A. J., Roman, S., Chapple, C. R., & MacNeil, S. (2013). Comparison of candidate scaffolds for tissue engineering for stress urinary incontinence and pelvic organ prolapse repair. *BJU International*, *112*(5), 674–685. DOI: 10.1111/bju.12186 PMID: 23773418

Mangır, N., Bullock, A. J., Roman, S., Osman, N., Chapple, C., & MacNeil, S. (2016). Production of ascorbic acid releasing biomaterials for pelvic floor repair. *Acta Biomaterialia*, *29*, 188–197. DOI: 10.1016/j.actbio.2015.10.019 PMID: 26478470

Marcheque, J., Bussolati, B., Csete, M., & Perin, L. (2019). Concise Reviews: Stem Cells and Kidney Regeneration: An Update. *Stem Cells Translational Medicine*, *8*(1), 82–92. DOI: 10.1002/sctm.18-0115 PMID: 30302937

Marra, G., Sturch, P., Oderda, M., Tabatabaei, S., Muir, G., & Gontero, P. (2016). Systematic review of lower urinary tract symptoms/benign prostatic hyperplasia surgical treatments on men's ejaculatory function: Time for a bespoke approach? *International journal of urology : official journal of the Japanese Urological Association, 23*(1), 22-35. DOI: 10.1111/iju.12866

Martínez-Salamanca, J. I., Zurita, M., Costa, C., Martínez-Salamanca, E., Fernández, A., Castela, A., Vaquero, J., Carballido, J., & Angulo, J. (2016). Dual Strategy With Oral Phosphodiesterase Type 5 Inhibition and Intracavernosal Implantation of Mesenchymal Stem Cells Is Superior to Individual Approaches in the Recovery of Erectile and Cavernosal Functions After Cavernous Nerve Injury in Rats. *Journal of Sexual Medicine*, *13*(1), 1–11. DOI: 10.1016/j.jsxm.2015.12.001 PMID: 26755080

Mauney, J. R., Cannon, G. M., Lovett, M. L., Gong, E. M., Di Vizio, D., Gomez, P.III, Kaplan, D. L., Adam, R. M., & Estrada, C. R.Jr. (2011). Evaluation of gel spun silk-based biomaterials in a murine model of bladder augmentation. *Biomaterials*, *32*(3), 808–818. DOI: 10.1016/j.biomaterials.2010.09.051 PMID: 20951426

McGee, E. A., & Hsueh, A. J. W. (2000). Initial and cyclic recruitment of ovarian follicles. *Endocrine Reviews*, *21*(2), 200–214. DOI: 10.1210/edrv.21.2.0394 PMID: 10782364

Medrano, J. V., Ramathal, C., Nguyen, H. A. N., Simon, C., & Reijo Pera, R. A. (2012). Divergent RNA-binding proteins, DAZL and VASA, induce meiotic progression in human germ cells derived in vitro. *Stem Cells (Dayton, Ohio)*, *30*(3), 441–451. DOI: 10.1002/stem.1012 PMID: 22162380

Meisel, R., Zibert, A., Laryea, M., Göbel, U., Däubener, W., & Dilloo, D. (2004). Human bone marrow stromal cells inhibit allogeneic T-cell responses by indoleamine 2,3-dioxygenase-mediated tryptophan degradation. *Blood*, *103*(12), 4619–4621. DOI: 10.1182/blood-2003-11-3909 PMID: 15001472

Metcalfe, S. M. (2020). Mesenchymal stem cells and management of COVID-19 pneumonia. *Medicine in Drug Discovery*, *5*, 100019. DOI: 10.1016/j.medidd.2020.100019 PMID: 32296777

Milenkovic, U., Albersen, M., & Castiglione, F. (2019). The mechanisms and potential of stem cell therapy for penile fibrosis. En *Nature Reviews Urology* (Vol. 16, Número 2, pp. 79-97). Nature Publishing Group. DOI: 10.1038/s41585-018-0109-7

Mirzaei, M., Bagherinasabsarab, M., Pakmanesh, H., Mohammadi, R., Teimourian, M., Farsinejad, A., & Jahani, Y. (2021). The Effect of Intracavernosal Injection of Stem Cell in the Treatment of Erectile Dysfunction in Diabetic Patients: A Randomized Single-blinded Clinical Trial. *Urology Journal*, *18*(6), 675–681. DOI: 10.22037/UJ.V18I.6503 PMID: 34655071

Mitsui, T., Fischer, I., Shumsky, J. S., & Murray, M. (2005). Transplants of fibroblasts expressing BDNF and NT-3 promote recovery of bladder and hindlimb function following spinal contusion injury in rats. *Experimental Neurology*, *194*(2), 410–431. DOI: 10.1016/j.expneurol.2005.02.022 PMID: 16022868

Mitsui, T., Kakizaki, H., Tanaka, H., Shibata, T., Matsuoka, I., & Koyanagi, T. (2003). Immortalized neural stem cells transplanted into the injured spinal cord promote recovery of voiding function in the rat. *The Journal of Urology*, *170*(4 Pt 1), 1421–1425. DOI: 10.1097/01.ju.0000075501.05758.33 PMID: 14501781

Mitsui, T., Neuhuber, B., & Fischer, I. (2011). Acute administration of AMPA/Kainate blocker combined with delayed transplantation of neural precursors improves lower urinary tract function in spinal injured rats. *Brain Research*, *1418*, 23–31. DOI: 10.1016/j.brainres.2011.08.023 PMID: 21937028

Mitterberger, M., Marksteiner, R., Margreiter, E., Pinggera, G. M., Frauscher, F., Ulmer, H., Fussenegger, M., Bartsch, G., & Strasser, H. (2008). Myoblast and Fibroblast Therapy for Post-Prostatectomy Urinary Incontinence: 1-Year Followup of 63 Patients. *The Journal of Urology*, *179*(1), 226–231. DOI: 10.1016/j.juro.2007.08.154 PMID: 18001790

Mitterberger, M., Pinggera, G. M., Marksteiner, R., Margreiter, E., Fussenegger, M., Frauscher, F., Ulmer, H., Hering, S., Bartsch, G., & Strasser, H. (2008). Adult Stem Cell Therapy of Female Stress Urinary Incontinence. *European Urology*, *53*(1), 169–175. DOI: 10.1016/j.eururo.2007.07.026 PMID: 17683852

Moghadasali, R., Hajinasrollah, M., Argani, H., Nassiri, S. M., Najarasl, M., Sodeifi, N., Baharvand, H., & Aghdami, N. (2015). Autologous transplantation of mesenchymal stromal cells tends to prevent progress of interstitial fibrosis in a rhesus Macaca mulatta monkey model of chronic kidney disease. *Cytotherapy*, *17*(11), 1495–1505. DOI: 10.1016/j.jcyt.2015.06.006 PMID: 26341479

Montorsi, F., Salonia, A., Centemero, A., Guazzoni, G., Nava, L., Da Pozzo, L. F., Cestari, A., Colombo, R., Barbagli, G., & Rigatti, P. (2002). Vestibular flap urethroplasty for strictures of the female urethra. Impact on symptoms and flow patterns. *Urologia Internationalis*, *69*(1), 12–16. DOI: 10.1159/000064353 PMID: 12119432

Moore, H., Udayashankar, R., & Aflatoonian, B. (2008). Stem cells for reproductive medicine. *Molecular and Cellular Endocrinology*, *288*(1-2), 104–110. DOI: 10.1016/j.mce.2008.02.019 PMID: 18406514

Morigi, M., Imberti, B., Zoja, C., Corna, D., Tomasoni, S., Abbate, M., Rottoli, D., Angioletti, S., Benigni, A., Perico, N., Alison, M., & Remuzzi, G. (2004). Mesenchymal stem cells are renotropic, helping to repair the kidney and improve function in acute renal failure. *Journal of the American Society of Nephrology*, *15*(7), 1794–1804. DOI: 10.1097/01.ASN.0000128974.07460.34 PMID: 15213267

Morigi, M., Rota, C., Montemurro, T., Montelatici, E., Lo Cicero, V., Imberti, B., Abbate, M., Zoja, C., Cassis, P., Longaretti, L., Rebulla, P., Introna, M., Capelli, C., Benigni, A., Remuzzi, G., & Lazzari, L. (2010). Life-sparing effect of human cord blood-mesenchymal stem cells in experimental acute kidney injury. *Stem Cells (Dayton, Ohio)*, *28*(3), 513–522. DOI: 10.1002/stem.293 PMID: 20049901

Morizane, R., Lam, A. Q., Freedman, B. S., Kishi, S., Valerius, M. T., & Bonventre, J. V. (2015). Nephron organoids derived from human pluripotent stem cells model kidney development and injury. *Nature Biotechnology*, *33*(11), 1193–1200. DOI: 10.1038/nbt.3392 PMID: 26458176

Mouravas, V., Filippopoulos, A., & Sfoungaris, D. (2014). Urethral plate grafting improves the results of tubularized incised plate urethroplasty in primary hypospadias. *Journal of Pediatric Urology*, *10*(3), 463–468. DOI: 10.1016/j.jpurol.2013.11.012 PMID: 24360521

Mukti, A. I., Ilyas, S., Warli, S. M., Putra, A., Rasyid, N., Munir, D., Siregar, K. B., & Ichwan, M. (2022). Umbilical Cord-Derived Mesenchymal Stem Cells Improve TGF-β, α-SMA and Collagen on Erectile Dysfunction in Streptozotocin-Induced Diabetic Rats. *Medical archives (Sarajevo, Bosnia and Herzegovina), 76*(1), 4-11. DOI: 10.5455/medarh.2022.76.4-11

Mulder, J., Sharmin, S., Chow, T., Rodrigues, D. C., Hildebrandt, M. R., D'Cruz, R., Rogers, I., Ellis, J., & Rosenblum, N. D. (2020). Generation of infant- and pediatric-derived urinary induced pluripotent stem cells competent to form kidney organoids. *Pediatric Research*, *87*(4), 647–655. DOI: 10.1038/s41390-019-0618-y PMID: 31629364

Mundy, A. R., & Andrich, D. E. (2011). Urethral trauma. Part I: Introduction, history, anatomy, pathology, assessment and emergency management. *BJU International*, *108*(3), 310–327. DOI: 10.1111/j.1464-410X.2011.10339.x PMID: 21771241

Murphy, C. M., O'Brien, F. J., Little, D. G., & Schindeler, A. (2013). Cell-scaffold interactions in the bone tissue engineering triad. *European Cells & Materials*, *26*, 120–132. DOI: 10.22203/eCM.v026a09 PMID: 24052425

Na, J., & Kim, G. J. (2020). Recent trends in stem cell therapy for premature ovarian insufficiency and its therapeutic potential: A review. *Journal of Ovarian Research*, *13*(1), 74. Advance online publication. DOI: 10.1186/s13048-020-00671-2 PMID: 32576209

Nakaki, F., Hayashi, K., Ohta, H., Kurimoto, K., Yabuta, Y., & Saitou, M. (2013). Induction of mouse germ-cell fate by transcription factors in vitro. *Nature*, *501*(7466), 222–226. DOI: 10.1038/nature12417 PMID: 23913270

Nappi, R. E., & Nijland, E. A. (2008). Women's perception of sexuality around the menopause: Outcomes of a European telephone survey. *European Journal of Obstetrics, Gynecology, and Reproductive Biology*, *137*(1), 10–16. DOI: 10.1016/j.ejogrb.2006.10.036 PMID: 17207567

National Kidney Foundation. (2002). K/DOQI clinical practice guidelines for chronic kidney disease: Evaluation, classification, and stratification. *American Journal of Kidney Diseases*, *39*(2, Suppl 1), S1–S266. PMID: 11904577

Nazm Bojnordi, M. (2017). The applications and recovery outcome of spermatogonia stem cells in regenerative medicine. *Middle East Fertility Society Journal*, *22*(4), 246–250. DOI: 10.1016/j.mefs.2017.05.008

Németh, K., Leelahavanichkul, A., Yuen, P. S. T., Mayer, B., Parmelee, A., Doi, K., Robey, P. G., Leelahavanichkul, K., Koller, B. H., Brown, J. M., Hu, X., Jelinek, I., Star, R. A., & Mezey, E. (2009). Bone marrow stromal cells attenuate sepsis via prostaglandin E(2)-dependent reprogramming of host macrophages to increase their interleukin-10 production. *Nature Medicine*, *15*(1), 42–49. DOI: 10.1038/nm.1905 PMID: 19098906

Nguyen, M. K., Jeon, O., Krebs, M. D., Schapira, D., & Alsberg, E. (2014). Sustained localized presentation of RNA interfering molecules from in situ forming hydrogels to guide stem cell osteogenic differentiation. *Biomaterials*, *35*(24), 6278–6286. DOI: 10.1016/j.biomaterials.2014.04.048 PMID: 24831973

NHI. (2023, noviembre 16). *The CRISPR Revolution*. https://www.nih.gov/about-nih/what-we-do/nih-turning-discovery-into-health/transformative-technologies/crispr-revolution#:~:text=NIH%20scientists%20are%20getting%20close,even%20removing%20the%20DNA%20as

Nicholas, C. R., Haston, K. M., Grewall, A. K., Longacre, T. A., & Reijo Pera, R. A. (2009). Transplantation directs oocyte maturation from embryonic stem cells and provides a therapeutic strategy for female infertility. *Human Molecular Genetics*, *18*(22), 4376–4389. DOI: 10.1093/hmg/ddp393 PMID: 19696121

Nicolaou, N., Renkema, K. Y., Bongers, E. M. H. F., Giles, R. H., & Knoers, N. V. A. M. (2015). Genetic, environmental, and epigenetic factors involved in CAKUT. *Nature Reviews. Nephrology*, *11*(12), 720–731. DOI: 10.1038/nrneph.2015.140 PMID: 26281895

NIH. (2024, agosto 16). CRISPR. National Human Genome Research Institute. https://www.genome.gov/genetics-glossary/CRISPR#:~:text=CRISPR%20(short%20for%20%E2%80%9Cclustered%20regularly,selectively%20modify%20the%20DNA%20of

Nikolic, A., Volarevic, V., Armstrong, L., Lako, M., & Stojkovic, M. (2016). Primordial Germ Cells: Current Knowledge and Perspectives. *Stem Cells International*, *2016*(1), 1741072. Advance online publication. DOI: 10.1155/2016/1741072 PMID: 26635880

Ning, H., Lin, G., Lue, T. F., & Lin, C. S. (2006). Neuron-like differentiation of adipose tissue-derived stromal cells and vascular smooth muscle cells. *Differentiation*, *74*(9-10), 510–518. DOI: 10.1111/j.1432-0436.2006.00081.x PMID: 17177848

Nishijima, S., Sugaya, K., Miyazato, M., Kadekawa, K., Oshiro, Y., Uchida, A., Hokama, S., & Ogawa, Y. (2007). Restoration of bladder contraction by bone marrow transplantation in rats with underactive bladder. *Biomedical Research*, *28*(5), 275–280. DOI: 10.2220/biomedres.28.275 PMID: 18000341

Nitta, M., Tamaki, T., Tono, K., Okada, Y., Masuda, M., Akatsuka, A., Hoshi, A., Usui, Y., & Terachi, T. (2010). Reconstitution of experimental neurogenic bladder dysfunction using skeletal muscle-derived multipotent stem cells. *Transplantation*, *89*(9), 1043–1049. DOI: 10.1097/TP.0b013e3181d45a7f PMID: 20150836

Oatley, J. M., & Brinster, R. L. (2008). Regulation of spermatogonial stem cell self-renewal in mammals. *Annual Review of Cell and Developmental Biology*, *24*(1), 263–286. DOI: 10.1146/annurev.cellbio.24.110707.175355 PMID: 18588486

Ochodnicky, P., Cruz, C. D., Yoshimura, N., & Cruz, F. (2012). Neurotrophins as regulators of urinary bladder function. *Nature Reviews. Urology*, *9*(11), 628–637. DOI: 10.1038/nrurol.2012.178 PMID: 23045265

Okamoto, M., & John, B. (2013). Synthetic biopolymer nanocomposites for tissue engineering scaffolds. *Progress in Polymer Science*, *38*(10), 1487–1503. DOI: 10.1016/j.progpolymsci.2013.06.001

Ortinski, P. I., O'Donovan, B., Dong, X., & Kantor, B. (2017). Integrase-Deficient Lentiviral Vector as an All-in-One Platform for Highly Efficient CRISPR/Cas9-Mediated Gene Editing. *Molecular Therapy. Methods & Clinical Development*, *5*, 153–164. DOI: 10.1016/j.omtm.2017.04.002 PMID: 28497073

Ory, J., Saltzman, R. G., Blachman-Braun, R., Dadoun, S., DiFede, D. L., Premer, C., Hurwitz, B., Hare, J. M., & Ramasamy, R. (2020). The Effect of Transendocardial Stem Cell Injection on Erectile Function in Men With Cardiomyopathy: Results From the TRIDENT, POSEIDON, and TAC-HFT Trials. *Journal of Sexual Medicine*, *17*(4), 695–701. DOI: 10.1016/j.jsxm.2020.01.003 PMID: 32059936

Osafune, K., Caron, L., Borowiak, M., Martinez, R. J., Fitz-Gerald, C. S., Sato, Y., Cowan, C. A., Chien, K. R., & Melton, D. A. (2008). Marked differences in differentiation propensity among human embryonic stem cell lines. *Nature Biotechnology*, *26*(3), 313–315. DOI: 10.1038/nbt1383

Packham, D. K., Fraser, I. R., Kerr, P. G., & Segal, K. R. (2016). Allogeneic Mesenchymal Precursor Cells (MPC) in Diabetic Nephropathy: A Randomized, Placebo-controlled, Dose Escalation Study. *EBioMedicine*, *12*, 263–269. DOI: 10.1016/j.ebiom.2016.09.011 PMID: 27743903

Palminteri, E., Berdondini, E., Fusco, F., De Nunzio, C., & Salonia, A. (2012). Long-term results of small intestinal submucosa graft in bulbar urethral reconstruction. *Urology*, *79*(3), 695–701. DOI: 10.1016/j.urology.2011.09.055 PMID: 22245298

Panula, S., Kurek, M., Kumar, P., Albalushi, H., Sánchez, S. P., Damdimopoulou, P., Olofsson, J. I., Hovatta, O., Lanner, F., & Stukenborg, J. B. (2019). Human induced pluripotent stem cells from two azoospermic patients with Klinefelter syndrome show similar X chromosome inactivation behavior to female pluripotent stem cells. *Human Reproduction (Oxford, England)*, *34*(11), 2297–2310. DOI: 10.1093/humrep/dez134 PMID: 31743397

Panula, S., Medrano, J. V., Kee, K., Bergström, R., Nguyen, H. N., Byers, B., Wilson, K. D., Wu, J. C., Simon, C., Hovatta, O., & Pera, R. A. R. (2011). Human germ cell differentiation from fetal- and adult-derived induced pluripotent stem cells. *Human Molecular Genetics*, *20*(4), 752–762. DOI: 10.1093/hmg/ddq520 PMID: 21131292

Pan, Z., & Ding, J. (2012). Poly(lactide-co-glycolide) porous scaffolds for tissue engineering and regenerative medicine. *Interface Focus*, *2*(3), 366–377. DOI: 10.1098/rsfs.2011.0123 PMID: 23741612

Papazova, D. A., Oosterhuis, N. R., Gremmels, H., van Koppen, A., Joles, J. A., & Verhaar, M. C. (2015). Cell-based therapies for experimental chronic kidney disease: A systematic review and meta-analysis. *Disease Models & Mechanisms*, *8*(3), 281–293. DOI: 10.1242/dmm.017699 PMID: 25633980

Pardo, B., Gómez-González, B., & Aguilera, A. (2009). DNA repair in mammalian cells: DNA double-strand break repair: how to fix a broken relationship. *Cellular and Molecular Life Sciences : CMLS*, *66*(6), 1039–1056. DOI: 10.1007/s00018-009-8740-3 PMID: 19153654

Park, G., Rim, Y. A., Sohn, Y., Nam, Y., & Ju, J. H. (2024). Replacing Animal Testing with Stem Cell-Organoids: Advantages and Limitations. *Stem Cell Reviews and Reports*, *20*(6), 1375–1386. DOI: 10.1007/s12015-024-10723-5 PMID: 38639829

Park, H. J., Jeong, H., Park, Y. H., Yim, H. W., Ha, U. S., Hong, S. H., Kim, S. W., Kim, N. J., & Lee, J. Y. (2019). Adipose Tissue-Derived Stem Cell Therapy for Cavernous Nerve Injury-Induced Erectile Dysfunction in the Rat Model: A Systematic Review and Meta-Analysis Using Methodological Quality Assessment. *International Journal of Stem Cells*, *12*(2), 206–217. DOI: 10.15283/ijsc18122 PMID: 31022999

Park, T. S., Galic, Z., Conway, A. E., Lindgren, A., Van Handel, B. J., Magnusson, M., Richter, L., Teitell, M. A., Mikkola, H. K. A., Lowry, W. E., Plath, K., & Clark, A. T. (2009). Derivation of primordial germ cells from human embryonic and induced pluripotent stem cells is significantly improved by coculture with human fetal gonadal cells. *Stem Cells (Dayton, Ohio)*, *27*(4), 783–795. DOI: 10.1002/stem.13 PMID: 19350678

Park, W. B., Kim, S. Y., Lee, S. H., Kim, H.-W., Park, J.-S., & Hyun, J. K. (2010). The effect of mesenchymal stem cell transplantation on the recovery of bladder and hindlimb function after spinal cord contusion in rats. *BMC Neuroscience*, *11*(1), 119. DOI: 10.1186/1471-2202-11-119 PMID: 20846445

Parreira, G. G., Ogawa, T., Avarbock, M. R., França, L. R., Brinster, R. L., & Russell, L. D. (1998). Development of germ cell transplants in mice. *Biology of Reproduction*, *59*(6), 1360–1370. DOI: 10.1095/biolreprod59.6.1360 PMID: 9828179

Patel, D. P., Pastuszak, A. W., & Hotaling, J. M. (2019). Emerging Treatments for Erectile Dysfunction: A Review of Novel, Non-surgical Options. En *Current Urology Reports* (Vol. 20, Número 8). Current Medicine Group LLC 1. DOI: 10.1007/s11934-019-0908-2

Peired, A. J., Sisti, A., & Romagnani, P. (2016). Mesenchymal Stem Cell-Based Therapy for Kidney Disease: A Review of Clinical Evidence. *Stem Cells International*, *4798639*(1), 4798639. Advance online publication. DOI: 10.1155/2016/4798639 PMID: 27721835

Pereira, E. M., Labilloy, A., Eshbach, M. L., Roy, A., Subramanya, A. R., Monte, S., Labilloy, G., & Weisz, O. A. (2016). Characterization and phosphoproteomic analysis of a human immortalized podocyte model of Fabry disease generated using CRISPR/Cas9 technology. *American Journal of Physiology. Renal Physiology*, *311*(5), F1015–F1024. DOI: 10.1152/ajprenal.00283.2016 PMID: 27681560

Pérez-Aizpurua, X., Garranzo-Ibarrola, M., Simón-Rodríguez, C., García-Cardoso, J. V., Chávez-Roa, C., López-Martín, L., Tufet i Jaumot, J. J., Alonso-Román, J., Maqueda-Arellano, J., Gómez-Jordana, B., Ruiz de Castroviejo-Blanco, J., Osorio-Ospina, F., González-Enguita, C., & García-Arranz, M. (2023). Stem Cell Therapy for Erectile Dysfunction: A Step towards a Future Treatment. *Life (Chicago, Ill.)*, *13*(2), 502. Advance online publication. DOI: 10.3390/life13020502 PMID: 36836859

Perin, L., Giuliani, S., Jin, D., Sedrakyan, S., Carraro, G., Habibian, R., Warburton, D., Atala, A., & De Filippo, R. E. (2007). Renal differentiation of amniotic fluid stem cells. *Cell Proliferation*, *40*(6), 936–948. DOI: 10.1111/j.1365-2184.2007.00478.x PMID: 18021180

Perin, L., Sedrakyan, S., Giuliani, S., Da Sacco, S., Carraro, G., Shiri, L., Lemley, K. V., Rosol, M., Wu, S., Atala, A., Warburton, D., & De Filippo, R. E. (2010). Protective effect of human amniotic fluid stem cells in an immunodeficient mouse model of acute tubular necrosis. *PLoS One*, *5*(2), e9357. DOI: 10.1371/journal.pone.0009357 PMID: 20195358

Pesce, M., Orlandi, A., Iachininoto, M. G., Straino, S., Torella, A. R., Rizzuti, V., Pompilio, G., Bonanno, G., Scambia, G., & Capogrossi, M. C. (2003). Myoendothelial differentiation of human umbilical cord blood-derived stem cells in ischemic limb tissues. *Circulation Research*, *93*(5). Advance online publication. DOI: 10.1161/01.RES.0000090624.04507.45 PMID: 12919944

Peters, K. M., Dmochowski, R. R., Carr, L. K., Robert, M., Kaufman, M. R., Sirls, L. T., Herschorn, S., Birch, C., Kultgen, P. L., & Chancellor, M. B. (2014). Autologous muscle derived cells for treatment of stress urinary incontinence in women. *The Journal of Urology*, *192*(2), 469–476. DOI: 10.1016/j.juro.2014.02.047 PMID: 24582537

Pleniceanu, O., Omer, D., Harari-Steinberg, O., & Dekel, B. (2018). Renal lineage cells as a source for renal regeneration. *Pediatric Research*, *83*(1-2), 267–274. DOI: 10.1038/pr.2017.255 PMID: 28985199

Pokrywczynska, M., Adamowicz, J., Czapiewska, M., Balcerczyk, D., Jundzill, A., Nowacki, M., Petros, P., & Drewa, T. (2016). Targeted therapy for stress urinary incontinence: A systematic review based on clinical trials. *Expert Opinion on Biological Therapy*, *16*(2), 233–242. DOI: 10.1517/14712598.2016.1118459 PMID: 26560683

Portilla Sanchez, R., Blanco, F. L., Santamarina, A., Casals Roa, J., Mata, J., & Kaufman, A. (1958). Vesical regeneration in the human after total cystectomy and implantation of a plastic mould. *British Journal of Urology*, *30*(2), 180–188. DOI: 10.1111/j.1464-410X.1958.tb06231.x PMID: 13546597

Preynat-Seauve, O., & Krause, K.-H. (2011). Stem cell sources for regenerative medicine: The immunological point of view. *Seminars in Immunopathology*, *33*(6), 519–524. DOI: 10.1007/s00281-011-0271-y PMID: 21598094

Protogerou, V., El Beshari, S., Michalopoulos, E., Mallis, P., Chrysikos, D., Samolis, A. A., Stavropoulos-Giokas, C., & Troupis, T. (2020a). The Combined Use of Stem Cells and Platelet Lysate Plasma for the Treatment of Erectile Dysfunction: A Pilot Study–6 Months Results. *Medicines (Basel, Switzerland)*, *7*(3), 14. DOI: 10.3390/medicines7030014 PMID: 32197323

Protogerou, V., Michalopoulos, E., Mallis, P., Gontika, I., Dimou, Z., Liakouras, C., Stavropoulos-Giokas, C., Kostakopoulos, N., Chrisofos, M., & Deliveliotis, C. (2019). Administration of Adipose Derived Mesenchymal Stem Cells and Platelet Lysate in Erectile Dysfunction: A Single Center Pilot Study. *Bioengineering (Basel, Switzerland)*, *6*(1), 21. DOI: 10.3390/bioengineering6010021 PMID: 30841525

Qin, D., Long, T., Deng, J., & Zhang, Y. (2014). Urine-derived stem cells for potential use in bladder repair. *Stem Cell Research & Therapy*, *5*(3), 69. DOI: 10.1186/scrt458 PMID: 25157812

Qiu, X., Villalta, J., Ferretti, L., Fandel, T. M., Albersen, M., Lin, G., Dai, Y., Lue, T. F., & Lin, C. S. (2012). Effects of Intravenous Injection of Adipose-Derived Stem Cells in a Rat Model of Radiation Therapy-Induced Erectile Dysfunction. *Journal of Sexual Medicine*, *9*(7), 1834–1841. DOI: 10.1111/j.1743-6109.2012.02753.x PMID: 22548750

Qi, X., Zhang, J., Yuan, H., Xu, Z., Li, Q., Niu, X., Hu, B., Wang, Y., & Li, X. (2016). Exosomes Secreted by Human-Induced Pluripotent Stem Cell-Derived Mesenchymal Stem Cells Repair Critical-Sized Bone Defects through Enhanced Angiogenesis and Osteogenesis in Osteoporotic Rats. *International Journal of Biological Sciences*, *12*(7), 836–849. DOI: 10.7150/ijbs.14809 PMID: 27313497

Quimby, J. M., Webb, T. L., Randall, E., Marolf, A., Valdes-Martinez, A., & Dow, S. W. (2016). Assessment of intravenous adipose-derived allogeneic mesenchymal stem cells for the treatment of feline chronic kidney disease: A randomized, placebo-controlled clinical trial in eight cats. *Journal of Feline Medicine and Surgery*, *18*(2), 165–171. DOI: 10.1177/1098612X15576980 PMID: 25784460

Qu-Petersen, Z., Deasy, B., Jankowski, R., Ikezawa, M., Cummins, J., Pruchnic, R., Mytinger, J., Cao, B., Gates, C., Wernig, A., & Huard, J. (2002). Identification of a novel population of muscle stem cells in mice: Potential for muscle regeneration. *The Journal of Cell Biology*, *157*(5), 851–864. DOI: 10.1083/jcb.200108150 PMID: 12021255

Radhakrishnan, G. (2017). Stem cells—The new agents in infertility treatment: The light at the end of the tunnel? *Fertility Science and Research*, *4*(2), 70. DOI: 10.4103/fsr.fsr_16_18

Ranghino, A., Bruno, S., Bussolati, B., Moggio, A., Dimuccio, V., Tapparo, M., Biancone, L., Gontero, P., Frea, B., & Camussi, G. (2017). The effects of glomerular and tubular renal progenitors and derived extracellular vesicles on recovery from acute kidney injury. *Stem Cell Research & Therapy*, *8*(1), 24. DOI: 10.1186/s13287-017-0478-5 PMID: 28173878

Ratajczak, M. Z., Ratajczak, J., & Kucia, M. (2019). Very Small Embryonic-Like Stem Cells (VSELs): An Update and Future Directions. *Circulation Research*, *124*(2), 208–210. DOI: 10.1161/CIRCRESAHA.118.314287 PMID: 30653438

Ratajczak, M. Z., Zuba-Surma, E., Wojakowski, W., Suszynska, M., Mierzejewska, K., Liu, R., Ratajczak, J., Shin, D. M., & Kucia, M. (2014). Very small embryonic-like stem cells (VSELs) represent a real challenge in stem cell biology: Recent pros and cons in the midst of a lively debate. *Leukemia*, *28*(3), 473–484. DOI: 10.1038/leu.2013.255 PMID: 24018851

Ratliff, B. B., & Goligorsky, M. S. (2013). Delivery of EPC embedded in HA-hydrogels for treatment of acute kidney injury. *Biomatter*, *3*(1), e23284. DOI: 10.4161/biom.23284 PMID: 23507925

Ravindran, S., Song, Y., & George, A. (2010). Development of three-dimensional biomimetic scaffold to study epithelial-mesenchymal interactions. *Tissue Engineering. Part A*, *16*(1), 327–342. DOI: 10.1089/ten.tea.2009.0110 PMID: 19712044

Raya-Rivera, A., Esquiliano, D. R., Yoo, J. J., Lopez-Bayghen, E., Soker, S., & Atala, A. (2011). Tissue-engineered autologous urethras for patients who need reconstruction: An observational study. *Lancet*, *377*(9772), 1175–1182. DOI: 10.1016/S0140-6736(10)62354-9 PMID: 21388673

Ribeiro-Filho, L. A., & Sievert, K.-D. (2015). Acellular matrix in urethral reconstruction. *Advanced Drug Delivery Reviews*, *82-83*, 38–46. DOI: 10.1016/j.addr.2014.11.019 PMID: 25477304

Rice, J. J., Martino, M. M., De Laporte, L., Tortelli, F., Briquez, P. S., & Hubbell, J. A. (2013). Engineering the regenerative microenvironment with biomaterials. *Advanced Healthcare Materials*, *2*(1), 57–71. DOI: 10.1002/adhm.201200197 PMID: 23184739

Rider, D. A., Dombrowski, C., Sawyer, A. A., Ng, G. H. B., Leong, D., Hutmacher, D. W., Nurcombe, V., & Cool, S. M. (2008). Autocrine fibroblast growth factor 2 increases the multipotentiality of human adipose-derived mesenchymal stem cells. *Stem Cells (Dayton, Ohio)*, *26*(6), 1598–1608. DOI: 10.1634/stemcells.2007-0480 PMID: 18356575

Rigaud, J., & Le Normand, L. (2004). Entérocystoplastie d'agrandissement. *Annales d'Urologie*, *38*(6), 298–310. DOI: 10.1016/j.anuro.2004.08.001 PMID: 15651484

Robinson, D., & Cardozo, L. (2011). Estrogens and the lower urinary tract. *Neurourology and Urodynamics*, *30*(5), 754–757. DOI: 10.1002/nau.21106 PMID: 21661025

Rocha, V., Wagner, J. E.Jr, Sobocinski, K. A., Klein, J. P., Zhang, M.-J., Horowitz, M. M., & Gluckman, E. (2000). Graft-Versus-Host Disease in Children Who Have Received a Cord-Blood or Bone Marrow Transplant from an HLA-Identical Sibling. *The New England Journal of Medicine*, *342*(25), 1846–1854. DOI: 10.1056/NEJM200006223422501 PMID: 10861319

Rodríguez, L. V., Alfonso, Z., Zhang, R., Leung, J., Wu, B., & Ignarro, L. J. (2006). Clonogenic multipotent stem cells in human adipose tissue differentiate into functional smooth muscle cells. *Proceedings of the National Academy of Sciences of the United States of America*, *103*(32), 12167–12172. DOI: 10.1073/pnas.0604850103 PMID: 16880387

Roman, S., Mangir, N., Bissoli, J., Chapple, C. R., & MacNeil, S. (2016). Biodegradable scaffolds designed to mimic fascia-like properties for the treatment of pelvic organ prolapse and stress urinary incontinence. *Journal of Biomaterials Applications*, *30*(10), 1578–1588. DOI: 10.1177/0885328216633373 PMID: 26896234

Roobrouck, V. D., Ulloa-Montoya, F., & Verfaillie, C. M. (2008). Self-renewal and differentiation capacity of young and aged stem cells. *Experimental Cell Research*, *314*(9), 1937–1944. DOI: 10.1016/j.yexcr.2008.03.006 PMID: 18439579

Rortveit, G., Hannestad, Y. S., Kjersti Daltveit, A., & Hunskaar, S. (2001). Age- and type-dependent effects of parity on urinary incontinence: The Norwegian EPINCONT study. *Obstetrics and Gynecology*, *98*(6), 1004–1010. DOI: 10.1097/00006250-200112000-00004 PMID: 11755545

Rossant, J. (2001). Stem cells from the Mammalian blastocyst. *Stem Cells (Dayton, Ohio)*, *19*(6), 477–482. DOI: 10.1634/stemcells.19-6-477 PMID: 11713338

Rossi, F., & van Griensven, M. (2014). Polymer functionalization as a powerful tool to improve scaffold performances. *Tissue Engineering. Part A*, *20*(15-16), 2043–2051. DOI: 10.1089/ten.tea.2013.0367 PMID: 24206079

Rota, C., Imberti, B., Pozzobon, M., Piccoli, M., De Coppi, P., Atala, A., Gagliardini, E., Xinaris, C., Benedetti, V., Fabricio, A. S. C., Squarcina, E., Abbate, M., Benigni, A., Remuzzi, G., & Morigi, M. (2012). Human amniotic fluid stem cell preconditioning improves their regenerative potential. *Stem Cells and Development*, *21*(11), 1911–1923. DOI: 10.1089/scd.2011.0333 PMID: 22066606

Rouet, P., Smih, F., & Jasin, M. (1994). Introduction of double-strand breaks into the genome of mouse cells by expression of a rare-cutting endonuclease. *Molecular and Cellular Biology*, *14*(12), 8096–8106. DOI: 10.1128/mcb.14.12.8096-8106.1994 PMID: 7969147

Rudin, N., Sugarman, E., & Haber, J. E. (1989). Genetic and physical analysis of double-strand break repair and recombination in Saccharomyces cerevisiae. *Genetics*, *122*(3), 519–534. DOI: 10.1093/genetics/122.3.519 PMID: 2668114

Ruggenenti, P., Perticucci, E., Cravedi, P., Gambara, V., Costantini, M., Sharma, S. K., Perna, A., & Remuzzi, G. (2008). Role of remission clinics in the longitudinal treatment of CKD. *Journal of the American Society of Nephrology*, *19*(6), 1213–1224. DOI: 10.1681/ASN.2007090970 PMID: 18354029

Ryu, C. M., Shin, J. H., Yu, H. Y., Ju, H., Kim, S., Lim, J., Heo, J., Lee, S., Shin, D. M., & Choo, M. S. (2019). N-acetylcysteine prevents bladder tissue fibrosis in a lipopolysaccharide-induced cystitis rat model. *Scientific Reports*, *9*(1), 8134. Advance online publication. DOI: 10.1038/s41598-019-44631-3 PMID: 31148586

Ryu, C. M., Yu, H. Y., Lee, H. Y., Shin, J. H., Lee, S., Ju, H., Paulson, B., Lee, S., Kim, S., Lim, J., Heo, J., Hong, K. S., Chung, H. M., Kim, J. K., Shin, D. M., & Choo, M. S. (2018). Longitudinal intravital imaging of transplanted mesenchymal stem cells elucidates their functional integration and therapeutic potency in an animal model of interstitial cystitis/bladder pain syndrome. *Theranostics*, *8*(20), 5610–5624. DOI: 10.7150/thno.27559 PMID: 30555567

Ryu, J. K., Kim, D. H., Song, K. M., Ryu, D. S., Kim, S. N., Shin, D. H., Yi, T., Suh, J. K., & Song, S. U. (2016). Intracavernous delivery of clonal mesenchymal stem cells rescues erectile function in the streptozotocin-induced diabetic mouse. *Andrology*, *4*(1), 172–184. DOI: 10.1111/andr.12138 PMID: 26711324

Saad, A., Dietz, A. B., Herrmann, S. M. S., Hickson, L. J., Glockner, J. F., McKusick, M. A., Misra, S., Bjarnason, H., Armstrong, A. S., Gastineau, D. A., Lerman, L. O., & Textor, S. C. (2017). Autologous Mesenchymal Stem Cells Increase Cortical Perfusion in Renovascular Disease. *Journal of the American Society of Nephrology*, *28*(9), 2777–2785. DOI: 10.1681/ASN.2017020151 PMID: 28461553

Sabatino, M., Ren, J., David-Ocampo, V., England, L., McGann, M., Tran, M., Kuznetsov, S. A., Khuu, H., Balakumaran, A., Klein, H. G., Robey, P. G., & Stroncek, D. F. (2012). The establishment of a bank of stored clinical bone marrow stromal cell products. *Journal of Translational Medicine*, *10*(1), 23. DOI: 10.1186/1479-5876-10-23 PMID: 22309358

Sack, B. S., Mauney, J. R., & Estrada, C. R.Jr. (2016). Silk Fibroin Scaffolds for Urologic Tissue Engineering. *Current Urology Reports*, *17*(2), 16. DOI: 10.1007/s11934-015-0567-x PMID: 26801192

Sadri-Ardekani, H., Mizrak, S. C., Van Daalen, S. K. M., Korver, C. M., Roepers-Gajadien, H. L., Koruji, M., Hovingh, S., De Reijke, T. M., De La Rosette, J. J. M. C. H., Van Der Veen, F., De Rooij, D. G., Repping, S., & Van Pelt, A. M. M. (2009). Propagation of human spermatogonial stem cells in vitro. *Journal of the American Medical Association*, *302*(19), 2127–2134. DOI: 10.1001/jama.2009.1689 PMID: 19920237

Saks, E. K., & Arya, L. A. (2009). Pharmacologic Management of Urinary Incontinence, Voiding Dysfunction, and Overactive Bladder. En *Obstetrics and Gynecology Clinics of North America* (Vol. 36, Número 3, pp. 493-507). Obstet Gynecol Clin North Am. DOI: 10.1016/j.ogc.2009.08.001

Sakuma, T., Matsumoto, T., Kano, K., Fukuda, N., Obinata, D., Yamaguchi, K., Yoshida, T., Takahashi, S., & Mugishima, H. (2009). Mature, adipocyte derived, dedifferentiated fat cells can differentiate into smooth muscle-like cells and contribute to bladder tissue regeneration. *The Journal of Urology*, *182*(1), 355–365. DOI: 10.1016/j.juro.2009.02.103 PMID: 19457498

Salehi-Pourmehr, H., Hajebrahimi, S., Rahbarghazi, R., Pashazadeh, F., Mahmoudi, J., Maasoumi, N., & Sadigh-Eteghad, S. (2020). Stem Cell Therapy for Neurogenic Bladder Dysfunction in Rodent Models: A Systematic Review. *International Neurourology Journal*, *24*(3), 241–257. DOI: 10.5213/inj.2040058.029 PMID: 33017895

Salehi-Pourmehr, H., Rahbarghazi, R., Mahmoudi, J., Roshangar, L., Chapple, C. R., Hajebrahimi, S., Abolhasanpour, N., & Azghani, M.-R. (2019). Intra-bladder wall transplantation of bone marrow mesenchymal stem cells improved urinary bladder dysfunction following spinal cord injury. *Life Sciences*, *221*, 20–28. DOI: 10.1016/j.lfs.2019.02.011 PMID: 30735734

Salem, H. K., & Thiemermann, C. (2010). Mesenchymal stromal cells: Current understanding and clinical status. *Stem Cells (Dayton, Ohio)*, *28*(3), 585–596. DOI: 10.1002/stem.269 PMID: 19967788

Samsonraj, R. M., Raghunath, M., Nurcombe, V., Hui, J. H., van Wijnen, A. J., & Cool, S. M. (2017). Concise Review: Multifaceted Characterization of Human Mesenchymal Stem Cells for Use in Regenerative Medicine. *Stem Cells Translational Medicine*, *6*(12), 2173–2185. DOI: 10.1002/sctm.17-0129 PMID: 29076267

Sanchez-Ramos, J. R., Song, S., Kamath, S. G., Zigova, T., Willing, A., Cardozo-Pelaez, F., Stedeford, T., Chopp, M., & Sanberg, P. R. (2001). Expression of neural markers in human umbilical cord blood. *Experimental Neurology*, *171*(1), 109–115. DOI: 10.1006/exnr.2001.7748 PMID: 11520125

Sandner, B., Ciatipis, M., Motsch, M., Soljanik, I., Weidner, N., & Blesch, A. (2016). Limited Functional Effects of Subacute Syngeneic Bone Marrow Stromal Cell Transplantation After Rat Spinal Cord Contusion Injury. *Cell Transplantation*, *25*(1), 125–139. DOI: 10.3727/096368915X687679 PMID: 25812176

Sangkum, P. (2016). Research highlights on stem cell therapy for the treatment of Peyronie's disease. *Translational Andrology and Urology*, *5*(3), 363–365. DOI: 10.21037/tau.2016.03.14 PMID: 27298783

Santeramo, I., Herrera Perez, Z., Illera, A., Taylor, A., Kenny, S., Murray, P., Wilm, B., & Gretz, N. (2017). Human Kidney-Derived Cells Ameliorate Acute Kidney Injury Without Engrafting into Renal Tissue. *Stem Cells Translational Medicine*, *6*(5), 1373–1384. DOI: 10.1002/sctm.16-0352 PMID: 28375556

Sato, T., Katagiri, K., Gohbara, A., Inoue, K., Ogonuki, N., Ogura, A., Kubota, Y., & Ogawa, T. (2011). In vitro production of functional sperm in cultured neonatal mouse testes. *Nature*, *471*(7339), 504–508. DOI: 10.1038/nature09850 PMID: 21430778

Sauerwein, D. (1990). [Surgical treatment of spastic bladder paralysis in paraplegic patients. Sacral deafferentation with implantation of a sacral anterior root stimulator]. *Der Urologe. Ausg. A*, *29*(4), 196–203. PMID: 2399626

Schäfer, F.-M., & Stehr, M. (2018). Tissue engineering in pediatric urology—A critical appraisal. *Innovative Surgical Sciences*, *3*(2), 107–118. DOI: 10.1515/iss-2018-0011 PMID: 31579774

Schlatt, S., Foppiani, L., Rolf, C., Weinbauer, G. F., & Nieschlag, E. (2002). Germ cell transplantation into X-irradiated monkey testes. *Human Reproduction (Oxford, England)*, *17*(1), 55–62. DOI: 10.1093/humrep/17.1.55 PMID: 11756362

Schwarzer, U., Sommer, F., Klotz, T., Braun, M., Reifenrath, B., & Engelmann, U. (2001). The prevalence of Peyronie's disease: Results of a large survey. *BJU International*, *88*(7), 727–730. DOI: 10.1046/j.1464-4096.2001.02436.x PMID: 11890244

Sèbe, P., Doucet, C., Cornu, J. N., Ciofu, C., Costa, P., De Medina, S. G. D., Pinset, C., & Haab, F. (2011). Intrasphincteric injections of autologous muscular cells in women with refractory stress urinary incontinence: A prospective study. *International Urogynecology Journal*, *22*(2), 183–189. DOI: 10.1007/s00192-010-1255-5 PMID: 20821309

Sedrakyan, S., Da Sacco, S., Milanesi, A., Shiri, L., Petrosyan, A., Varimezova, R., Warburton, D., Lemley, K. V., De Filippo, R. E., & Perin, L. (2012). Injection of amniotic fluid stem cells delays progression of renal fibrosis. *Journal of the American Society of Nephrology*, *23*(4), 661–673. DOI: 10.1681/ASN.2011030243 PMID: 22302195

Sedrakyan, S., Villani, V., Da Sacco, S., Tripuraneni, N., Porta, S., Achena, A., Lavarreda-Pearce, M., Petrosyan, A., Soloyan, H., Filippo, R. E. D., Bussolati, B., & Perin, L. (2017). Amniotic fluid stem cell-derived vesicles protect from VEGF-induced endothelial damage. *Scientific Reports*, *7*(1), 16875. DOI: 10.1038/s41598-017-17061-2 PMID: 29203902

Semedo, P., Correa-Costa, M., Antonio Cenedeze, M., Maria Avancini Costa Malheiros, D., Antonia dos Reis, M., Shimizu, M. H., Seguro, A. C., Pacheco-Silva, A., & Saraiva Camara, N. O. (2009). Mesenchymal stem cells attenuate renal fibrosis through immune modulation and remodeling properties in a rat remnant kidney model. *Stem Cells (Dayton, Ohio)*, *27*(12), 3063–3073. DOI: 10.1002/stem.214 PMID: 19750536

Semedo, P., Wang, P. M., Andreucci, T. H., Cenedeze, M. A., Teixeira, V. P. A., Reis, M. A., Pacheco-Silva, A., & Câmara, N. O. S. (2007). Mesenchymal stem cells ameliorate tissue damages triggered by renal ischemia and reperfusion injury. *Transplantation Proceedings*, *39*(2), 421–423. DOI: 10.1016/j.transproceed.2007.01.036 PMID: 17362746

Semmelink, H. J., de Wilde, P. C., van Houwelingen, J. C., & Vooijs, G. P. (1990). Histomorphometric study of the lower urogenital tract in pre- and post-menopausal women. *Cytometry, 11*(6), 700–707. DOI: 10.1002/cyto.990110607 PMID: 2384029

Semmens, J. P., & Wagner, G. (1982). Estrogen deprivation and vaginal function in postmenopausal women. *Journal of the American Medical Association, 248*(4), 445–448. DOI: 10.1001/jama.1982.03330040033027 PMID: 6283189

Seval, M. M., & Koyuncu, K. (2022). Current status of stem cell treatments and innovative approaches for stress urinary incontinence. *Frontiers in Medicine, 9*(December 2009). DOI: 10.3389/fmed.2022.1073758

Shah, J. S., Sabouni, R., Cayton Vaught, K. C., Owen, C. M., Albertini, D. F., & Segars, J. H. (2018). Biomechanics and mechanical signaling in the ovary: A systematic review. *Journal of Assisted Reproduction and Genetics, 35*(7), 1135–1148. DOI: 10.1007/s10815-018-1180-y PMID: 29691711

Shan, H. T., Zhang, H. B., Chen, W. T., Chen, F. Z., Wang, T., Luo, J. T., Yue, M., Lin, J. H., & Wei, A. Y. (2017). Combination of low-energy shock-wave therapy and bone marrow mesenchymal stem cell transplantation to improve the erectile function of diabetic rats. *Asian Journal of Andrology, 19*(1), 26–33. DOI: 10.4103/1008-682X.184271 PMID: 27427555

Sharifiaghdas, F., Tajalli, F., Taheri, M., Naji, M., Moghadasali, R., Aghdami, N., Baharvand, H., Azimian, V., & Jaroughi, N. (2016). Effect of autologous muscle-derived cells in the treatment of urinary incontinence in female patients with intrinsic sphincter deficiency and epispadias: A prospective study. *International journal of urology : official journal of the Japanese Urological Association, 23*(7), 581-586. DOI: 10.1111/iju.13097

Sharma, S., Wistuba, J., Pock, T., Schlatt, S., & Neuhaus, N. (2019). Spermatogonial stem cells: Updates from specification to clinical relevance. *Human Reproduction Update, 25*(3), 275–297. DOI: 10.1093/humupd/dmz006 PMID: 30810745

Sheashaa, H., Lotfy, A., Elhusseini, F., Aziz, A. A., Baiomy, A., Awad, S., Alsayed, A., El-Gilany, A.-H., Saad, M.-A. A. A., Mahmoud, K., Zahran, F., Salem, D. A., Sarhan, A., Ghaffar, H. A., & Sobh, M. (2016). Protective effect of adipose-derived mesenchymal stem cells against acute kidney injury induced by ischemia-reperfusion in Sprague-Dawley rats. *Experimental and Therapeutic Medicine, 11*(5), 1573–1580. DOI: 10.3892/etm.2016.3109 PMID: 27168775

Shi, J.-G., Fu, W.-J., Wang, X.-X., Xu, Y.-D., Li, G., Hong, B.-F., Hu, K., Cui, F.-Z., Wang, Y., & Zhang, X. (2012). Transdifferentiation of human adipose-derived stem cells into urothelial cells: Potential for urinary tract tissue engineering. *Cell and Tissue Research*, *347*(3), 737–746. DOI: 10.1007/s00441-011-1317-0 PMID: 22290635

Shi, M., Liu, Z.-W., & Wang, F.-S. (2011). Immunomodulatory properties and therapeutic application of mesenchymal stem cells. *Clinical and Experimental Immunology*, *164*(1), 1–8. DOI: 10.1111/j.1365-2249.2011.04327.x PMID: 21352202

Shin, J. H., Ryu, C. M., Yu, H. Y., Shin, D. M., & Choo, M. S. (2020a). Current and Future Directions of Stem Cell Therapy for Bladder Dysfunction. En *Stem Cell Reviews and Reports* (Vol. 16, Número 1, pp. 82-93). Springer. DOI: 10.1007/s12015-019-09922-2

Shin, J. H., Park, J. H., Ryu, C.-M., Shin, D.-M., & Choo, M.-S. (2024). Stem cell therapy for interstitial cystitis/bladder pain syndrome. *Lower Urinary Tract Symptoms*, *16*(4), e12527. DOI: 10.1111/luts.12527 PMID: 38867432

Shirvan, M. K., Alamdari, D. H., Mahboub, M. D., Ghanadi, A., Rahimi, H. R., & Seifalian, A. M. (2013). A novel cell therapy for stress urinary incontinence, short-term outcome. *Neurourology and Urodynamics*, *32*(4), 377–382. DOI: 10.1002/nau.22301 PMID: 22972395

Siegel, N., Valli, A., Fuchs, C., Rosner, M., & Hengstschläger, M. (2009). Induction of mesenchymal/epithelial marker expression in human amniotic fluid stem cells. *Reproductive Biomedicine Online*, *19*(6), 838–846. DOI: 10.1016/j.rbmo.2009.09.015 PMID: 20031026

Simoni, M., & Taylor, H. S. (2018). Therapeutic strategies involving uterine stem cells in reproductive medicine. *Current Opinion in Obstetrics & Gynecology*, *30*(3), 209–216. DOI: 10.1097/GCO.0000000000000457 PMID: 29652725

Singh, M., Kapoor, R., Kapoor, D., Kapoor, R., Srivastav, A., & Chipde, S. (2013). Dorsal onlay vaginal graft urethroplasty for female urethral stricture. *Indian Journal of Urology : IJU : Journal of the Urological Society of India*, *29*(2), 124–128. DOI: 10.4103/0970-1591.21041 PMID: 23956514

Sivaraman, S., Amoroso, N., Gu, X., Purves, J. T., Hughes, F. M.Jr, Wagner, W. R., & Nagatomi, J. (2019). Evaluation of Poly (Carbonate-Urethane) Urea (PCUU) Scaffolds for Urinary Bladder Tissue Engineering. *Annals of Biomedical Engineering*, *47*(3), 891–901. DOI: 10.1007/s10439-018-02182-0 PMID: 30542784

Smolar, J., Salemi, S., Horst, M., Sulser, T., & Eberli, D. (2016). Stem Cells in Functional Bladder Engineering. *Transfusion Medicine and Hemotherapy; Offizielles Organ der Deutschen Gesellschaft für Transfusionsmedizin und Immunhämatologie*, *43*(5), 328–335. DOI: 10.1159/000447977 PMID: 27781020

Soler, R., Füllhase, C., Hanson, A., Campeau, L., Santos, C., & Andersson, K.-E. (2012). Stem cell therapy ameliorates bladder dysfunction in an animal model of Parkinson disease. *The Journal of Urology*, *187*(4), 1491–1497. DOI: 10.1016/j.juro.2011.11.079 PMID: 22341818

Song, B., Smink, A. M., Jones, C. V., Callaghan, J. M., Firth, S. D., Bernard, C. A., Laslett, A. L., Kerr, P. G., & Ricardo, S. D. (2012). The directed differentiation of human iPS cells into kidney podocytes. *PLoS One*, *7*(9), e46453. DOI: 10.1371/journal.pone.0046453 PMID: 23029522

Song, H.-W., & Wilkinson, M. F. (2012). In vitro spermatogenesis: A long journey to get tails. *Spermatogenesis*, *2*(4), 238–244. DOI: 10.4161/spmg.22069 PMID: 23248764

Song, K. M., Jin, H. R., Park, J. M., Choi, M. J., Kwon, M. H., Kwon, K. D., Batbold, D., Yin, G. N., Kim, W. J., Koh, G. Y., Ryu, J. K., & Suh, J. K. (2014). Intracavernous delivery of stromal vascular fraction restores erectile function through production of angiogenic factors in a mouse model of cavernous nerve injury. *Journal of Sexual Medicine*, *11*(8), 1962–1973. DOI: 10.1111/jsm.12597 PMID: 24902866

Song, M., Heo, J., Chun, J. Y., Bae, H. S., Kang, J. W., Kang, H., Cho, Y. M., Kim, S. W., Shin, D. M., & Choo, M. S. (2014). The paracrine effects of mesenchymal stem cells stimulate the regeneration capacity of endogenous stem cells in the repair of a bladder-outlet-obstruction-induced overactive bladder. *Stem Cells and Development*, *23*(6), 654–663. DOI: 10.1089/scd.2013.0277 PMID: 24192209

Song, M., Lim, J., Yu, H. Y., Park, J., Chun, J. Y., Jeong, J., Heo, J., Kang, H., Kim, Y., Cho, Y. M., Kim, S. W., Oh, W., Choi, S. J., Jang, S. W., Park, S., Shin, D. M., & Choo, M. S. (2015). Mesenchymal Stem Cell Therapy Alleviates Interstitial Cystitis by Activating Wnt Signaling Pathway. *Stem Cells and Development*, *24*(14), 1648–1657. DOI: 10.1089/scd.2014.0459 PMID: 25745847

Song, M., Yu, H. Y., Chun, J. Y., Shin, D. M., Song, S. H., Choo, M. S., & Song, Y. S. (2016). The fibrosis of ketamine, a noncompetitive N-methyl-D-aspartic acid receptor antagonist dose-dependent change in a ketamine-induced cystitis rat model. *Drug and Chemical Toxicology*, *39*(2), 206–212. DOI: 10.3109/01480545.2015.1079916 PMID: 26324997

Song, Y., Peng, C., Lv, S., Cheng, J., Liu, S., Wen, Q., Guan, G., & Liu, G. (2017). Adipose-derived stem cells ameliorate renal interstitial fibrosis through inhibition of EMT and inflammatory response via TGF-β1 signaling pathway. *International Immunopharmacology*, *44*, 115–122. DOI: 10.1016/j.intimp.2017.01.008 PMID: 28092863

Spaggiari, G. M., Capobianco, A., Becchetti, S., Mingari, M. C., & Moretta, L. (2006). Mesenchymal stem cell-natural killer cell interactions: Evidence that activated NK cells are capable of killing MSCs, whereas MSCs can inhibit IL-2-induced NK-cell proliferation. *Blood*, *107*(4), 1484–1490. DOI: 10.1182/blood-2005-07-2775 PMID: 16239427

Stagg, J. (2007). Immune regulation by mesenchymal stem cells: Two sides to the coin. *Tissue Antigens*, *69*(1), 1–9. DOI: 10.1111/j.1399-0039.2006.00739.x PMID: 17212702

Stanasel, I., Mirzazadeh, M., & Smith, J. J. (2010). Bladder tissue engineering. En *Urologic Clinics of North America* (Vol. 37, Número 4, pp. 593-599). Urol Clin North Am. DOI: 10.1016/j.ucl.2010.06.008

Stangel-Wojcikiewicz, K., Jarocha, D., Piwowar, M., Jach, R., Uhl, T., Basta, A., & Majka, M. (2014). Autologous muscle-derived cells for the treatment of female stress urinary incontinence: A 2-year follow-up of a polish investigation. *Neurourology and Urodynamics*, *33*(3), 324–330. DOI: 10.1002/nau.22404 PMID: 23606303

Steers, W. D., & De Groat, W. C. (1988). Effect of bladder outlet obstruction on micturition reflex pathways in the rat. *The Journal of Urology*, *140*(4), 864–871. DOI: 10.1016/S0022-5347(17)41846-5 PMID: 3418824

Stimpfel, M., Skutella, T., Cvjeticanin, B., Meznaric, M., Dovc, P., Novakovic, S., Cerkovnik, P., Vrtacnik-Bokal, E., & Virant-Klun, I. (2013). Isolation, characterization and differentiation of cells expressing pluripotent/multipotent markers from adult human ovaries. *Cell and Tissue Research*, *354*(2), 593–607. DOI: 10.1007/s00441-013-1677-8 PMID: 23820736

Strasser, H., Marksteiner, R., Margreiter, E., Mitterberger, M., Pinggera, G. M., Frauscher, F., Fussenegger, M., Kofler, K., & Bartsch, G. (2007). Transurethral ultrasonography-guided injection of adult autologous stem cells versus transurethral endoscopic injection of collagen in treatment of urinary incontinence. *World Journal of Urology*, *25*(4), 385–392. DOI: 10.1007/s00345-007-0190-7 PMID: 17701044

Suman, S., Domingues, A., Ratajczak, J., & Ratajczak, M. Z. (2019). Potential clinical applications of stem cells in regenerative medicine. In *Advances in Experimental Medicine and Biology* (Vol. 1201, pp. 1–22). Springer., DOI: 10.1007/978-3-030-31206-0_1

Sun, D., Bu, L., Liu, C., Yin, Z., Zhou, X., Li, X., & Xiao, A. (2013). Therapeutic effects of human amniotic fluid-derived stem cells on renal interstitial fibrosis in a murine model of unilateral ureteral obstruction. *PLoS One*, *8*(5), e65042. DOI: 10.1371/journal.pone.0065042 PMID: 23724119

Sung, P.-H., Chiang, H.-J., Chen, C.-H., Chen, Y.-L., Huang, T.-H., Zhen, Y.-Y., Chang, M.-W., Liu, C.-F., Chung, S.-Y., Chen, Y.-L., Chai, H.-T., Sun, C.-K., & Yip, H.-K. (2016). Combined Therapy With Adipose-Derived Mesenchymal Stem Cells and Ciprofloxacin Against Acute Urogenital Organ Damage in Rat Sepsis Syndrome Induced by Intrapelvic Injection of Cecal Bacteria. *Stem Cells Translational Medicine*, *5*(6), 782–792. DOI: 10.5966/sctm.2015-0116 PMID: 27075767

Sun, X., Luo, L. H., Feng, L., Li, D. S., & Zhong, K. Z. (2017). B Cell Lymphoma-2-Modified Bone Marrow-Derived Mesenchymal Stem Cells Transplantation for the Treatment of Diabetes Mellitus-Induced Erectile Dysfunction in a Rat Model. *Urologia Internationalis*, *98*(3), 358–366. DOI: 10.1159/000452253 PMID: 27894122

Swaminathan, M., Stafford-Smith, M., Chertow, G. M., Warnock, D. G., Paragamian, V., Brenner, R. M., Lellouche, F., Fox-Robichaud, A., Atta, M. G., Melby, S., Mehta, R. L., Wald, R., Verma, S., & Mazer, C. D.ACT-AKI investigators. (2018). Allogeneic Mesenchymal Stem Cells for Treatment of AKI after Cardiac Surgery. *Journal of the American Society of Nephrology*, *29*(1), 260–267. DOI: 10.1681/ASN.2016101150 PMID: 29038286

Taguchi, A., Kaku, Y., Ohmori, T., Sharmin, S., Ogawa, M., Sasaki, H., & Nishinakamura, R. (2014). Redefining the in vivo origin of metanephric nephron progenitors enables generation of complex kidney structures from pluripotent stem cells. *Cell Stem Cell*, *14*(1), 53–67. DOI: 10.1016/j.stem.2013.11.010 PMID: 24332837

Taguchi, H., Ishizuka, E., & Saito, K. (1977). Cystoplasty by regeneration of the bladder. *The Journal of Urology*, *118*(5), 752–756. DOI: 10.1016/S0022-5347(17)58181-1 PMID: 916093

Takahashi, K., Tanabe, K., Ohnuki, M., Narita, M., Ichisaka, T., Tomoda, K., & Yamanaka, S. (2007). Induction of pluripotent stem cells from adult human fibroblasts by defined factors. *Cell*, *131*(5), 861–872. DOI: 10.1016/j.cell.2007.11.019 PMID: 18035408

Takasato, M., Er, P. X., Becroft, M., Vanslambrouck, J. M., Stanley, E. G., Elefanty, A. G., & Little, M. H. (2014). Directing human embryonic stem cell differentiation towards a renal lineage generates a self-organizing kidney. *Nature Cell Biology*, *16*(1), 118–126. DOI: 10.1038/ncb2894 PMID: 24335651

Takasato, M., Er, P. X., Chiu, H. S., Maier, B., Baillie, G. J., Ferguson, C., Parton, R. G., Wolvetang, E. J., Roost, M. S., Chuva de Sousa Lopes, S. M., & Little, M. H. (2015). Kidney organoids from human iPS cells contain multiple lineages and model human nephrogenesis. *Nature*, *526*(7574), 564–568. DOI: 10.1038/nature15695 PMID: 26444236

Takata, M., Sasaki, M. S., Sonoda, E., Morrison, C., Hashimoto, M., Utsumi, H., Yamaguchi-Iwai, Y., Shinohara, A., & Takeda, S. (1998). Homologous recombination and non-homologous end-joining pathways of DNA double-strand break repair have overlapping roles in the maintenance of chromosomal integrity in vertebrate cells. *The EMBO Journal*, *17*(18), 5497–5508. DOI: 10.1093/emboj/17.18.5497 PMID: 9736627

Tamadon, A., Zhan-byrbekuly, U., Kairgaliyev, I., Khoradmehr, A., Tamadon, A., Zhan-byrbekuly, U., Kairgaliyev, I., & Khoradmehr, A. (2019). *Mesenchymal Stem Cell Therapy of Male Infertility*. Male Reproductive Health., DOI: 10.5772/intechopen.88343

Tanavde, V., Vaz, C., Rao, M. S., Vemuri, M. C., & Pochampally, R. R. (2015). Research using Mesenchymal Stem/Stromal Cells: Quality metric towards developing a reference material. *Cytotherapy*, *17*(9), 1169–1177. DOI: 10.1016/j.jcyt.2015.07.008 PMID: 26276001

Tang, X., Zhang, X., Wu, Y., Yin, H., Du, Y., Zhang, X., Li, Q., Liu, S., & Xu, T. (2020). The Clinical Effects of Utilizing Allogeneic Acellular Dermal Matrix in the Surgical Therapy of Anterior Urethral Stricture. *Urologia Internationalis*, *104*(11-12), 933–938. DOI: 10.1159/000510317 PMID: 33022676

Tanigawa, S., Islam, M., Sharmin, S., Naganuma, H., Yoshimura, Y., Haque, F., Era, T., Nakazato, H., Nakanishi, K., Sakuma, T., Yamamoto, T., Kurihara, H., Taguchi, A., & Nishinakamura, R. (2018). Organoids from Nephrotic Disease-Derived iPSCs Identify Impaired NEPHRIN Localization and Slit Diaphragm Formation in Kidney Podocytes. *Stem Cell Reports*, *11*(3), 727–740. DOI: 10.1016/j.stemcr.2018.08.003 PMID: 30174315

Tan, J., Li, P., Wang, Q., Li, Y., Li, X., Zhao, D., Xu, X., & Kong, L. (2016). Autologous menstrual blood-derived stromal cells transplantation for severe Asherman's syndrome. *Human Reproduction (Oxford, England), 31*(12), 2723–2729. DOI: 10.1093/humrep/dew235 PMID: 27664218

Tarng, D.-C., Tseng, W.-C., Lee, P.-Y., Chiou, S.-H., & Hsieh, S.-L. (2016). Induced Pluripotent Stem Cell-Derived Conditioned Medium Attenuates Acute Kidney Injury by Downregulating the Oxidative Stress-Related Pathway in Ischemia-Reperfusion Rats. *Cell Transplantation, 25*(3), 517–530. DOI: 10.3727/096368915X688542 PMID: 26132529

Tedesco, M., Bellei, B., Garelli, V., Caputo, S., Latini, A., Giuliani, M., Cota, C., Chichierchia, G., Romani, C., Foddai, M. L., Cristaudo, A., Morrone, A., & Migliano, E. (2020). Adipose tissue stromal vascular fraction and adipose tissue stromal vascular fraction plus platelet-rich plasma grafting: New regenerative perspectives in genital lichen sclerosus. *Dermatologic Therapy, 33*(6), e14277. DOI: 10.1111/dth.14277 PMID: 32888384

Temeltas, G., Dagci, T., Evren, V., & Lekili, M. (2009). Effects of neuronal and glial restricted precursor cells transplantation on erectile function after experimentally induced spinal cord injury. *Journal of Sexual Medicine, 6*(12), 3265–3273. DOI: 10.1111/j.1743-6109.2009.01376.x PMID: 19570038

Thacker, J., & Wilkinson, R. E. (1991). The genetic basis of resistance to ionising radiation damage in cultured mammalian cells. *Mutation Research, 254*(2), 135–142. DOI: 10.1016/0921-8777(91)90004-9 PMID: 2002810

Thatava, T., Armstrong, A. S., De Lamo, J. G., Edukulla, R., Khan, Y. K., Sakuma, T., Ohmine, S., Sundsbak, J. L., Harris, P. C., Kudva, Y. C., & Ikeda, Y. (2011). Successful disease-specific induced pluripotent stem cell generation from patients with kidney transplantation. *Stem Cell Research & Therapy, 2*(6), 48. DOI: 10.1186/scrt89 PMID: 22142803

The Hinxton Group: World Stem Cell Policies. (s. f.). Recuperado 25 de agosto de 2024, de http://www.hinxtongroup.org/wp.html

Thiel, A., Yavanian, G., Nastke, M.-D., Morales, P., Kouris, N. A., Kimbrel, E. A., & Lanza, R. (2015). Human embryonic stem cell-derived mesenchymal cells preserve kidney function and extend lifespan in NZB/W F1 mouse model of lupus nephritis. *Scientific Reports, 5*(1), 17685. DOI: 10.1038/srep17685 PMID: 26628350

Tian, H., Bharadwaj, S., Liu, Y., Ma, P. X., Atala, A., & Zhang, Y. (2010). Differentiation of human bone marrow mesenchymal stem cells into bladder cells: Potential for urological tissue engineering. *Tissue Engineering. Part A, 16*(5), 1769–1779. DOI: 10.1089/ten.tea.2009.0625 PMID: 20020816

Tilgner, K., Atkinson, S. P., Golebiewska, A., Stojković, M., Lako, M., & Armstrong, L. (2008). Isolation of primordial germ cells from differentiating human embryonic stem cells. *Stem Cells (Dayton, Ohio), 26*(12), 3075–3085. DOI: 10.1634/stemcells.2008-0289 PMID: 18802037

Tilly, J. L., Niikura, Y., & Rueda, B. R. (2009). The current status of evidence for and against postnatal oogenesis in mammals: A case of ovarian optimism versus pessimism? *Biology of Reproduction, 80*(1), 2–12. DOI: 10.1095/biolreprod.108.069088 PMID: 18753611

Torres Zambrano, G. M., Valverde, C. A. V., Hernandez, A. B., Hadi, L. A., Rivero, R. A., & Carmenate, Y. V. (2020). Renal Involvement in Patients with COVID-19 Pneumonia and Outcomes After Stem Cell Nebulization. *medRxiv*, 2020.12.16.20236877. DOI: 10.1101/2020.12.16.20236877

Torres-zambrano, G. M., Villegas-valverde, C. A., Bencomo-hernandez, A. A., Hadi, L. A., Adukkadukkam, D., Rivero-jimenez, R. A., Ahmed, Y. M., Castillo-aleman, Y. M., Ventura-carmenate, Y., Torres-zambrano, G. M., Villegas-valverde, C. A., Bencomo-hernandez, A. A., Hadi, L. A., Rivero-jimenez, R. A., Ahmed, Y. M., Castillo-aleman, Y. M., & Renal, Y. V. (2021). *Renal Involvement and Outcomes in Severe COVID-19 Patients After Stem Cell Jet-Nebulization To cite this article : 10*(6), 231-237. DOI: 10.11648/j.cmr.20211006.19

Tournaye, H., & Goossens, E. (2011). Spermatogonial stem cells: What does the future hold? *Facts, Views & Vision in ObGyn, 3*(1), 36–40. PMID: 24753846

Toyohara, T., Mae, S.-I., Sueta, S.-I., Inoue, T., Yamagishi, Y., Kawamoto, T., Kasahara, T., Hoshina, A., Toyoda, T., Tanaka, H., Araoka, T., Sato-Otsubo, A., Takahashi, K., Sato, Y., Yamaji, N., Ogawa, S., Yamanaka, S., & Osafune, K. (2015). Cell Therapy Using Human Induced Pluripotent Stem Cell-Derived Renal Progenitors Ameliorates Acute Kidney Injury in Mice. *Stem Cells Translational Medicine, 4*(9), 980–992. DOI: 10.5966/sctm.2014-0219 PMID: 26198166

Tremolada, C.Image Regenerative Clinic. (2019). Mesenchymal Stromal Cells and Micro Fragmented Adipose Tissue: New Horizons of Effectiveness of Lipogems. *Journal of Stem Cells Research. HSOA Journal of Stem Cells Research, Development & Therapy, 5*(1), 1–7. DOI: 10.24966/SRDT-2060/100017

Tremp, M., Salemi, S., Largo, R., Andersson, K. E. A., Plock, J., Aboushwareb, T., Sulser, T., & Eberli, D. (2014). Adipose-derived stem cells (ADSCs) and muscle precursor cells (MPCs) for the treatment of bladder voiding dysfunction. *World Journal of Urology*, *32*(5), 1241–1248. DOI: 10.1007/s00345-013-1200-6 PMID: 24217741

Tresa, V., Yaseen, A., Lanewala, A. A., Hashmi, S., Khatri, S., Ali, I., & Mubarak, M. (2017). Etiology, clinical profile and short-term outcome of acute kidney injury in children at a tertiary care pediatric nephrology center in Pakistan. *Renal Failure*, *39*(1), 26–31. DOI: 10.1080/0886022X.2016.1244074 PMID: 27767356

Trounson, A., & DeWitt, N. D. (2016). Pluripotent stem cells progressing to the clinic. *Nature Reviews. Molecular Cell Biology*, *17*(3), 194–200. DOI: 10.1038/nrm.2016.10 PMID: 26908143

Tseng, L. H., Lin, Y.-H., Chang, S.-D., Lim, C.-K., Chiang, C.-C., & Huang, C.-H. (s. f.). *Three-dimensional printing vaginal pessary to treat pelvic organ prolapse: A pilot study*. Recuperado 6 de julio de 2025, de https://www.authorea.com/users/517755/articles/592178-three-dimensional-printing-vaginal-pessary-to-treat-pelvic-organ-prolapse-a-pilot-study

Tsuji, I., Kuroda, K., Fujieda, J., Shiraishi, Y., Kassai, T., & Shida, H. (1963). A clinical and experimental study on cystoplasty not using the intestine. *The Journal of Urology*, *89*(2), 214–225. DOI: 10.1016/S0022-5347(17)64531-2 PMID: 13994538

Tsuji, I., Kuroda, K., Fujieda, J., Shiraishi, Y., Kunishima, K., & Orikasa, S. (1967). Clinical experiences of bladder reconstruction using preserved bladder and gelatin sponge bladder in the case of bladder cancer. *The Journal of Urology*, *98*(1), 91–92. DOI: 10.1016/S0022-5347(17)62828-3 PMID: 6067880

Tsulukidze, A., Murvanidze, D., Dvali, R., & Ivashchenko, G. (1964). FORMATION OF A BLADDER BY A PLASTIC SHELL AFTER TOTAL CYSTECTOMY. *British Journal of Urology*, *36*(1), 102–105. DOI: 10.1111/j.1464-410X.1964.tb09488.x PMID: 14126062

Uebersax, L., Merkle, H. P., & Meinel, L. (2009). Biopolymer-based growth factor delivery for tissue repair: From natural concepts to engineered systems. *Tissue Engineering. Part B, Reviews*, *15*(3), 263–289. DOI: 10.1089/ten.teb.2008.0668 PMID: 19416020

Uhlén, M., Fagerberg, L., Hallström, B. M., Lindskog, C., Oksvold, P., Mardinoglu, A., Sivertsson, Å., Kampf, C., Sjöstedt, E., Asplund, A., Olsson, I. M., Edlund, K., Lundberg, E., Navani, S., Szigyarto, C. A. K., Odeberg, J., Djureinovic, D., Takanen, J. O., Hober, S., & Pontén, F. (2015). Tissue-based map of the human proteome. *Science*, *347*(6220), 1260419. Advance online publication. DOI: 10.1126/science.1260419 PMID: 25613900

Urogenital tract fistulas in females—UpToDate. (s. f.). Recuperado 11 de agosto de 2024, de https://www.uptodate.com/contents/urogenital-tract-fistulas-in-females?search=urogenital-tractfistulas_in-women&source=search_result&selectedTitle=1~150&usage_type=default&display_rank=1

Vacanti, N. M., Cheng, H., Hill, P. S., Guerreiro, J. D. T., Dang, T. T., Ma, M., Watson, S., Hwang, N. S., Langer, R., & Anderson, D. G. (2012). Localized delivery of dexamethasone from electrospun fibers reduces the foreign body response. *Biomacromolecules*, *13*(10), 3031–3038. DOI: 10.1021/bm300520u PMID: 22920794

Vaegler, M., Lenis, A. T., Daum, L., Amend, B., Stenzl, A., Toomey, P., Renninger, M., Damaser, M. S., & Sievert, K. D. (2012). Stem cell therapy for voiding and erectile dysfunction. En *Nature Reviews Urology* (Vol. 9, Número 8, pp. 435-447). DOI: 10.1038/nrurol.2012.111

Vakalopoulos, I., Memmos, D., Mykoniatis, I., Toutziaris, C., & Dimitriadis, G. (2018). Stem cell therapy in erectile dysfunction: Science fiction or realistic treatment option? En *Hormones* (Vol. 17, Número 3, pp. 315-320). Springer. DOI: 10.1007/s42000-018-0050-4

Valentin, J. E., Freytes, D. O., Grasman, J. M., Pesyna, C., Freund, J., Gilbert, T. W., & Badylak, S. F. (2009). Oxygen diffusivity of biologic and synthetic scaffold materials for tissue engineering. *Journal of Biomedical Materials Research. Part A*, *91*(4), 1010–1017. DOI: 10.1002/jbm.a.32328 PMID: 19097154

van den Berg, C. W., Ritsma, L., Avramut, M. C., Wiersma, L. E., van den Berg, B. M., Leuning, D. G., Lievers, E., Koning, M., Vanslambrouck, J. M., Koster, A. J., Howden, S. E., Takasato, M., Little, M. H., & Rabelink, T. J. (2018). Renal Subcapsular Transplantation of PSC-Derived Kidney Organoids Induces Neo-vasculogenesis and Significant Glomerular and Tubular Maturation In Vivo. *Stem Cell Reports*, *10*(3), 751–765. DOI: 10.1016/j.stemcr.2018.01.041 PMID: 29503086 van der Spoel, T. I. G., Jansen of Lorkeers, S. J., Agostoni, P., van Belle, E., Gyongyosi, M., Sluijter, J. P. G., Cramer, M. J., Doevendans, P. A., & Chamuleau, S. A. J. (2011). Human relevance of pre-clinical studies in stem cell therapy: Systematic review and meta-analysis of large animal models of ischaemic heart disease. *Cardiovascular Research*, *91*(4), 649–658. DOI: 10.1093/cvr/cvr113 PMID: 21498423

Vashaghian, M., Ruiz-Zapata, A. M., Kerkhof, M. H., Zandieh-Doulabi, B., Werner, A., Roovers, J. P., & Smit, T. H. (2017). Toward a new generation of pelvic floor implants with electrospun nanofibrous matrices: A feasibility study. *Neurourology and Urodynamics*, *36*(3), 565–573. DOI: 10.1002/nau.22969 PMID: 26840206

Vassena, R., Eguizabal, C., Heindryckx, B., Sermon, K., Simon, C., Van Pelt, A. M. M., Veiga, A., & Zambelli, F. (2015). Stem cells in reproductive medicine: Ready for the patient? *Human Reproduction (Oxford, England)*, *30*(9), 2014–2021. DOI: 10.1093/humrep/dev181 PMID: 26202914

Vigneau, C., Polgar, K., Striker, G., Elliott, J., Hyink, D., Weber, O., Fehling, H.-J., Keller, G., Burrow, C., & Wilson, P. (2007). Mouse embryonic stem cell-derived embryoid bodies generate progenitors that integrate long term into renal proximal tubules in vivo. *Journal of the American Society of Nephrology*, *18*(6), 1709–1720. DOI: 10.1681/ASN.2006101078 PMID: 17475814

Vij, S. C., Sabanegh, E.Jr, & Agarwal, A. (2018). Biological therapy for non-obstructive azoospermia. *Expert Opinion on Biological Therapy*, *18*(1), 19–23. DOI: 10.1080/14712598.2018.1380622 PMID: 28927307

Viswanathan, S., Keating, A., Deans, R., Hematti, P., Prockop, D., Stroncek, D. F., Stacey, G., Weiss, D. J., Mason, C., & Rao, M. S. (2014). Soliciting strategies for developing cell-based reference materials to advance mesenchymal stromal cell research and clinical translation. *Stem Cells and Development*, *23*(11), 1157–1167. DOI: 10.1089/scd.2013.0591 PMID: 24422625

Vizzard, M. A. (1999). Alterations in growth-associated protein (GAP-43) expression in lower urinary tract pathways following chronic spinal cord injury. *Somatosensory & Motor Research*, *16*(4), 369–381. DOI: 10.1080/08990229970429 PMID: 10632033

Volarevic, V., Bojic, S., Nurkovic, J., Volarevic, A., Ljujic, B., Arsenijevic, N., Lako, M., & Stojkovic, M. (2014). Stem cells as new agents for the treatment of infertility: Current and future perspectives and challenges. *BioMed Research International*, *2014*, 1–8. Advance online publication. DOI: 10.1155/2014/507234 PMID: 24826378

Wang, B., Zhou, J., Banie, L., Reed-Maldonado, A. B., Ning, H., Lu, Z., Ruan, Y., Zhou, T., Wang, H. S., Oh, B. S., Wang, G., Qi, S. L., Lin, G., & Lue, T. F. (2018). Low-intensity extracorporeal shock wave therapy promotes myogenesis through PERK/ATF4 pathway. *Neurourology and Urodynamics*, *37*(2), 699–707. DOI: 10.1002/nau.23380 PMID: 28763567

Wang, D., Li, M., Huang, W., Lu, M., Hu, C., Li, K., Qiu, J., & Gao, X. (2015). Repair of urethral defects with polylactid acid fibrous membrane seeded with adipose-derived stem cells in a rabbit model. *Connective Tissue Research*, *56*(6), 434–439. DOI: 10.3109/03008207.2015.1035376 PMID: 25943462

Wang, D., Mou, H., Li, S., Li, Y., Hough, S., Tran, K., Li, J., Yin, H., Anderson, D. G., Sontheimer, E. J., Weng, Z., Gao, G., & Xue, W. (2015). Adenovirus-Mediated Somatic Genome Editing of Pten by CRISPR/Cas9 in Mouse Liver in Spite of Cas9-Specific Immune Responses. *Human Gene Therapy*, *26*(7), 432–442. DOI: 10.1089/hum.2015.087 PMID: 26086867

Wang, G., Lin, G., Zhang, H., Qiu, X., Ning, H., Banie, L., Fandel, T., Albersen, M., Lue, T. F., & Lin, C. S. (2011). Effects of prolonged vaginal distension and β-aminopropionitrile on urinary continence and urethral structure. *Urology*, *78*(4), 968.e13–968.e19. DOI: 10.1016/j.urology.2011.07.1381 PMID: 21982017

Wang, H. J., Chuang, Y. C., & Chancellor, M. B. (2011). Development of cellular therapy for the treatment of stress urinary incontinence. *International Urogynecology Journal*, *22*(9), 1075–1083. DOI: 10.1007/s00192-011-1432-1 PMID: 21505907

Wang, J., & Sauer, M. V. (2006). In vitro fertilization (IVF): A review of 3 decades of clinical innovation and technological advancement. *Therapeutics and Clinical Risk Management*, *2*(4), 355–364. DOI: 10.2147/tcrm.2006.2.4.355 PMID: 18360648

Wang, S., Li, Y., Zhao, J., Zhang, J., & Huang, Y. (2013). Mesenchymal stem cells ameliorate podocyte injury and proteinuria in a type 1 diabetic nephropathy rat model. *Biology of Blood and Marrow Transplantation*, *19*(4), 538–546. DOI: 10.1016/j.bbmt.2013.01.001 PMID: 23295166

Wang, X. Y., Liu, C. L., Li, S. D., Xu, Y., Chen, P., Liu, Y., Ding, Q., Wahafu, W., Hong, B. F., & Yang, M. H. (2015). Hypoxia precondition promotes adipose-derived mesenchymal stem cells based repair of diabetic erectile dysfunction via augmenting angiogenesis and neuroprotection. *PLoS One*, *10*(3), e0118951. Advance online publication. DOI: 10.1371/journal.pone.0118951 PMID: 25790284

Wang, X., Liu, C., Xu, Y., Chen, P., Shen, Y., Xu, Y., Zhao, Y., Chen, W., Zhang, X., Ouyang, Y., Wang, Y., Xie, C., Zhou, M., & Liu, C. (2017). Combination of mesenchymal stem cell injection with icariin for the treatment of diabetes-associated erectile dysfunction. *PLoS One*, *12*(3), e0174145. Advance online publication. DOI: 10.1371/journal.pone.0174145 PMID: 28350842

Wang, Y., & Liao, L. (2014). Histologic and functional outcomes of small intestine submucosa-regenerated bladder tissue. *BMC Urology*, *14*(1), 69. DOI: 10.1186/1471-2490-14-69 PMID: 25148849

Wankhade, U. D., Shen, M., Kolhe, R., & Fulzele, S. (2016). Advances in Adipose-Derived Stem Cells Isolation, Characterization, and Application in Regenerative Tissue Engineering. *Stem Cells International*, *3206807*(1), 3206807. Advance online publication. DOI: 10.1155/2016/3206807 PMID: 26981130

Wen, L., & Tang, F. (2016). Single-cell sequencing in stem cell biology. *Genome Biology*, *17*(1), 71. Advance online publication. DOI: 10.1186/s13059-016-0941-0 PMID: 27083874

Wen, Y., Wani, P., Zhou, L., Baer, T., Phadnis, S. M., Reijo Pera, R. A., & Chen, B. (2013). Reprogramming of fibroblasts from older women with pelvic floor disorders alters cellular behavior associated with donor age. *Stem Cells Translational Medicine*, *2*(2), 118–128. DOI: 10.5966/sctm.2012-0092 PMID: 23341439

West, J. A., Park, I. H., Daley, G. Q., & Geijsen, N. (2006). In vitro generation of germ cells from murine embryonic stem cells. *Nature Protocols*, *1*(4), 2026–2036. DOI: 10.1038/nprot.2006.303 PMID: 17487192

Wever, K. E., Menting, T. P., Rovers, M., van der Vliet, J. A., Rongen, G. A., Masereeuw, R., Ritskes-Hoitinga, M., Hooijmans, C. R., & Warlé, M. (2012). Ischemic preconditioning in the animal kidney, a systematic review and meta-analysis. *PLoS One*, *7*(2), e32296. DOI: 10.1371/journal.pone.0032296 PMID: 22389693

White, Y. A. R., Woods, D. C., Takai, Y., Ishihara, O., Seki, H., & Tilly, J. L. (2012). Oocyte formation by mitotically active germ cells purified from ovaries of reproductive-age women. *Nature Medicine*, *18*(3), 413–421. DOI: 10.1038/nm.2669 PMID: 22366948

Wiafe, B., Metcalfe, P. D., & Adesida, A. B. (2015). Stem Cell Therapy: Current Applications and Potential for Urology. *Current Urology Reports*, *16*(11), 77. DOI: 10.1007/s11934-015-0551-5 PMID: 26385812

Woo, J. C., Bae, W. J., Kim, S. J., Kim, S. D., Sohn, D. W., Hong, S. H., Lee, J. Y., Hwang, T. K., Sung, Y. C., & Kim, S. W. (2011). Transplantation of muscle-derived stem cells into the corpus avernosum restores erectile function in a rat model of cavernous nerve injury. *Korean Journal of Urology*, *52*(5), 359–363. DOI: 10.4111/kju.2011.52.5.359 PMID: 21687398

Woo, L. L., Tanaka, S. T., Anumanthan, G., Pope, J. C.IV, Thomas, J. C., Adams, M. C., Brock, J. W., & Bhowmick, N. A. (2011). Mesenchymal stem cell recruitment and improved bladder function after bladder outlet obstruction: Preliminary data. *The Journal of Urology*, *185*(3), 1132–1138. DOI: 10.1016/j.juro.2010.10.033 PMID: 21255803

Wu, J. M., Hundley, A. F., Fulton, R. G., & Myers, E. R. (2009). Forecasting the prevalence of pelvic floor disorders in U.S. Women: 2010 to 2050. *Obstetrics and Gynecology*, *114*(6), 1278–1283. DOI: 10.1097/AOG.0b013e3181c2ce96 PMID: 19935030

Wu, R., Liu, G., Fan, Y., Rohozinski, J., Lu, X., Rodriguez, G., Farney, A., Atala, A., & Zhang, Y. (2013). 249 HUMAN URINE-DERIVED STEM CELLS ORIGINATE FROM PARIETAL STEM CELLS. *The Journal of Urology*, *189*(4S). Advance online publication. DOI: 10.1016/j.juro.2013.02.1631

Wu, X., Goodyear, S. M., Abramowitz, L. K., Bartolomei, M. S., Tobias, J. W., Avarbock, M. R., & Brinster, R. L. (2012). Fertile offspring derived from mouse spermatogonial stem cells cryopreserved for more than 14 years. *Human Reproduction (Oxford, England)*, *27*(5), 1249–1259. DOI: 10.1093/humrep/des077 PMID: 22416011

Wu, X., Jia, Y., Sun, X., & Wang, J. (2020). Tissue engineering in female pelvic floor reconstruction. *Engineering in Life Sciences*, *20*(7), 275–286. DOI: 10.1002/elsc.202000003 PMID: 32647506

Wu, X., Wang, Y., Zhu, C., Tong, X., Yang, M., Yang, L., Liu, Z., Huang, W., Wu, F., Zong, H., Li, H., & He, H. (2016). Preclinical animal study and human clinical trial data of co-electrospun poly(L-lactide-co-caprolactone) and fibrinogen mesh for anterior pelvic floor reconstruction. *International Journal of Nanomedicine*, *11*, 389–397. DOI: 10.2147/IJN.S88803 PMID: 26893556

Wu, Z., Zhang, S., Zhou, L., Cai, J., Tan, J., Gao, X., Zeng, Z., & Li, D. (2017). Thromboembolism Induced by Umbilical Cord Mesenchymal Stem Cell Infusion: A Report of Two Cases and Literature Review. *Transplantation Proceedings*, *49*(7), 1656–1658. DOI: 10.1016/j.transproceed.2017.03.078 PMID: 28838459

Wyndaele, J. J., Madersbacher, H., & Kovindha, A. (2001). Conservative treatment of the neuropathic bladder in spinal cord injured patients. *Spinal Cord*, *39*(6), 294–300. DOI: 10.1038/sj.sc.3101160 PMID: 11438850

Xiao, Y., Song, Y. J., Song, B., Huang, C. B., Ling, Q., & Yu, X. (2017). TGF-β/MAPK signaling mediates the effects of bone marrow mesenchymal stem cells on urinary control and interstitial cystitis after urinary bladder transplantation. *American Journal of Translational Research*, *9*(3), 1193.

Xia, Y., Nivet, E., Sancho-Martinez, I., Gallegos, T., Suzuki, K., Okamura, D., Wu, M.-Z., Dubova, I., Esteban, C. R., Montserrat, N., Campistol, J. M., & Izpisua Belmonte, J. C. (2013). Directed differentiation of human pluripotent cells to ureteric bud kidney progenitor-like cells. *Nature Cell Biology*, *15*(12), 1507–1515. DOI: 10.1038/ncb2872 PMID: 24240476

Xie, J., Liu, B., Chen, J., Xu, Y., Zhan, H., Yang, F., Li, W., & Zhou, X. (2018). Umbilical cord-derived mesenchymal stem cells alleviated inflammation and inhibited apoptosis in interstitial cystitis via AKT/mTOR signaling pathway. *Biochemical and Biophysical Research Communications*, *495*(1), 546–552. DOI: 10.1016/j.bbrc.2017.11.072 PMID: 29137981

Xie, M., Song, L., Wang, J., Fan, S., Zhang, Y., & Xu, Y. (2013). Evaluation of stretched electrospun silk fibroin matrices seeded with urothelial cells for urethra reconstruction. *The Journal of Surgical Research*, *184*(2), 774–781. DOI: 10.1016/j.jss.2013.04.016 PMID: 23706393

Xu, K., Han, Y., Huang, Y., Wei, P., Yin, J., & Jiang, J. (2022). The application of 3D bioprinting in urological diseases. *Materials Today. Bio*, *16*, 100388. DOI: 10.1016/j.mtbio.2022.100388 PMID: 35967737

Xu, Y., Guan, R., Lei, H., Li, H., Wang, L., Gao, Z., Song, W., & Xin, Z. (2014). Therapeutic potential of adipose-derived stem cells-based micro-tissues in a rat model of postprostatectomy erectile dysfunction. *Journal of Sexual Medicine*, *11*(10), 2439–2448. DOI: 10.1111/jsm.12636 PMID: 25042722

Xu, Y.-M., Fu, Q., Sa, Y.-L., Zhang, J., Song, L.-J., & Feng, C. (2013). Outcome of small intestinal submucosa graft for repair of anterior urethral strictures. *International Journal of Urology*, *20*(6), 622–629. DOI: 10.1111/j.1442-2042.2012.03230.x PMID: 23131085

Yamamoto, T., Gotoh, M., Kato, M., Majima, T., Toriyama, K., Kamei, Y., Iwaguro, H., Matsukawa, Y., & Funahashi, Y. (2012). Periurethral injection of autologous adipose-derived regenerative cells for the treatment of male stress urinary incontinence: Report of three initial cases. *International Journal of Urology*, *19*(7), 652–659. DOI: 10.1111/j.1442-2042.2012.02999.x PMID: 22435469

Yan, H., Ding, Y., & Lu, M. (2020). Current Status and Prospects in the Treatment of Erectile Dysfunction by Adipose-Derived Stem Cells in the Diabetic Animal Model. En *Sexual Medicine Reviews* (Vol. 8, Número 3, pp. 486-491). Elsevier B.V. DOI: 10.1016/j.sxmr.2019.09.006

Yang, Q., Chen, X., Zheng, T., Han, D., Zhang, H., Shi, Y., Bian, J., Sun, X., Xia, K., Liang, X., Liu, G., Zhang, Y., & Deng, C. (2016). Transplantation of human urine-derived stem cells transfected with pigment epithelium-derived factor to protect erectile function in a rat model of cavernous nerve injury. *Cell Transplantation*, *25*(11), 1987–2001. DOI: 10.3727/096368916X691448 PMID: 27075964

Yang, R., Fang, F., Wang, J., & Guo, H. (2015). Adipose-derived stem cells ameliorate erectile dysfunction after cavernous nerve cryoinjury. *Andrology*, *3*(4), 694–701. DOI: 10.1111/andr.12047 PMID: 26198799

Yang, X., Yu, Y., Xu, J., Shu, H., Xia, J., Liu, H., Wu, Y., Zhang, L., Yu, Z., Fang, M., Yu, T., Wang, Y., Pan, S., Zou, X., Yuan, S., & Shang, Y. (2020). Clinical course and outcomes of critically ill patients with SARS-CoV-2 pneumonia in Wuhan, China: A single-centered, retrospective, observational study. *The Lancet. Respiratory Medicine*, *8*(5), 475–481. DOI: 10.1016/S2213-2600(20)30079-5 PMID: 32105632

Yao, C., Hedrick, M., Pareek, G., Renzulli, J., Haleblian, G., & Webster, T. J. (2013). Nanostructured polyurethane-poly-lactic-co-glycolic acid scaffolds increase bladder tissue regeneration: An in vivo study. *International Journal of Nanomedicine*, *8*, 3285–3296. DOI: 10.2147/IJN.S44901 PMID: 24039415

Ye, H., Zheng, T., Li, W., Li, X., Fu, X., Huang, Y., Hu, C., Li, J., Huang, J., Liu, Z., Zheng, L., & Zheng, Y. (2017). Ovarian Stem Cell Nests in Reproduction and Ovarian Aging. *Cellular physiology and biochemistry : international journal of experimental cellular physiology, biochemistry, and pharmacology, 43*(5), 1917-1925. DOI: 10.1159/000484114

Yiou, R., Hamidou, L., Birebent, B., Bitari, D., Le Corvoisier, P., Contremoulins, I., Rodriguez, A. M., Augustin, D., Roudot-Thoraval, F., de la Taille, A., & Rouard, H. (2017). Intracavernous Injections of Bone Marrow Mononucleated Cells for Postradical Prostatectomy Erectile Dysfunction: Final Results of the INSTIN Clinical Trial. *European Urology Focus*, *3*(6), 643–645. DOI: 10.1016/j.euf.2017.06.009 PMID: 28753830

Yiou, R., Hamidou, L., Birebent, B., Bitari, D., Lecorvoisier, P., Contremoulins, I., Khodari, M., Rodriguez, A. M., Augustin, D., Roudot-Thoraval, F., De La Taille, A., & Rouard, H. (2016). Safety of Intracavernous Bone Marrow-Mononuclear Cells for Postradical Prostatectomy Erectile Dysfunction: An Open Dose-Escalation Pilot Study. *European Urology*, *69*(6), 988–991. DOI: 10.1016/j.eururo.2015.09.026 PMID: 26439886

Yokoyama, T., Huard, J., Pruchnic, R., Yoshimura, N., Qu, Z., Cao, B., de Groat, W. C., Kumon, H., & Chancellor, M. B. (2001). Muscle-derived cell transplantation and differentiation into lower urinary tract smooth muscle. *Urology*, *57*(4), 826–831. DOI: 10.1016/S0090-4295(00)01083-9 PMID: 11306423

Yoshida, M., & Yamaguchi, O. (2014). Detrusor Underactivity: The Current Concept of the Pathophysiology. *Lower Urinary Tract Symptoms*, *6*(3), 131–137. DOI: 10.1111/luts.12070 PMID: 26663593

You, D., Jang, M. J., Kim, B. H., Song, G., Lee, C., Suh, N., Jeong, I. G., Ahn, T. Y., & Kim, C.-S. (2015). Comparative Study of Autologous Stromal Vascular Fraction and Adipose-Derived Stem Cells for Erectile Function Recovery in a Rat Model of Cavernous Nerve Injury. *Stem Cells Translational Medicine*, *4*(4), 351–358. DOI: 10.5966/sctm.2014-0161 PMID: 25792486

You, D., Jang, M. J., Song, G., Shin, H. C., Suh, N., Kim, Y. M., Ahn, T. Y., & Kim, C. S. (2021). Safety of autologous bone marrow-derived mesenchymal stem cells in erectile dysfunction: An open-label phase 1 clinical trial. *Cytotherapy*, *23*(10), 931–938. DOI: 10.1016/j.jcyt.2021.06.001 PMID: 34326007

Zakrzewski, W., Dobrzyński, M., Szymonowicz, M., & Rybak, Z. (2019). Stem cells: Past, present, and future. *Stem Cell Research & Therapy*, *10*(1), 68. Advance online publication. DOI: 10.1186/s13287-019-1165-5 PMID: 30808416

Zambon, J. P., Batezini, N. S. S., Pinto, E. R. S., Skaff, M., Girotti, M. E., & Almeida, F. G. (2010). Do we need new surgical techniques to repair vesico-vaginal fistulas? *International Urogynecology Journal*, *21*(3), 337–342. DOI: 10.1007/s00192-009-1040-5 PMID: 19949771

Zambon, J. P., Williams, K. J., Bennington, J., & Badlani, G. H. (2019). Applicability of regenerative medicine and tissue engineering for the treatment of stress urinary incontinence in female patients. *Neurourology and Urodynamics, 38*(S4), S76–S83. DOI: 10.1002/nau.24033 PMID: 31099087

Zaret, K. S. (2009). Using small molecules to great effect in stem cell differentiation. *Cell Stem Cell, 4*(5), 373–374. DOI: 10.1016/j.stem.2009.04.012 PMID: 19427285

Zasieda, Y. (2020). ERECTILE DYSFUNCTION TREATMENT WITH COMBINATION OF MESENCHYMAL STEM CELL DERIVED EXOSOMES AND FOCUSED LOW-INTENSIVE SHOCK WAVE THERAPY. *Men's Health. Gender and Psychosomatic Medicine, 1-2*, 70–78. DOI: 10.37321/UJMH.2020.1-2-08

Zegers-Hochschild, F., Adamson, G. D., De Mouzon, J., Ishihara, O., Mansour, R., Nygren, K., Sullivan, E., & Van Der Poel, S. (2009). The International Committee for Monitoring Assisted Reproductive Technology (ICMART) and the World Health Organization (WHO) Revised Glossary on ART Terminology, 2009. *Human Reproduction (Oxford, England), 24*(11), 2683–2687. DOI: 10.1093/humrep/dep343 PMID: 19801627

Zhang, S., Li, J., & Li, C. XuminXie, null, He, J., Ling, F., & Liu, G. (2022). Isolation and identification of a mesenchymal stem/stromal cell-like population from pediatric urethral tissue. *In Vitro Cellular & Developmental Biology. Animal, 58*(6), 503-511. DOI: 10.1007/s11626-022-00697-4

Zhang, H., Qiu, X., Shindel, A. W., Ning, H., Ferretti, L., Jin, X., Lin, G., Lin, C. S., & Lue, T. F. (2012a). Adipose tissue-derived stem cells ameliorate diabetic bladder dysfunction in a type II diabetic rat model. *Stem Cells and Development, 21*(9), 1391–1400. DOI: 10.1089/scd.2011.0244 PMID: 22008016

Zhang, H., Zhao, Y., Wang, M., Song, W., Sun, P., & Jin, X. (2019). A promising therapeutic option for diabetic bladder dysfunction: Adipose tissue-derived stem cells pretreated by defocused low-energy shock wave. *Journal of Tissue Engineering and Regenerative Medicine, 13*(6), 986–996. DOI: 10.1002/term.2844 PMID: 30811857

Zhang, H., Zheng, W., Shen, Y., Adhikari, D., Ueno, H., & Liu, K. (2012). Experimental evidence showing that no mitotically active female germline progenitors exist in postnatal mouse ovaries. *Proceedings of the National Academy of Sciences of the United States of America, 109*(31), 12580–12585. DOI: 10.1073/pnas.1206600109 PMID: 22778414

Zhang, M., Peng, Y., Zhou, Z., Zhou, J., Wang, Z., & Lu, M. (2013). Differentiation of human adipose-derived stem cells co-cultured with urothelium cell line toward a urothelium-like phenotype in a nude murine model. *Urology*, *81*(2), 465.e15–465.e22. DOI: 10.1016/j.urology.2012.10.030 PMID: 23374843

Zhang, P. Y., Fan, Y., Tan, T., & Yu, Y. (2020). Generation of Artificial Gamete and Embryo From Stem Cells in Reproductive Medicine. *Frontiers in Bioengineering and Biotechnology*, *8*, 781. Advance online publication. DOI: 10.3389/fbioe.2020.00781 PMID: 32793569

Zhang, S., Li, J., Li, C., Xie, X., He, J., Ling, F., Li, B., Wu, H., Li, Z., & Zheng, J. (2022). CD73-Positive Small Extracellular Vesicles Derived From Umbilical Cord Mesenchymal Stem Cells Promote the Proliferation and Migration of Pediatric Urethral Smooth Muscle Cells Through Adenosine Pathway. *Frontiers in Bioengineering and Biotechnology*, *10*, 895998. DOI: 10.3389/fbioe.2022.895998 PMID: 35573239

Zhang, S., Li, J., Li, C., Xie, X., He, J., Ling, F., Li, B., Wu, H., Li, Z., Zhen, J., & Liu, G. (2024). CD73-positive pediatric urethral mesenchymal stem-like cell-derived small extracellular vesicles stimulate angiogenesis. *Regenerative Therapy*, *25*, 77–84. DOI: 10.1016/j.reth.2023.12.002 PMID: 38111468

Zhang, Y., Ma, Y., Chen, J., Wang, M., Cao, Y., Li, L., Yang, H., Liu, X., Li, Y., & Zhu, L. (2021). Mesenchymal stem cell transplantation for vaginal repair in an ovariectomized rhesus macaque model. *Stem Cell Research & Therapy*, *12*(1), 406. DOI: 10.1186/s13287-021-02488-2 PMID: 34266489

Zhang, Y., McNeill, E., Tian, H., Soker, S., Andersson, K. E., Yoo, J. J., & Atala, A. (2008). Urine derived cells are a potential source for urological tissue reconstruction. *The Journal of Urology*, *180*(5), 2226–2233. DOI: 10.1016/j.juro.2008.07.023 PMID: 18804817

Zhang, Y., Shi, L., Lin, X., Zhou, F., Xin, L., Xu, W., Yu, H., Li, J., Pan, M., Pan, Y., Dai, Y., Zhang, Y., Shen, J., Zhao, L., Lu, M., & Zhang, S. (2021). Unresponsive thin endometrium caused by Asherman syndrome treated with umbilical cord mesenchymal stem cells on collagen scaffolds: A pilot study. *Stem Cell Research & Therapy*, *12*(1), 420. DOI: 10.1186/s13287-021-02499-z PMID: 34294152

Zhan, T., Li, X., Liu, J., & Ye, C. (2024). CRISPR-based gene expression platform for precise regulation of bladder cancer. *Cellular & Molecular Biology Letters*, *29*(1), 66. DOI: 10.1186/s11658-024-00569-7 PMID: 38724931

Zhao, Y. X., Chen, S. R., Su, P. P., Huang, F. H., Shi, Y. C., Shi, Q. Y., & Lin, S. (2019). Using Mesenchymal Stem Cells to Treat Female Infertility: An Update on Female Reproductive Diseases. *Stem Cells International*, *2019*, 1–10. Advance online publication. DOI: 10.1155/2019/9071720 PMID: 31885630

Zhao, Y., Luo, Q., Zhang, X., Qin, Y., Hao, J., Kong, D., Wang, H., Li, G., Gu, X., & Wang, H. (2020). Clinical Efficacy and Safety of Stem Cell-Based Therapy in Treating Asherman Syndrome: A System Review and Meta-Analysis. *Stem Cells International*, *2020*, 1–11. Advance online publication. DOI: 10.1155/2020/8820538 PMID: 33414830

Zhou, F., Hui, Y., Xin, H., De Xu, Y., Lei, H. E., Yang, B. C., Guan, R. L., Li, M., Hou, J. Q., & Xin, Z. C. (2017). Therapeutic effects of adipose-derived stem cells-based microtissues on erectile dysfunction in streptozotocin-induced diabetic rats. *Asian Journal of Andrology*, *19*(1), 91–97. DOI: 10.4103/aja.aja_29_20 PMID: 27345005

Zhou, F., Hui, Y., Xu, Y., Lei, H., Yang, B., Guan, R., Gao, Z., Xin, Z., & Hou, J. (2016). Effects of adipose-derived stem cells plus insulin on erectile function in streptozotocin-induced diabetic rats. *International Urology and Nephrology*, *48*(5), 657–669. DOI: 10.1007/s11255-016-1221-3 PMID: 26820518

Zhou, L., Song, Q., Shen, J., Xu, L., Xu, Z., Wu, R., Ge, Y., Zhu, J., Wu, J., Dou, Q., & Jia, R. (2017). Comparison of human adipose stromal vascular fraction and adipose-derived mesenchymal stem cells for the attenuation of acute renal ischemia/reperfusion injury. *Scientific Reports*, *7*(1), 44058. DOI: 10.1038/srep44058 PMID: 28276451

Zhou, Q., Wang, M., Yuan, Y., Wang, X., Fu, R., Wan, H., Xie, M., Liu, M., Guo, X., Zheng, Y., Feng, G., Shi, Q., Zhao, X. Y., Sha, J., & Zhou, Q. (2016). Complete Meiosis from Embryonic Stem Cell-Derived Germ Cells In Vitro. *Cell Stem Cell*, *18*(3), 330–340. DOI: 10.1016/j.stem.2016.01.017 PMID: 26923202

Zhou, S., Zhang, K., Atala, A., Khoury, O., Murphy, S. V., Zhao, W., & Fu, Q. (2016). Stem Cell Therapy for Treatment of Stress Urinary Incontinence: The Current Status and Challenges. *Stem Cells International*, *2016*(1), 7060975. Advance online publication. DOI: 10.1155/2016/7060975 PMID: 26880983

Zhou, Y., Xu, H., Xu, W., Wang, B., Wu, H., Tao, Y., Zhang, B., Wang, M., Mao, F., Yan, Y., Gao, S., Gu, H., Zhu, W., & Qian, H. (2013). Exosomes released by human umbilical cord mesenchymal stem cells protect against cisplatin-induced renal oxidative stress and apoptosis in vivo and in vitro. *Stem Cell Research & Therapy*, *4*(2), 34. DOI: 10.1186/scrt194 PMID: 23618405

Zhu, J. (2010). Bioactive modification of poly(ethylene glycol) hydrogels for tissue engineering. *Biomaterials*, *31*(17), 4639–4656. DOI: 10.1016/j.biomaterials.2010.02.044 PMID: 20303169

Zhu, W.-D., Xu, Y.-M., Feng, C., Fu, Q., Song, L.-J., & Cui, L. (2010). Bladder reconstruction with adipose-derived stem cell-seeded bladder acellular matrix grafts improve morphology composition. *World Journal of Urology*, *28*(4), 493–498. DOI: 10.1007/s00345-010-0508-8 PMID: 20091038

Zhu, X.-Y., Urbieta-Caceres, V., Krier, J. D., Textor, S. C., Lerman, A., & Lerman, L. O. (2013). Mesenchymal stem cells and endothelial progenitor cells decrease renal injury in experimental swine renal artery stenosis through different mechanisms. *Stem Cells (Dayton, Ohio)*, *31*(1), 117–125. DOI: 10.1002/stem.1263 PMID: 23097349

Zoja, C., Garcia, P. B., Rota, C., Conti, S., Gagliardini, E., Corna, D., Zanchi, C., Bigini, P., Benigni, A., Remuzzi, G., & Morigi, M. (2012). Mesenchymal stem cell therapy promotes renal repair by limiting glomerular podocyte and progenitor cell dysfunction in adriamycin-induced nephropathy. *American Journal of Physiology. Renal Physiology*, *303*(9), F1370–F1381. DOI: 10.1152/ajprenal.00057.2012 PMID: 22952284

Zou, J., Yang, W., Cui, W., Li, C., Ma, C., Ji, X., Hong, J., Qu, Z., Chen, J., Liu, A., & Wu, H. (2023). Therapeutic potential and mechanisms of mesenchymal stem cell-derived exosomes as bioactive materials in tendon-bone healing. *Journal of Nanobiotechnology*, *21*(1), 14. DOI: 10.1186/s12951-023-01778-6 PMID: 36642728

Zou, K., Yuan, Z., Yang, Z., Luo, H., Sun, K., Zhou, L., Xiang, J., Shi, L., Yu, Q., Zhang, Y., Hou, R., & Wu, J. (2009). Production of offspring from a germline stem cell line derived from neonatal ovaries. *Nature Cell Biology*, *11*(5), 631–636. DOI: 10.1038/ncb1869 PMID: 19363485

Zuk, P. A., Zhu, M., Mizuno, H., Huang, J., Futrell, J. W., Katz, A. J., Benhaim, P., Lorenz, H. P., & Hedrick, M. H. (2001). Multilineage cells from human adipose tissue: Implications for cell-based therapies. *Tissue Engineering*, *7*(2), 211–228. DOI: 10.1089/107632701300062859 PMID: 11304456

About the Authors

Gina Torres Zambrano has a degree in Medicine from the University of Rosario, Bogotá, Colombia, in 2002. She is a European Qualified Urological Surgeon, with a Master's Degree in urology from the Hospital of Getafe, Spain (from 2005 to 2010), endorse for the European Board of Urology certification (EBU). She has studied and been trained at different reference centers in Europe. She made an observationalship in the pelvic floor units and reconstructive surgery at the Leuven Hospital, Belgium, and at the Royal Hallamshire Hospital in Sheffield, England, in 2013. Her doctoral thesis is the contribution of the Spanish branch to the Randomized European Study Screening for Cancer De Prostate (ERSPC), in 2015, at the University of Alcalá, Spain. She worked at the Abu Dhabi Stem Cell Center (ADSCC) in the study of stem cell treatments in the urological field. Also she worked at Medeor Al Ain and RAK Hospital in Ras AL Khaima as a Consultant Urologist. Currently working at Sheik Khalifa Medical City (SKMC) in Abu Dhabi. Dr. Torres is a member of several societies, including the European Association of Urology (EAU), the EAU Section of Female and Functional Urology (ESFFU), Arab Association of Urology (AAU), the Endourological Society,and the Association of Spanish Scientists and Researchers in the United Arab Emirates (ACIEAU), Boar member of the Gender and Investigation commission (IgR) from Raicex (Network Of Spanish Scientists And Researchers Abroad).

Juan Uria Gonzalez Tova has a PhD in Urology with 22 years of clinical experience in Hospital Care and Primary Care and with a broad formation and development in clinical management and laparoscopy. He has 3 years of experience in university teaching. His is affiliation with NMC Royal in the United Arab Emirates.

Braulio Peramo Moyaa, an Obstetrician, Gynecologist and IVF physician, is the Medical Director and the General Manager of the Al Ain Fertility Center (AAFC). Dr. Peramo has been internationally practicing in the field of reproductive medicine, specializing in Assisted Reproductive Techniques (ARTs), gynecology and andrology for almost two decades in the USA, UK, mainland Europe and UAE. Dr. Peramo obtained his Medical and Surgery degree from the University of Madrid in 1991. Following this, he received his obstetrics and gynecology training in Doce de Octubre Hospital Universitario, where his work won the XVIII Prize of the Gynecological Society of Madrid in 1995. Later in 1997, he became a Clinical Fellow of Obstetrics and Gynecology during his visit at the Florida Hospital in Orlando, USA. In 2004, Dr. Peramo was awarded a Masters in Reproductive Medicine from the Universidad Complutense de Madrid, and in 2007, he attended advanced workshops on IVF and infertility in the New York University, USA. In January 2008, he received the European Diploma of Gynecological Endoscopy from the University of Auvergne in Clermont-Ferrand, France. Dr. Peramo has been the Medical Director at both the Clinica Tambre in Madrid and the Marques Institute in Barcelona, Spain. Under his lead, Clinica Tambre became a finalist for the European Foundation's Quality Management twice, in 2005 and 2008. He has also been IVF Consultant at the London Women's Clinic in Harley Street, UK. His expertise and research in the field of reproductive medicine has contributed in the development of infertility treatments and optimization of a number of ART protocols. Dr. Peramo has produced over 100 publications in short communications, full-length research papers, abstracts, posters and book chapters. His work in standardizing assisted reproduction aspects, from ovarian stimulation to the successful management of difficult andrology cases, has been internationally recognized. Dr. Peramo's outlook is to reduce the burden of infertility by doing everything possible to help couples in achieving their dream of creating a healthy family. He has particular interest in several aspects of infertility

Mariano Rosello Gaya, Experienced Medical Doctor with a demonstrated history of working in the hospital & health care industry. Skilled surgeon in penile implantation and Peyronies disease patients as well as all andrological procedures, together with both female and male genital reconstructive surgery. Team Building, and Public Speaking. Strong healthcare services professional with a Master's Degree focused in Master in Clinical Andrology from Fundació Puigvert, Barcelona, together with active participation European Society for Sexual Medicine. FECSM Fellow of the European Committee for Sexual Medicine since 2014.

Fathima Mohammed Ahamed is a researcher of Medical Biotechnology and Human Genetics (Rare Disorders), affiliated with the Al Ain Fertility Center of Al Ain Hospital.

Index

A

B

C

D

E

F

M

N

O

P

R

S

T

www.ingramcontent.com/pod-product-compliance
Lightning Source LLC
LaVergne TN
LVHW080902300926
849162LV00003B/514